Use of Probiotics in Aquaculture

Use of Probiotics in Aquaculture

Editor:

Preethi Kartan

Delve Publishing

www.delvepublishing.com

Use of Probiotics in Aquaculture

Editor: Preethi Kartan

708 3rd Avenue, 6th Floor
New York, NY 10017
United States of America

www.delvepublishing.com

ISBN: 978-1-68095-845-4

Library of Congress Control Number: 2016951598

Delve Publishing LLC publishes wide variety of books and eBooks. For more information about Delve Publishing and its products, visit our website at **www.delvepublishing.com**

About the Editor

Preethi Kartan

Preethi is a postgraduate in Biotechnology from University of Leeds, UK. She is currently working as a Scientific Associate at one of the CRO's in India and her interest lie in life sciences related writing.

Contents

List of Contributors

Monique Mancuso
Institute for Coastal Marine Environment (IAMC), National Research Council (CNR), Section of Messina, Italy

Patricia Martınez Cruz
Departamento de Sistemas Biologicos, Universidad Aut ´ onoma Metropolitana-Xochimilco, Calzada del Hueso 1100, ´ 04960 Mexico City, Mexico
Universidad Autonoma Metropolitana-Xochimilco, Calzada del Hueso 1100, 04960 Mexico City, Mexico

Ana L. Iba´nez
Departamento de Hidrobiolog´ıa, Universidad Autonoma Metropolitana-Iztapalapa, Av. San Rafael Atlixco 186, ´ 09340 Mexico City, Mexico

Oscar A. Monroy Hermosillo
Departamento de Biotecnolog´ıa, Universidad Autonoma Metropolitana-Iztapalapa, Av. San Rafael Atlixco 186, ´ 09340 Mexico City, Mexico

Hugo C. Ramırez Saad
Departamento de Sistemas Biologicos, Universidad Aut ´ onoma Metropolitana-Xochimilco, Calzada del Hueso 1100, ´ 04960 Mexico City, Mexico

Ekundayo Taiye Michael
Modibbo Adama University of Technology, Yola, Adamawa State, Nigeria

Sogbesan Olukayode Amos
Modibbo Adama University of Technology, Yola, Adamawa State, Nigeria

Lauratu Tahir Hussaini
Modibbo Adama University of Technology, Yola, Adamawa State, Nigeria

Jorge Olmos
Molecular Microbiology Laboratory, Department of Marine Biotechnology, Centro de Investigación Científica y de Educación Superior de Ensenada (CICESE), Ensenada, Baja California, México

J Paniagua-Michel
Bioactive Compounds and Bioremediation Laboratory, Department of Marine Biotechnology, Centro de Investigación Científica y de Educación Superior de Ensenada (CICESE), Ensenada, Baja California, México

Suxu He
Key Laboratory for Feed Biotechnology of the Ministry of Agriculture, Feed Research Institute, Chinese Academy of Agricultural Sciences, Beijing 100081, P. R. China

Wenshu Liu
Key Laboratory for Feed Biotechnology of the Ministry of Agriculture, Feed Research Institute, Chinese Academy of Agricultural Sciences, Beijing 100081, P. R. China

Zhigang Zhou
Key Laboratory for Feed Biotechnology of the Ministry of Agriculture, Feed Research Institute, Chinese Academy of Agricultural Sciences, Beijing 100081, P. R. China

Wei Mao
Key Laboratory for Feed Biotechnology of the Ministry of Agriculture, Feed Research Institute, Chinese Academy of Agricultural Sciences, Beijing 100081, P. R. China

Pengfei Ren
Key Laboratory for Feed Biotechnology of the Ministry of Agriculture, Feed Research Institute, Chinese Academy of Agricultural Sciences, Beijing 100081, P. R. China

Toshihiro Marubashi
Calpis Co., Ltd, Tokyo, Japan

Einar Ringø
Norwegian College of Fishery Science, Faculty of Biosciences, Fisheries and Economics, University of Tromsø, 9037 Tromsø, Norway

Loh Teng-Hern Tan
Biomedical Research Laboratory, Jeffrey Cheah School of Medicine and Health Sciences, Monash University Malaysia, Bandar Sunway, Malaysia

Kok-Gan Chan
Division of Genetics and Molecular Biology, Institute of Biological Sciences, Faculty of Science, University of Malaya, Kuala Lumpur, Malaysia

Learn-Han Lee
Biomedical Research Laboratory, Jeffrey Cheah School of Medicine and Health Sciences, Monash University Malaysia, Bandar Sunway, Malaysia

Bey-Hing Goh
Biomedical Research Laboratory, Jeffrey Cheah School of Medicine and Health Sciences, Monash University Malaysia, Bandar Sunway, Malaysia

Paola Navarrete
Biotechnology Laboratory, Institute of Nutrition and Food Technology (INTA), University of Chile, Santiago, Chile

Dariel Tovar-Ramírez
Physiology and Functional Genomics Laboratory, The Center for Biological Research of NW (CIBNOR), La Paz, B.C.S., México

ArupTewary
Aquaculture Research Unit, Department of Zoology, Vidyasagar University, Midnapore- 721 102, West Bengal, India

Bidhan C Patra
Aquaculture Research Unit, Department of Zoology, Vidyasagar University, Midnapore- 721 102, West Bengal, India

Thomas L. Welker
Agricultural Research Service, United States Department of Agriculture, Hagerman Fish Culture Station, 3059F National Fish Hatchery Road, Hagerman, ID, USA

Chhorn Lim
Agricultural Research Service, United States Department of Agriculture, Aquatic Animal Health Research Unit, 990 Wire Road, Auburn, AL 36832, USA

Mohamed S Hassaan
Fish Nutrition Research Laboratory, National Institute of Oceanography and Fisheries (NIOF), Egypt

Magdy A Soltan
Animal Production Department, Faculty of Agriculture, Benha University, Egypt

Bestha Lakshmi
Department of Virology, Sri Venkateswara University, Tirupati 517502, India

Buddolla Viswanath
Department of Virology, Sri Venkateswara University, Tirupati 517502, India

D. V. R. Sai Gopal
Department of Virology, Sri Venkateswara University, Tirupati 517502, India

Ponpisit Utiswannakul
Department of Microbiology, Bangkok 10330, Thailand

Siripen Sangchai
Department of Microbiology, Bangkok 10330, Thailand

Sirirat Rengpipat
Department of Microbiology, Bangkok 10330, Thailand
The Center of Excellence for Marine Biotechnology (CEMB), Faculty of Science, Chulalongkorn University, 254 Phayathai Road, Bangkok 10330, Thailand

Meeran Mohideen
Centre for Aquaculture Research and Extension, St. Xavier's (Autonomous) College, Palayamkottai, 627002, Tamil Nadu, India
Institute for Research in Molecular Medicine, University Sains Malaysia, Pulau Penang, 11800, Malaysia

Haniffa MA
Centre for Aquaculture Research and Extension, St. Xavier's (Autonomous) College, Palayamkottai, 627002, Tamil Nadu, India

Ambas I
Sustainable Aquatic Resources and Biotechnology, Department of Environment and Agriculture, Curtin University, Bentley 6102, Australia
Faculty of Marine Science and Fishery, Department of Fishery, Hasanuddin University, Makassar, 90225, Indonesia

Fotedar R
Sustainable Aquatic Resources and Biotechnology, Department of Environment and Agriculture, Curtin University, Bentley 6102, Australia

Buller N
Department of Agriculture and Food Western Australia, Animal Health Laboratories, South Perth, WA, Australia

Fengxing Xie
Tianjin Research Center of Agricultural Biotechnology, Tianjin 300192, People's Republic of China

Taicheng Zhu
Institute of Microbiology, Chinese Academy of Sciences, Beijing 100101, People's Republic of China

Fengfeng Zhang
Tianjin Research Center of Agricultural Biotechnology, Tianjin 300192, People's Republic of China

Ke Zhou
Tianjin Research Center of Agricultural Biotechnology, Tianjin 300192, People's Republic of China

Yujie Zhao
Tianjin Research Center of Agricultural Biotechnology, Tianjin 300192, People's Republic of China

Zhenghua Li
Tianjin Institute of Industrial Biotechnology, Chinese Academy of Sciences, 32XiQiDao, Tianjin Airport Economic Park, Tianjin 300308, People's Republic of China

Rebeca Cerezuela
Fish Innate Immune System Group, Department of Cell Biology and Histology, Faculty of Biology, University of Murcia. 30100 Murcia, Spain

Jose Meseguer
Fish Innate Immune System Group, Department of Cell Biology and Histology, Faculty of Biology, University of Murcia. 30100 Murcia, Spain

Ma Angeles Esteban
Fish Innate Immune System Group, Department of Cell Biology and Histology, Faculty of Biology, University of Murcia. 30100 Murcia, Spain

Estefanía Muñoz-Atienza
Grupo de Seguridad y Calidad de los Alimentos por Bacterias Lácticas, Bacteriocinas y Probióticos (Grupo SEGABALBP) Departamento de Nutrición, Bromatología y Tecnología de los Alimentos, Facultad de Veterinaria, Universidad Complutense de Madrid, Madrid 28040, Spain

Beatriz Gómez-Sala
Grupo de Seguridad y Calidad de los Alimentos por Bacterias Lácticas, Bacteriocinas y Probióticos (Grupo SEGABALBP) Departamento de Nutrición, Bromatología y Tecnología de los Alimentos, Facultad de Veterinaria, Universidad Complutense de Madrid, Madrid 28040, Spain

Carlos Araújo
Grupo de Seguridad y Calidad de los Alimentos por Bacterias Lácticas, Bacteriocinas y Probióticos (Grupo SEGABALBP) Departamento de Nutrición, Bromatología y Tecnología de los Alimentos, Facultad de Veterinaria, Universidad Complutense de Madrid, Madrid 28040, Spain
Centro de Genética e Biotecnologia, Universidade de Trás-os-Montes e Alto Douro, Vila Real, Portugal

Cristina Campanero
Grupo de Seguridad y Calidad de los Alimentos por Bacterias Lácticas, Bacteriocinas y Probióticos (Grupo SEGABALBP) Departamento de Nutrición, Bromatología y Tecnología de los Alimentos, Facultad de Veterinaria, Universidad Complutense de Madrid, Madrid 28040, Spain

Rosa del Campo
Servicio de Microbiología, Hospital Universitario Ramón y Cajal, Madrid 28034, Spain

Pablo E Hernández
Grupo de Seguridad y Calidad de los Alimentos por Bacterias Lácticas, Bacteriocinas y Probióticos (Grupo SEGABALBP) Departamento de Nutrición, Bromatología y Tecnología de los Alimentos, Facultad de Veterinaria, Universidad Complutense de Madrid, Madrid 28040, Spain

Carmen Herranz
Grupo de Seguridad y Calidad de los Alimentos por Bacterias Lácticas, Bacteriocinas y Probióticos (Grupo SEGABALBP) Departamento de Nutrición, Bromatología y Tecnología de los Alimentos, Facultad de Veterinaria, Universidad Complutense de Madrid, Madrid 28040, Spain

Luis M Cintas
Grupo de Seguridad y Calidad de los Alimentos por Bacterias Lácticas, Bacteriocinas y Probióticos (Grupo SEGABALBP) Departamento de Nutrición, Bromatología y Tecnología de los Alimentos, Facultad de Veterinaria, Universidad Complutense de Madrid, Madrid 28040, Spain

Einar Ringø
Norwegian College of Fishery Science, Faculty of Biosciences, Fisheries and Economics, University of Tromsø, Tromsø, Norway

Rolf Erik Olsen
Institute of Marine Research, Matre Research Station, Matredal, Norway

Jose L. Gonzalez Vecino
EWOS Innovation AS, Dirdal Norway

Simon Wadsworth
EWOS Innovation AS, Dirdal Norway

Seong Kyu Song
School of BioScience, Handong University, Pohang, South Korea

Preface

Aquaculture is one of the fastest developing growth sectors in the world. Disease outbreaks are constraint to aquaculture production thereby affects both economic development of the country and socio-economic status of the local people in many countries. Disease control in aquaculture industry has been achieved by different methods using traditional ways, synthetic chemicals and antibiotics. However, the use of such expensive chemotherapeutants for controlling diseases has been widely criticized for their negative impacts like accumulation of residues, development of drug resistance, and reduced consumer preference for aqua products treated with antibiotics and traditional methods are ineffective against controlling new diseases in large aquaculture systems. Therefore, alternative methods need to be developed to maintain a healthy environment in the aquaculture systems. Use of probiotics is one of such method that is gaining importance. Probiotics are defined as live cells or a substrate that provides benefits through stimulation of growth, improved digestion, improved immune response and ingested with the aim of promoting good health. Probiotics have been widely used in livestock rearing, have also recently been applied to aquaculture to elevate the production. Probiotics can also improve water quality and pond management. This article describes some of the applications of probiotics in aquaculture.

Editor
Preethi Kartan

INTRODUCTION

Probiotics are microorganisms that offer some form of health benefit to the host they can be found in various different foods. Probiotics are believed to play very important roles in regulating proper intestinal function and digestion by balancing intestinal microflora.

Probiotics are well known and routinely used additives in the main livestock species. They claim to improve gut health by stabilizing gut flora being their effect reflected in a better overall health status, welfare and performance of the animals.

In aquaculture, probiotics are administered by feed and/or as a water additive. The supplementation of probiotics through feed is a better method to ensuring the efficiency of the probiotic bacteria in the GI tract of fish. However, their use in fish feed production is still scarce.

Different factors might be behind the lack of this type of products for aquaculture: (i) The digestive system of fishes differs between cultured species, and gut microbiota may vary considerably among individuals of one specie, under the same rearing conditions;(ii) Probiotics used in monogastric and ruminants available nowadays are mainly based on bacteria or yeast that need enough warm temperature to develop in the animals' gut. It can be difficult to reach taking into account that fish are poikiloterms and in some specific productions water temperature is extremely low; (iii) Up today, it is not well known if the microorganisms from probiotics can develop, multiply as well as modify fish gut flora in these environmental and gut conditions and, (iv) Feed fish processing is extremely hard in terms of temperature and pressure so, how to apply this alive microorganism to fish pellets is still being studied. Their inclusion by coating after pelleting can be the solution, although the stability of this microorganism in this oily solutions as well as once reach water in tanks or sea need still to be demonstrated.

Despite all this, the application of probiotics in aquaculture practices has already gained momentum, and nowadays, numerous microorganisms, both from indigenous and exogenous sources, are used as probiotics. The commonly used probiotics in fish culture practices belong to Saccharomyces, Clostridium, Bacillus, Enteroccus, Lactobacillus, Shewanella, Leuconostoc, Lactococcus, Carnobacterium, Aeromonas and several other genera.

In the European Union, there is still only one probiotic based on the lactic acid bacteria strain Pediococcus acidilactici that has been agreed as aquafeed supplement, after demonstration of beneficial effects on salmonids and shrimps. In some other countries, especially in Asia, many other probiotics are available for aquaculture.

The expected benefits may vary. For example, increased survival and production of channel catfish (Ictalurus punctatus) by Bacillus species, improved growth and immunity of Nile tilapia (Oreochromis niloticus) by Bacillus subtilis and Rhodopseudomonas, and protection of rainbow trout (Oncorhynchus mykiss) against Vibrio anguillarum by Pseudomonas fluorescens have been demonstrated. More generally, probiotics have shown that they can help to increase fish growth by enhancing the feed conversion efficiency, as well as confer protection against harmful bacteria by competitive exclusion, production of organic acids (formic acid, acetic acid, lactic acid), hydrogen peroxide and several other compounds. They can also effectively trigger the piscine immune system. For all this reasons, their use will definitively increase in the near future.

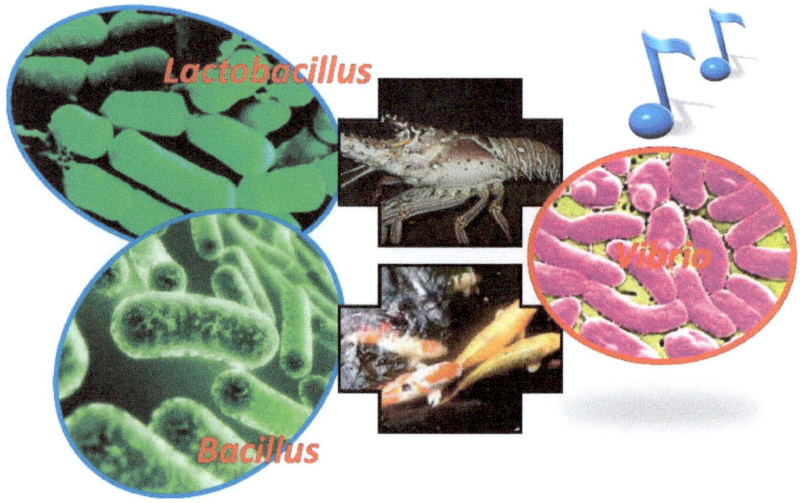

Prebiotics and Synbiotics

The prebiotic comes before and helps the probiotic, and then the two can combine to have a synergistic effect, known as synbiotics. A prebiotic is actually a nondigestible carbohydrate that acts as food for the probiotics and bacteria in your gut. The definition of the effect of prebiotics is the selective stimulation of growth and/or activity of one or a limited number of microbial genus/species in the gut microbiota that confer(s) health benefits to the host. The health benefits have been suggested to include acting as a remedy for gastrointestinal (GI) complications such as enteritis, constipation, and irritable bowel disease; prevention and treatment of various cancers; decreasing allergic inflammation; treatment of nonalcoholic fatty liver disease (NAFLD), and fighting immune deficiency diseases. There has also been research showing that the dietary intake of particular food products with a prebiotic effect has been shown, especially in adolescents, but also tentatively in postmenopausal women, to increase calcium absorption as well as bone calcium accretion and bone mineral density. The benefits for obesity and type 2 diabetes are growing as recent data, both from experimental models and from human studies, have shown particular food products with prebiotics have influences on energy homeostasis, satiety regulation, and body weight gain.

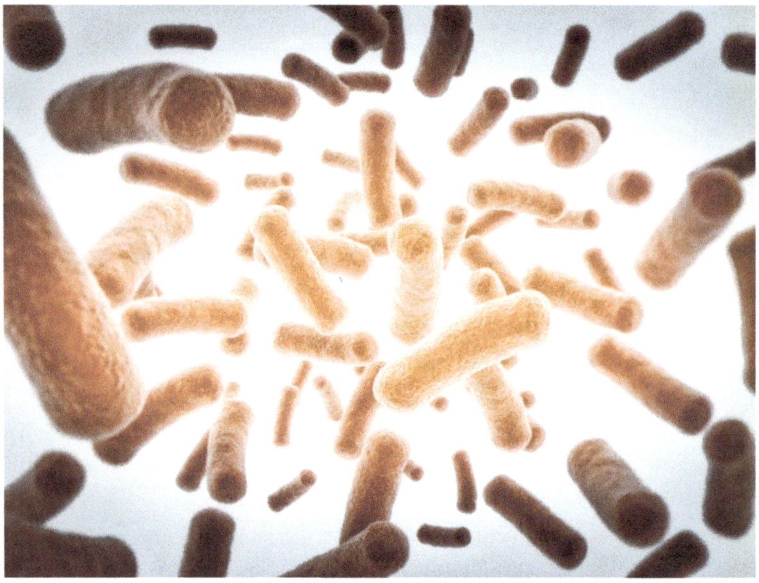

Most of the prebiotics identified are oligosaccharides. They are resistant to the human digestive enzymes that work on all other carbohydrates. This means that they pass through the upper GI system without being digested. They then get fermented in the lower colon and produce short-chain fatty acids that will then nourish the beneficial microbiota that live there. Oligosaccharides can be synthesized or obtained from natural sources. These sources include asparagus, artichoke, bamboo shoots, banana, barley, chicory, leeks, garlic, honey, lentils, milk, mustards, onion, rye, soybean, sugar beet, sugarcane juice, tomato, wheat, and yacón. The health benefits from these oligosaccharides are a topic of ongoing research.

Microbes and their role in our Health

Microbes are tiny organisms (bacteria, viruses, parasites, or fungi) -- so tiny that millions can fit into the eye of a needle -- that are so powerful that an imbalance in the body is related to numerous diseases. These microorganisms can be found in almost every part of the human body, living on the skin, in the nose, and in the gut. There are trillions of these microorganisms in our bodies. They outnumber human cells by 10 to one, but due to their small size, they only make up 1%-3% of a body's total mass.

The National Institute of Health (NIH) is sponsoring the Human Microbiome Project (HMP), which is developing research resources to enable the study of the microbial communities that live in and on our bodies and the roles they play in human health and disease. The NIH has funded many more medical studies using HMP data and techniques, including the role of the gut microbiome in Crohn's disease, ulcerative colitis, and esophageal cancer; skin microbiome in psoriasis, atopic dermatitis, and immunodeficiency; urogenital microbiome in reproductive and sexual history and circumcision; and a number of childhood disorders, including pediatric abdominal pain, intestinal inflammation, and necrotizing enterocolitis, a severe condition in premature infants in which the intestine tissue dies due to lack of oxygen.

Health Benefits of Probiotics

Probiotics may seem new to the food and supplement industry, but they have been with us from our first breath. During a delivery through the birth canal, a newborn picks up the bacteria Bacteroides, Bifidobacterium, Lactobacillus, and Escherichia coli from his/her mother. These good bacteria are not transmitted when a Cesarean section is performed and have been shown to be the reason why some infants born by C-section have allergies, less than optimal immune systems, and lower levels of gut microflora.

What exactly do probiotics do? They are believed to protect us in two ways. The first is the role that they play in our digestive tract. We know that our digestive tract needs a healthy balance between the good and bad bacteria, so what gets in the way of this? It looks like our lifestyle is both the problem and the solution. Poor food choices, emotional stress, lack of sleep, antibiotic overuse, other drugs, and environmental influences can all shift the balance in favor of the bad bacteria.

When the digestive tract is healthy, it filters out and eliminates things that can damage it, such as harmful bacteria, toxins, chemicals, and other waste products. The healthy balance of bacteria assists with the regulation of gastrointestinal motility and maintenance of gut barrier function. Research has shown some benefits for the use of probiotics for infectious diarrhea, antibiotic-associated diarrhea, gut transit, IBS, abdominal pain and bloating, ulcerative colitis, Helicobacter pylori infection, nonalcoholic fatty liver disease (NAFLD), and necrotizing enterocolitis.

The other way that probiotics help is the impact that they have on our immune system. Some believe that this role is the most important. Our immune system is our protection against germs. When it doesn't function properly, we can suffer from allergic reactions, autoimmune disorders (for example, ulcerative colitis, Crohn's disease, and rheumatoid arthritis), and infections (for example, infectious diarrhea, H. pylori, skin infections, and vaginal infections). By maintaining the correct balance from birth, the hope would be to prevent these ailments. Our immune system can benefit anytime that balanced is restored, so it's never too late.

Research into the benefits of probiotics has been branching out, and new areas are emerging. Preliminary research has linked them to supporting the health of the reproductive tract, oral cavity, lungs, skin and gut-brain axis, and the prevention and treatment of obesity and type 1 and type 2 diabetes.

Types of Probiotics

Probiotic foods, beverages, and supplements products contain bacteria and/or yeasts. Up until the 1960s, the only gut microflora that they were able to identify were clostridia, lactobacilli, enterococci, and E. coli. Since then, innovative techniques have discovered many more bacteria.

There are several different kinds of probiotics, and their health benefits are determined by the job that they do in the gut. They must be identified by their genus, species, and strain level. Here is a list of probiotics and their possible health benefits.

Lactobacillus

There are more than 50 species of lactobacilli. They are naturally found in the digestive, urinary, and genital systems. Foods that are fermented, like yogurt, and dietary supplements also contain these bacteria. Lactobacillus has been used for treating and preventing a wide variety of diseases and conditions.

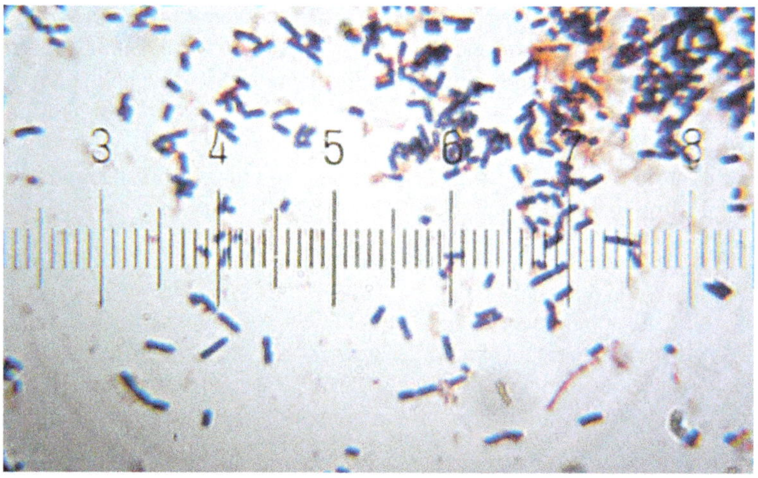

Some of the lactobacilli found in foods and supplements are Lactobacillus acidophilus, L. acidophilus DDS-1, Lactobacillus bulgaricus, Lactobacillus rhamnosus GG, Lactobacillus plantarium, Lactobacillus reuteri, Lactobacillus salivarius, Lactobacillus casei, Lactobacillus johnsonii, and Lactobacillus gasseri.

Studies have shown some benefits linked to Lactobacillus and treating and/or preventing yeast infections, bacterial vaginosis, urinary tract infection, irritable bowel syndrome, antibiotic-related diarrhea, traveler's diarrhea, diarrhea resulting from Clostridium difficile, treating lactose intolerance, skin disorders (fever blisters, eczema, acne, and canker sores), and prevention of respiratory infections.

Bifidobacteria

There are approximately 30 species of bifidobacteria. They make up most of the healthy bacteria in the colon. They appear in the intestinal tract within days of birth, especially in breastfed infants and are thought to be the best marker of intestinal health.

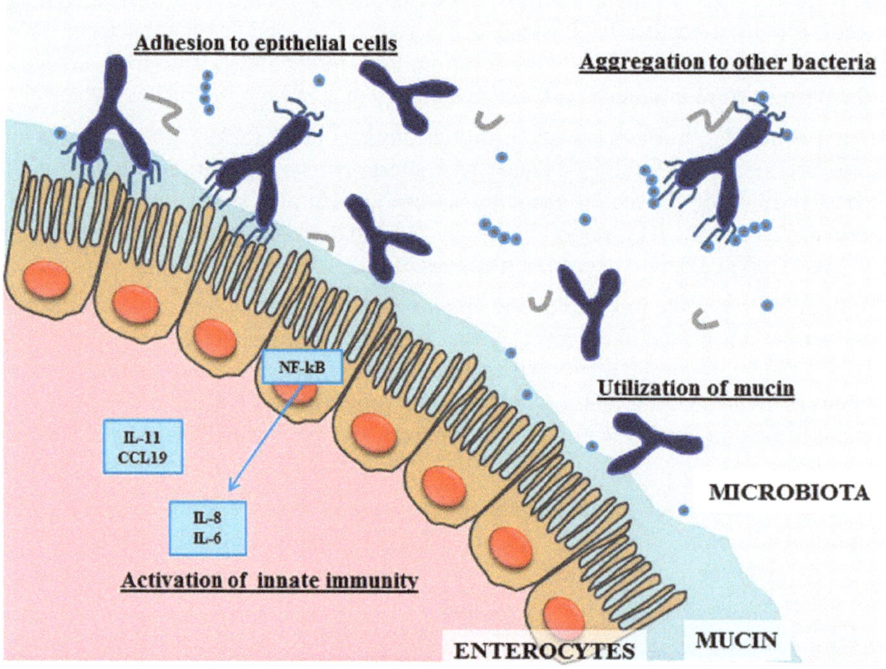

Some of the bifidobacteria used as probiotics are Bifidobacterium bifidum, Bifidobacterium lactis, Bifidobacterium longum, Bifidobacterium breve, Bifidobacterium infantis, Bifidobacterium thermophilum, and Bifidobacterium pseudolongum.

Studies have shown that bifidobacteria can help with improving blood lipids and glucose tolerance. Bifidobacteria have been shown to effectively alleviate IBS and significantly improve IBS symptoms like pain/discomfort, distension/bloating, urgency, and digestive disorders.

Saccharomyces boulardii

This is also known as S. boulardii and is the only yeast probiotic. Some studies have shown that it is effective in preventing and treating diarrhea associated with the use of antibiotics and traveler's diarrhea. It has also been reported to prevent the reoccurrence of C. difficile, to treat acne, and to reduce side effects of treatment for H. pylori.

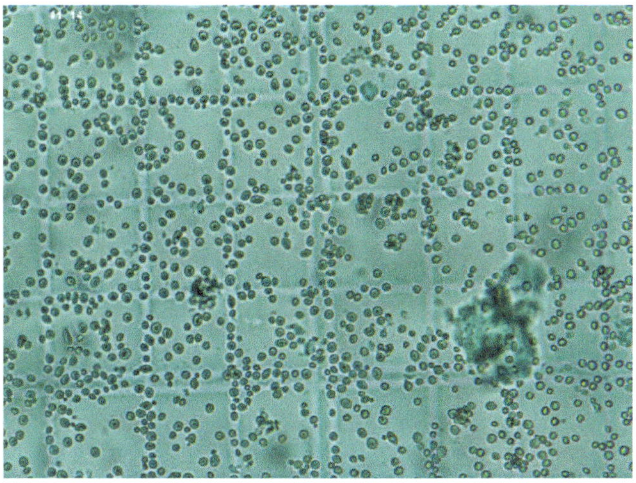

Streptococcus thermophilus

This produces large quantities of the enzyme lactase, making it effective, according to some reports, in the prevention of lactose intolerance.

Enterococcus faecium

This is normally found in the intestinal tract of humans and animals.

Leuconostoc

This has been used extensively in food processing throughout human history, and ingestion of foods containing live bacteria, dead bacteria, and metabolites of these microorganisms has taken place for a long time.

Foods Contain Probiotics

The discovery of the benefits of probiotics began with sour milk. Today we have many other options to get various bacteria from our foods, although it's not as simple as just adding them to the food. For there to be health benefits, the microorganism has to be able to survive the passage through the gastrointestinal tract, survive the food manufacturing process, and grow and survive during the ripening or storage period. Also, the bacteria must not negatively affect product quality and be included on the Generally Recognized as Safe (GRAS) list.

Most bacteria are included through the fermentation process. Fermentation helps extend the shelf life of perishable foods. It is a slow decomposition process of organic substances induced by microorganisms or enzymes that essentially convert carbohydrates to alcohols or organic acids. The lactic acid supplies the bacteria that then add the health benefits to the food. You can purchase foods that are fermented or ferment them yourself.

Kefir: This could be the most ideal probiotic dairy product because it contains both bacteria and yeast working together to provide the numerous health benefits. In a recent eight-week study, people with diabetes were given kefir milk containing Lactobacillus casei, Lactobacillus acidophilus, and bifidobacteria vs. conventional fermented milk. The hemoglobin A1C levels were significantly lower in the group consuming the kefir.

Kimchi: This fermented vegetable is made from Chinese cabbage (beachu), radish, green onion, red pepper powder, garlic, ginger, and fermented seafood (jeotgal). Many bacteria have been found to be present and can include any of the following: Leuconostoc mesenteroides and Lactobacillus plantarum, L. mesenteroides, L. citreum, L. gasicomitatum, L. brevis, L. curvatus, L. plantarum, L. sakei, L. lactis, Pediococcus pentosaceus, Weissella confusa, and W. koreensis. A recent review linked the health benefits of kimchi to anticancer, antiobesity, anticonstipation, colon health promotion, cholesterol reduction, antioxidative and antiaging properties, brain health promotion, immune promotion, and skin health promotion.

Yogurt: It can contain Streptococcus thermophilus, Lactobacillus bulgaricus, L. acidophilus, and Bifidobacterium bifidum. Research has shown links with yogurt to have positive effects on the gut microbiota and is associated with a reduced risk for gastrointestinal disease and improvement of lactose intolerance (especially among children), type 2 diabetes, cardiovascular disease, allergies and respiratory diseases, as well as improved dental and bone health.

Sweet acidophilus milk: L. acidophilus and L. acidophilus plus bifidobacteria are added to make this milk.

Other foods without substantial research: miso (fermented soybean paste); tempeh; sauerkraut; aged soft cheese; sourdough bread; sour pickles; gundruk (nonsalted, fermented, and acidic vegetable product); sinki (indigenous fermented radish tap root food); khalpi (fermented cucumber); inziangsang (traditional fermented leafy vegetable product prepared from mustard leaves); soidonis (widespread fermented product prepared from the tip of mature bamboo shoots)

Side Effects and Risks of Probiotics

Supplements play an important role when the diet is not adequate to supply our needs. In the case of probiotics, one's diet is the ideal source for probiotics. These are live bacteria and need to be carefully monitored, stored, and combined for the health benefits that one would be taking them for. At this time, supplements are not monitored in the U.S. the way that food or medication is. They fall under the Dietary Supplement Health and Education Act of 1994 (DSHEA). This requires that the dietary supplement or dietary ingredient manufacturer be responsible for ensuring that a dietary supplement or ingredient is safe before it is marketed. The only time that the U.S. Food and Drug Administration (FDA) may get involved is if action is needed to be taken against a manufacturer after the supplement is marketed and then found to be unsafe. This means that as much as we may know about probiotics, we can't be certain of the safety or content of the supplements available to us.

There is one Voluntary Certification Program by which a supplement manufacturer can choose to be evaluated. ConsumerLab.com (CL) is the leading provider of independent test results and information to help consumers and health-care professionals identify the best quality health and nutrition products. Products that have passed their testing for identity, strength, purity, and disintegration can print the CL Seal of Approval on their product. This is one step toward being confident that one is getting the amount and type of probiotic promised by the manufacturer.

Caution needs to be taken by everyone who chooses to take these supplements, but this is especially true for children, pregnant women, elderly people, and people with compromised immune systems. For people with compromised immune systems due to disease or treatment for a disease (such as cancer chemotherapy), taking probiotics may actually increase one's chances of getting sick. It has been shown that the use of various probiotics for immunocompromised patients or patients with a leaky gut has resulted in infections and sepsis (infection of the bloodstream). One case of bacteremia (bacteria in the bloodstream) was recently found when someone with active severe inflammatory bowel diseases with mucosal disruption was given Lactobacillus GG. Always speak with a doctor before taking any supplement under these circumstances.

CHAPTER 1

PROBIOTICS IN AQUACULTURE

Monique Mancuso

Keywords: Aquaculture, Marine fish, Probiotics

INTRODUCTION

In 2010, global production of farmed food fish was 59.9 million tones [1]. Over the years, aquaculture sector has undergone a sea change in order to meet the increasing demand. The production is maximized through intensification with addition of commercial diets, growth promoters, antibiotics, and several other additives; all these practices create stressful conditions that cause problems related to diseases and deterioration of environmental conditions often occur and result in serious economic losses [2].

The prevention and the control of diseases have led during recent decades to a substantial increase in the use of veterinary medicines. Probiotics are originally defined as the organisms and substances which contribute to the intestinal microbial balance [3]. The term probiotic means promoter of life infact help, in a natural way, to improve the overall health status of the host organism [4]. Probiotics open a new era in health management strategy for fish/shellfish gaining increasing scientific and commercial interest. The obvious potential advantages of such approaches are that they promote specific microbes in the intestine for restoring the intestinal microbial balance and exerting numerous beneficial effects in host Oliva-Teles 2012 [5].

Probiotics are live microorganisms which when administered in adequate amounts confer a health benefits on host. They are used for aquaculture practices include: Lactobacillus, Lactococcus, Leuconostoc, Enterococcus, Carnobacterium, Shewanella, Bacillus, Aeromonas, Vibrio, Enterobacter, Pseudomonas, Clostridium, and Saccharomyces species. The involvement of probiotics in nutrition, disease resistance and other beneficial activities in fish has proven [6]. Among the numerous health benefits attributed to probiotics, modulation of immune system is one of the most commonly purported benefits of the probiotics and their potency to stimulate the systemic and local immunity under in vitro and in vivo conditions is noteworthy. For this reason there is a close link between the welfare and the use of probiotics. Recently the probiotics are used as alternative measures to control the fish diseases; infact they inhibit pathogenic microorganisms and have been used therapeutically to treat a variety of gastrointestinal and even systemic disorders. Probiotics transiently colonize the bowel and except when used to treat an acute disorder, must be regularly consumed to maintain benefit. Use of microbial probiotics to promote health maintenance and disease prevention and control is now widely accepted as the new ecofriendly alternative measures for sustainable aquaculture [7]. Probiotics are also known to play an important role in developing innate immunity among the fishes, and hence help them to fight against any pathogenic bacterias as well as against environmental stressors [8].

CONCLUSIONS

The current global food crisis and increasing production costs has put pressure on governments and the international community to ensure sufficient food supply for a growing population. Thus, aquaculture is presented as a way to meet the growing demand for fresh water food. Improving aquaculture practices will become an important alternative to overexploitation and modification of aquatic ecosystems caused by capture fisheries. The use of probiotics can potentiate the benefits of this activity because it offers viable alternatives for the generation of a higher-quality livestock product in terms of size, production time, and health. It is necessary to conduct studies relating to probiotics resistance to antibiotics,

Citation: Mancuso M (2013) Probiotics in Aquaculture. J Fisheries Livest Prod 1:e107. doi: 10.4172/2332-2608.1000e107.

and the chances of transmission of genetic elements to other microorganisms in the fish Gastro Intestinal Tract (GIT), and thus to humans when consuming the aquaculture product. And on the other hand, there is a need to strengthen studies of microbial ecology in aquaculture systems, correlating microbial communities with animal growth and its relationship to the water quality.

REFERENCES

1. FAO (2012) The State of World Fisheries and Aquaculture.
2. Mancuso M (2012) Photobacteriosis exchange between wild and farmed fish in the Mediterranean area- Editorial - Journal of aquaculture research and development (JARD) 3: 3-6.
3. Parker RB, Probiotics, the other half of the antibiotic story, Anim. Nutr. Health 29: 4-8
4. Gismondo MR, Drago L, Lombardi A (1999) Review of probiotics available to modify gastrointestinal flora. Int J Antimicrob Agents 12: 287-292
5. Oliva-Teles A (2012) Nutrition and health of acquaculture fish. Journal of Fish Dis. 35: 83-108
6. Nayak SK (2010) Probiotics and immunity: a fish perspective. Fish Shellfish Immunol 29: 2-14.
7. Ram C Sihag, Parvati Sharma (2012) Probiotics: The New Ecofriendly Alternative Measures of Disease Control for Sustainable Aquaculture. Journal of Fisheries and Aquatic Science 7:72-103.
8. Mohapatra S, Chakraborty T, Kumar V, Deboeck G, Mohanta KN (2013) Aquaculture and stress management: a review of probiotic intervention. J Anim Physiol Anim Nutr (Berl) 97: 405-430.

CHAPTER 2

USE OF PROBIOTICS IN AQUACULTURE

Patricia Martınez Cruz Ana L. Iba´nez, Oscar A. Monroy Hermosillo and Hugo C. Ramırez Saad

INTRODUCTION

Aquaculture is the farming of aquatic organisms by intervention in the rearing process to enhance production and private ownership of the stock being cultivated. Compared to fishing, this activity allows a selective increase in the production of species used for human consumption, industry or sport fishing. Due to overfishing of wild populations, aquaculture has become an economic activity of great importance around the world. Aquaculture's contribution to world food production, raw materials for industrial and pharmaceutical use, and aquatic organisms for stocking or ornamental trade has increased dramatically in recent decades. The report World Aquaculture 2012 found that global production of fish from aquaculture grew more than 30 percent between 2006 and 2011, from 47.3 million tons to 63.6 million tons. It also forecasts that by 2012 more than 50 percent of the world's food fish consumption will come from aquaculture, so it is expected to overtaking capture fisheries as a source of edible fish. This growth rate is due to several factors: (1) many fisheries have reached their maximum sustainable exploitation, (2) consumer concerns about security and safety of their food, (3) the market demand for high-quality, healthy, low-calorie, and high-protein aquatic products, and (4) aquatic breeding makes only a minimum contribution to carbon dioxide emission [1, 2].

Aquaculture has a long history, originating at least in the year 475 B.C. in China [3], but became important in the late nineteen-forties, since the methods of aquaculture could be used to restock the waters as a complement to natural spawning. Nowadays, aquaculture is a lucrative industry [2, 4]. However, the intensification of aquaculture practices requires cultivation at high densities, which has caused significant damage to the environment due to discharges of concentrated organic wastes, that deplete dissolved oxygen in ponds, giving rise to toxic metabolites (such as hydrogen sulfide, methane, ammonia, and nitrites), that often are responsible for mortality. Additionally, aquaculture has appropriated of water bodies used for recreational purpose, and sometimes makes a water's waste because this natural resource is not reused in extensive aquaculture systems [5, 6]. Moreover, under these conditions of intensive production, aquatic species are subjected to high-stress conditions, increasing the incidence of diseases and causing a decrease in productivity [7].

Outbreaks of viral, bacterial, and fungal infections have caused devastating economic losses worldwide, that is, China reported disease-associated losses of $750 million in 1993, while India reported $210 million losses from 1995 to 1996. Added to this, significant stock mortality has been reported due to poor environmental conditions on farms, unbalanced nutrition, generation of toxins, and genetic factors [8]. In recent decades, prevention and control of animal diseases has focused on the use of chemical additives and veterinary medicines, especially antibiotics, which generate significant risks to public health by promoting the selection, propagation, and persistence of bacterial-resistant strains [9–11].

Resistance to antibiotics is acquired in two ways: chromosomal mutation or acquisition of plasmids. Chromosomal mutations are not transferred laterally to other bacteria but resistance plasmids can be transferred very rapidly, producing a high percentage of pathogenic bacteria that develop mediated plasmid-resistance in a short period of time [12]. This can be an important public health issue as well. For example, transfer of multidrug resistance occurred in Ecuador during the cholera epizootic happened between 1991 and 1994, which originated from shrimp farm workers. Although the original epidemic strain Vibrio cholerae 01 was

Citation: Patricia Martınez Cruz Ana L. Iba´nez, Oscar A. Monroy Hermosillo and Hugo C. Ramırez Saad, Use of Probiotics in Aquaculture, doi:10.5402/2012/916845.

susceptible to 12 antimicrobial agents tested, on the coast of Ecuador it acquired multidrug resistance due to transfer of resistance genes from other Vibrio species that were pathogenic to shrimp [13].

Probiotic is a relatively new term which is used to name microorganisms that are associated with the beneficial effects for the host. Kozasa made the first empirical application of probiotics in aquaculture [14], considering the benefits exerted by the use of probiotics on humans and poultry. He used spores of Bacillus toyoi as feed additive to increase the growth rate of yellow tail, Seriola quinqueradiata. In 1991, Porubcan [15, 16] documented the use of Bacillus spp, to test its ability to increase productivity of Penaeus monodon farming and to improve water quality by decreasing the concentrations of ammonia and nitrite. In order to avoid or reduce the use of certain antimicrobials, biological control was tested, described as the use of natural enemies to reduce the damage caused by harmful organisms. Strictly speaking, a probiotic should not be classified as a biological control agent because it is not necessarily a natural enemy of the pathogen [17]. However, certain probiotics have the ability to inhibit the growth of pathogenic bacteria. Moriarty determined the ability of Bacillus spp. to decrease the proportion of Vibrio spp. in shrimp ponds, especially in sediments [18]. Further studies have stressed probiotics ability to stimulate appetite, improve absorption of nutrients, and strengthen the host immune system [19, 20]. The aim of this paper is to present an updated concept of probiotics, current uses in aquaculture, commercial presentations, and a final approach to prospect applications in this area.

DEFINITION OF PROBIOTIC

The term "probiotic" comes from Greek pro and bios meaning "prolife" [56], having different meanings over the years. In 1905, Dr. Elie Metchnikoff was the first to describe the positive role played by some bacteria among farmers who consumed pathogen-containing milk and that "reliance on gut microbes for food makes it possible to take steps to change the flora of our bodies and to replace harmful microbes by beneficial microbes" [57]. However, the term probiotic was introduced until 1965 by Lilly and Stillwell [58] as a modification of the original word "probiotika." It was used to describe substances produced by a microorganism that prolong the logarithmic growth phase in other species. It was described as an agent which has the opposite function of antibiotics. Later, Sperti [59] modified the concept of "tissue extracts that stimulate microbial growth."

The first use of the term to describe a microbial feed/food supplement was by Parker in 1974 [60]. He defined it as "organisms and substances that contribute to intestinal microbial balance." Fuller [61] expanded the definition to "live microbial food supplement that benefits the host (human or animal) by improving the microbial balance of the body" and said that it would be effective in a range of extreme temperatures and salinity variations. Afterwards, it was suggested that probiotics were "monocultures or mixed cultures of microorganisms applied to animals or humans, that benefit the host by improving properties of indigenous microflora" [62]. In 1998, Guarner and Schaafsma assumed that probiotics are live microorganisms which, when consumed in adequate amounts, confer health benefits to the host [63]. Gatesoupe in 1999, defined them as "microbial cells administered in a certain way, which reaches the gastrointestinal tract and remain alive with the aim of improving health" [25]. In the same year, studies were carried out on the inhibition of pathogens using probiotics, this expanded the definition to "... live microbial supplement which benefits the host by improving its microbial balance" [34].

Knowledge of probiotics has increased, currently it is known that these microorganisms have an antimicrobial effect through modifying the intestinal microbiota, secreting antibacterial substances (bacteriocins and organic acids), competing with pathogens to prevent their adhesion to the intestine, competing for nutrients necessary for pathogen survival, and producing an antitoxin effect. Probiotics are also capable of modulating the immune system, regulating allergic response of the body, and reducing proliferation of cancer in mammals. Because of this, when provided at certain concentration and viability, probiotics favorably affect host health [64]. In fact, terms such as "friendly bacteria," "friendly," or "healthy" are commonly used to describe probiotics [19].

For many years, studies focused on microorganisms characteristic from intestinal microbiota, and the term "probiotic" was mainly restricted to gram-positive lactic acid bacteria [65], particularly representative of the genera Bifidobacterium, Lactobacillus, and Streptococcus [66]. In contrast to terrestrial animals, gastrointestinal microbiota of aquatic species is particularly dependent on the external environment due to the flow of water passing through the digestive tract. Thus, the majority of bacteria are transient in the intestine, due to constant intake of water and food, together with microorganisms present in them. Although in the gastrointestinal tract (GIT) of aquatic animals have been reported potentially pathogenic bacteria such as Salmonella, Listeria, and Escherichia coli, probiotic bacteria and other microorganisms have also been identified. These include gram-positive bacteria such as Bacillus, Carnobacterium, Enterococcus, and several species of Lactobacillus; gramnegative, facultative anaerobic such as Vibrio and Pseudomonas, as well as certain fungi, yeasts, and algae of the genera Debaryomyces, Saccharomyces, and Tetraselmis, respectively [20, 67, 68]. Due to the increasing interest of probiotics in aquaculture, Moriarty [18] proposed extending the definition of these to "living microbial additives that benefit the health of hydrobionts and therefore increase productivity."

A more general and common concept of probiotic is "one or more microorganisms with beneficial effects for the host, able to persist in the digestive tract because of its tolerance to acid and bile salts" [20]. Although the use of probiotics in aquaculture is relatively recent, interest in them has increased due to their potential in disease control [19]; however, gradually other applications have been proposed (summarized in Table 1), that will be addressed in detail later.

COMMERCIAL PREPARATIONS

The interest in probiotics as an environmentally friendly alternative is increasing and its application is both empirical and scientific. According to Soccol et al. [69], the global market for probiotic ingredients, supplements and foods, reached US $15,900 million in 2008 and is projected to increase to US $19,600 million in 2013, representing an annual growth rate of 4.3%. At present, there are several commercial preparations of probiotics that contain one or more live microorganisms, which have been introduced to improve the cultivation of aquatic organisms. Probiotics can be used as a food additive added directly to the culture tank or mixed with food.

Apart from laboratory preparation of bacteria, some commercially available products are now available. One of the first evaluations of commercial products focussed on a bacterial preparation called Biostart that is derived from Bacillus isolates. It was used during the production of cultured catfish studying the effect of inoculum concentration [22]. In 1998, Moriarty reported that the use of commercial probiotic strains of Bacillus spp. increased the quality and viability of pond-raised shrimp [18]. Meanwhile, Chang and Liu [31] evaluated the effect of Enterococcus faecium SF68 and Bacillus toyoi isolates present in Cernivet LBC and Toyocerin, respectively, to decrease the mortality of the European eel because of the edwardsielosis, ensuring greater efficiency with E. faecium SF68. It is relevant to note that E. faecium has long been known as a probiotic for humans, whereas B. toyoi has been used with terrestrial animals. Moreover, a B. subtilis strain combined with hydrolytic enzymes to produce Biogen, was used to supplement the feed of Oreochromis niloticus, obtaining significant increases in productivity [70].

The lactic acid-producing bacteria have been the focus of much interest. The human probiotic, Lactobacillus rhamnosus ATCC (American Type Culture Collection, Rockville, MD, USA), was used in rainbow trout for 51 days to reduce mortality by Aeromonas salmonicida, the causative agent of the fish disease “furunculosis” (one of the major fish diseases in many parts of world). Mortality was reduced from 52.6 to 18.9% when 109 cells g−1 were administered with feed, when probiotic dose was increased to 1012 cells g−1 of feed the mortality reached 46.3% [32]. Apparently, increasing dosage does not necessarily improve protection. Abasali and Mohamad [55] increased the gonadosomatic index and the production of fingerlings in females of reproductive age, using mixed cultures consisting of L. acidophilus, L. casei, E. faecium, and B. thermophilum (Primalac).

Studies with Penaeus vannamei showed that using mixed cultures of probiotics increases survival, feed conversion, and the final production of farmed shrimp [71]. Meanwhile, Taoka et al. [72] used Alchem Poseidon and Alchem Korea CO and Wonju Korea CO, which have mixed cultures of bacteria (Bacillus subtilis, Lactobacillus acidophilus, and Clostridium butyricum) and yeast (Saccharomyces cerevisiae), enhanced nonspecific immune parameters of tilapia Oreochromis niloticus such as lysozyme activity, migration onf neutrophils, and plasma bactericidal activity, resulting in improvement of resistance to Edwardsiella tarda infection. Previous studies in humans and land animals using prebiotics (nondigestible ingredients of the diet that stimulates the growth of microorganisms) showed their ability to stimulate the activity of probiotic bacteria in the colon [73, 74]. Some commercial aquaculture products included prebiotics in their formulation, such as mannans, glucans, and yucca extract that further increase the beneficial effects of the product [75, 76].

Currently, commercial products are available in liquid or powder presentations, and various technologies have been developed for improvement. on the case of fermentation processes, the interest has been focused on optimizing the fermentation conditions to increase the viability and functionality of probiotics, improving performance [77]. Generally, the production is carried out in batch cultures due to the difficulty of industrial scale operation of continuous systems [69]. More recently, systems have been developed for immobilization of probiotics, especially using microencapsulation. Microbial cells at high density are encapsulated in a colloidal matrix using alginate, chitosan, carboxymethylcellulose, or pectin to physically and chemically protect the microorganisms. The methods commonly used for microencapsulation of probiotics are the emulsion, extrusion, spray drying, and adhesion to starch [78]. Focused on the application to aquaculture, Rosas et al. [79] have effectively encapsulated cells of Shewanella putrefaciens in calcium alginate, demonstrating the survival of encapsulated probiotic cells through the gastrointestinal tract of sole (Solea senegalensis). Encapsulation in alginate matrices protects bacteria from low pH and digestive enzymes; this protection helps to release the probiotic into the intestine without any significant damage [80]. Currently, the lyophilized commercial preparations have advantages for storage and transport. However, conditions for reconstitution of these preparations such as temperature, degree of hydration, and osmolarity of the solution are vital to ensure the viability of bacteria [81]. It is important to emphasize that these products must provide a health benefit to the host; for this, it is necessary that contained microorganisms have the ability to survive storage conditions, and after that in the digestive tract of aquatic species, remaining viable and stable, and finally improving production [20]. According to the opinion of the producers, these preparations are safe to use and effective in preserving the health of aquatic animals [19].

Table 1: Different applications of probiotics in aquaculture

Application	Identity of the probiotic	Applied to aquatic species	Reference
Growth promoter	*Bacillus* sp. S11	*Penaeus monodon*	[21]
	Bacillus sp.	Catfish	[22]
	Carnobacterium divergens	*Gadus morhua*	[23]
	Alteromonas CA2	*Crassostrea gigas*	[24]
	Lactobacillus helveticus	*Scophthalmus maximus*	[25]
	Lactobacillus lactis AR21	*Brachionus plicatilis*	[26]
	Streptococcus thermophilus	*Scophthalmus maximus*	[25]
	Streptomyces	*Xiphophorus helleri*	[27]
	L. casei	*Poeciliopsis gracilis*	[28]
	Bacillus NL 110, *Vibrio* NE 17	*Macrobrachium rosenbergii*	[29]
	Bacillus coagulans	*Cyprinus carpio* koi	[30]
Pathogen inhibition	*Bacillus* sp.	Penaeids	[18]
	Enterococcus faecium SF 68	*Anguilla anguilla*	[31]
	L. rhamnosus ATCC53103	*Oncorhynchus mykiss*	[32]
	Micrococcus luteus A1-6	*Oncorhynchus mykiss*	[33]
	Pseudomonas fluorescens	*Oncorhynchus mykiss*	[34]
	P. fluorescens AH2	*Oncorhynchus mykiss*	[35]
	*Pseudomonas*sp.	*Oncorhynchus mykiss*	[36]
	Roseobacter sp. BS. 107	Scallop larvae	[37]
	Saccharomyces cerevisiae, S. exiguous, Phaffia rhodozyma	*Litopenaeus vannamei*	[38]
	Vibrio alginolyticus	Salmonids	[39]
	V. fluvialis	*Oncorhynchus mykiss*	[33]
	Tetraselmis suecica	*Salmo salar*	[40]
	Carnobacterium sp. Hg4-03	*Hepialus gonggaensis* larvae	[41]
	Lactobacillus acidophilus	*Clarias gariepinus*	[42]
	Bacillus spp., *Enterococcus*sp.	*Farfantepenaeus brasiliensis*	[43]
	Lactococcus lactis	*Epinephelus coioides*	[44]
Nutrient digestibility	*L. helveticus*	*Scophthalmus maximus*	[25]
	Bacillus NL 110, *Vibrio* NE 17	*Macrobrachium rosenbergii*	[29]
	Carnobacterium sp. Hg4-03	*Hepialus gonggaensis* larvae	[41]
	Lactobacillus acidophilus	*Clarias gariepinus*	[45]
	Shewanella putrefaciens Pdp11	*Solea senegalensis*	[46]
Water quality	*Bacillus* sp. 48	*Penaeus monodon*	[19]
	Bacillus NL 110, *Vibrio* sp. NE 17	*Macrobrachium rosenbergii*	[29]
	Lactobacillus acidophilus	*Clarias gariepinus*	[45]
	B. coagulans SC8168	*Pennaeus vannamei*	[47]
	Bacillus sp., *Saccharomyces* sp.	*Penaeus monodon*	[48]
Stress tolerance	*Lactobacillus delbrueckii*	*Dicentrarchus labrax*	[49]
	Alteromonas sp.	*Sparus auratus*	[50]
	B. subtilis, L. acidophilus, S. cerevisiae	*Paralichthys olivaceus*	[51]
	L. casei	*Poecilopsis gracilis*	[28]
	Pediococcus acidilactici	*Litopenaeus stylirostris*	[52]
	Shewanella putrefaciens Pdp11	Makimaki	[46]
Reproduction improvement	*Bacillus subtilis*	*Poecilia reticulata, Xiphophorus maculatus*	[53]
	L. rhamnosus	*Danio rerio*	[54]
	L. acidophilus, L. casei, Enterococcus faecium, Bifidobacterium thermophilum	*Xiphophorus helleri*	[55]

APPLICATIONS OF PROBIOTICS IN AQUACULTURE

The need for sustainable aquaculture has promoted research into the use of probiotics on aquatic organisms. The initial interest was focused on their use as growth promoters and to improve the health of animals; however, new areas have been found, such as their effect on reproduction or stress tolerance, although this requires a more scientific development.

Growth Promoter

Probiotics have been used in aquaculture to increase the growth of cultivated species, in reality it is not known whether these products increase the appetite, or if, by their nature, improve digestibility. Some people are inclined to think that it could be both factors; furthermore, it would be important to determine whether probiotics actually taste good for aquaculture species [20]. According to Balcazar et al. [´ 13], probiotic microorganisms are able to colonize gastrointestinal tract when administered over a long period of time because they have a higher multiplication rate than the rate of expulsion, so as probiotics constantly added to fish cultures, they adhere to the intestinal mucosa of them, developing and exercising their multiple benefits. This also depends on factors such as hydrobionts species, body temperature, enzyme levels, genetic resistance, and water quality.

The effect of probiotics has been tested on phytoplankton (microalgae), which forms the basis of aquatic food chains, due to its nutrient-producing photosynthetic machinery that in most cases, higher organisms are unable to synthesize such is the case of polyunsaturated fatty acids and vitamins. Within groups of microalgae used in aquaculture are distinguished central diatoms as Chaetoceros spp., which have proven to be a good live food; however, production has limitations due to the complexity of their nutritional requirements [82]. Gomez et al. [´ 83] assessed the growth of Vibrio alginolyticus C7b probiotic in the presence of the microalgae Chaetoceros muelleri, proving that these organisms can be grown together to achieve high density and fed to shrimp.

Rotifers are indispensable as the first live feed for larvae of most cultured aquatic species, due to their small size they are more accessible to larvae, for example, the nauplii of brine shrimp, which is a very common live feed. Planas et al. [84] used lactic acid bacteria to increase the growth of the rotifer Brachionus plicatilis and obtained best results with the addition of Lactococcus casei ssp. casei, Pediococcus acidilactici, and Lactobacillus lactis ssp. lactis.

The use of probiotics as growth promoters of edible fishes has been reported. Diet of Nile tilapia (Oreochromis niloticus) was amended with a probiotic Streptococcus strain, increasing significantly the content of crude protein and crude lipid in the fish, also weight has increased from 0.154 g to 6.164 g in 9 weeks of culture [85]. Due to the commercial importance of this species, the effect of supplementing diet with probiotics produced an increase of 115.3% when commercial formulation was used at a concentration of 2% [70].

Examples of growth improvement of ornamental fishes include swordtail (Xiphophorus helleri, X. maculatus) and guppy, (Poecilia reticulate, P. sphenops), their feed was supplemented with Bacillus subtilis and Streptomyces, finding significative increases in growth and survival of Xiphophorus and Poecilia after 90 and 50 days of administration, respectively [27, 86].

Probiotics also have been tested successfully in shellfish culture. Macey and Coyne [87] isolated two yeasts and one bacterial strain (designated SS1, AY1, and SY9, resp.) from the digestive tract of abalone (Haliotis midae). A diet was formulated with a mixture of the three putative probiotics. Each probiont was added to the feed to achieve a final concentration of approximately 107 cells g−1 of dry feed. The growth rate of small (20 mm) and large (67 mm) abalone was improved by 8% and 34%, respectively, in eightmonth cultures. Furthermore, abalones supplemented with probiotics had a survival rate of 62% to the pathogenic bacterium Vibrio anguillarum compared to 25% survival of untreated animals.

Inhibition of Pathogens

Antibiotics were used for a long time in aquaculture to prevent diseases in the crop. However, this caused various problems such as the presence of antibiotic residues in animal tissues, the generation of bacterial resistance mechanisms, as well as an imbalance in the gastrointestinal microbiota of aquatic species, which affected their health [88]. In fact, the European Union has regulated the use of antibiotics in organisms for human consumption [89]. Today, consumers demand natural products, free of additives such as antibiotics; moreover, there is a tendency for preventing diseases rather than treating them. Thus, the use of probiotics is a viable alternative for the inhibition of pathogens and disease control in aquaculture species.

Probiotic microorganisms have the ability to release chemical substances with bactericidal or bacteriostatic effect on pathogenic bacteria that are in the intestine of the host, thus constituting a barrier against the proliferation of opportunistic pathogens. In general, the antibacterial effect is due to one or more of the following factors: production of antibiotics, bacteriocins, siderophores, enzymes (lysozymes, proteases) and/or hydrogen peroxide, as well as alteration of the intestinal pH due to the generation of organic acids [65].

Taoka et al. [51] showed that viable probiotics administered to tilapia Oreochromis niloticus, increased nonspecific immune response, determined by parameters such as lysozyme activity, neutrophile migration, and bactericidal activity, which improved the resistance of fish to infection by Edwardsiella tarda. In turn, Robertson et al. [90] isolated a strain of Carnobacterium sp. from salmon bowel and administered alive to rainbow trout and Atlantic salmon, demonstrating in vitro antagonism against known fish pathogens: Aeromonas hydrophila, A. salmonicida, Flavobacterium psychrophilum, Photobacterium damselae, and Vibrio species. There is also evidence on the effect of dead probiotic cultures consisting on a mixture of Vibrio fluvialis A3-47S, Aeromonas hydrophila A3-51, and Carnobacterium BA211, in the control of furunculosis in rainbow trout. For this specific case, the number of leukocytes was greater than with live cells, in fact, the data suggest that cellular immunity more than humoral factors was involved in the benefits of these preparations of inactivated bacterial cells [91]. In the case of shrimp, studies have focused on the evaluation of probiotics such as Bacillus cereus, Paenibacillus polymyxa, and Pseudomonas sp. PS-102 as biocontrol agents against pathogens of various Vibrio species [92, 93].

Probiotic strains isolated from the gastrointestinal tract of clownfish (Amphiprion percula) have been used to inactivate several pathogens such as Aeromonas hydrophila and Vibrio alginolyticus among others. It has been observed that probiotics in vivo generate a density such that allow the production of antimicrobial metabolites therefore, the bacteria isolated from adult clownfish have the potential to colonize the intestinal mucus and therefore can be used as prophylactic agent and/or therapeutic [94, 95]. Furthermore, it has been found that concentrations of 106 to 108 cells g−1 of probiotic promote the development of healthy microbiota in the gastrointestinal tract of ornamental fishes from the genera Poecilia and Xiphophorus, decreasing the amount of heterotrophic microorganisms [86].

Gomez et al. [´ 96] reported the use of Vibrio alginolyticus strains as probiotics to increase survival and growth of white shrimp (Litopenaeus vannamei), also by using probiotics in Ecuadorian shrimp hatcheries, production increased by 35%, while with the use of antimicrobials it decreased by 94%.

Improvement in Nutrient Digestion

A study has suggested that probiotics have a beneficial effect on the digestive processes of aquatic animals because probiotic strains synthesize extracellular enzymes such as proteases, amylases, and lipases as well as provide growth factors such as vitamins, fatty acids, and aminoacids [13]. Therefore, nutrients are absorbed more efficiently when the feed is supplemented with probiotics [70].

In this sense, probiotics have been used in edible fishes, as in the case of larvae of European bass (Dicentrarchus labrax). It has been reported that the probiotic yeast Debaryomyces hansenii HF1 has the ability to produce spermine and spermidine, two polyamines involved in the differentiation and maturation of the gastrointestinal tract in mammals. In addition this yeast secretes amylase and trypsin, enzymes that aid digestion in sea bass larvae [97]. Studies in juvenile common dentex Dentex dentex L. showed that when diet is supplemented with 0.5 g of Bacillus cereus strain E kg−1 of food, increased fish growth due to more efficient use of the food [98]. In the case of rainbow trout, similar results were obtained using B. subtilis, B. licheniformis, and Enterococcus faecium, when these probiotics were provided for 10 weeks along with the diet of fish [99]. In some trials, diet of European sea bass larvae (Dicentrarchus labrax) has supplemented with probiotic yeast (Saccharomyces cerevisiae strain X2180), assessing fish growth and activity and expression of antioxidative key enzymes (catalase, glutathione peroxidase, and superoxide dismutase), finding differences in enzyme activity and gene expression patterns between probiotic supplemented and nonsupplemented treatments, which was attributed to the presence of the yeast [100].

In white shrimp Litopenaeus vannamei Boone and Fenneropenaeus indicus, various strains of Bacillus have been used as probiotics to increase apparent digestibility of dry matter, crude protein, and phosphorus. Results showed higher sizes when the diet is supplemented with 50 g of probiotic kg−1 of food [101]. Other research has suggested the importance of managing the probiotic in all ontogenetic stages of the shrimp to generate a constant effect on the production of digestive enzymes [102].

In guppies (Poecilia reticulata, P. sphenops), and swordtail (Xiphophorus helleri, X. maculatus), the effect of incorporating Bacillus subtilis, isolated from the intestine of Cirrhinus mrigala into their diet has been evaluated. The results show an increase in the length and weight of the ornamental fishes as well as the specific activity of proteases and amylases in the digestive tract [86]. According to Moriarty [103], Bacillus secretes a wide range of exoenzymes that complement the activities of the fish and increases enzymatic digestion. In fact, the bacteria isolated from the digestive tract of aquatic animals have shown chitinases, proteases, cellulases, lipases, and trypsin [67].

Improvement of Water Quality

In several studies, water quality was recorded during the addition of probiotic strains especially of the gram-positive genus Bacillus. Probably since this bacterial group is more efficient than gram-negative in transforming organic matter to CO_2. It is suggested that maintaining high levels of probiotics in production ponds, fish farmers can minimize the accumulation of dissolved and particulate organic carbon during the growing season. In addition, this can balance the production of phytoplankton [13]. However, this hypothesis could not be confirmed on tests carried out during cultivation of shrimps or channel catfish, using one or more species of Bacillus, Nitrobacter, Pseudomonas, Enterobacter, Cellulomonas, and Rhodopseudomonas. Thus published evidence for improving water quality is limited, except for the nitrification [65].

In the case of edible fish, trout production farms generate high concentrations of nitrogen ranging from 0.05– 3.3 mg L^{-1} of total Kjeldahl nitrogen and up to 6.4 mg L^{-1} after 7 months of monitoring [104]. For tilapia production in recirculating systems, concentrations of total ammonia (NH_4 + NH_3) increased from 4.73 to 14.87 mg L^{-1} in a 21-dayexperiment, while the nitrite concentration increased from 3.75 to 9.77 mg L^{-1} [105]. Due to the high concentrations of produced nitrogen compounds, especially the highly toxic total ammonia; the use of probiotics is recommended, as they may improve water quality. Haroun et al. supplemented the food of Nile tilapia Oreochromis niloticus L. with a commercial probiotic made from Bacillus licheniformis and B. subtilis in 17 weeks of culture. Assessment of water quality parameters showed an acceptable range for fish cultivation: 5.7–6.3 mg L^{-1} for dissolved oxygen concentration, 0.36– 0.42 mg L^{-1} for ammonia concentration, and pH between 6.3 and 8.2 [70].

Lalloo et al. [106] isolated several strains of Bacillus from Cyprinus carpio and carried out tests to improve water quality in ornamental fish culture and to inhibit the growth of Aeromonas hydrophila. Three out of nine isolates showed high capacity to inhibit the pathogen in 78 of relative incidence rate; moreover, concentrations of ammonia, nitrate, and phosphate were lowered in rates of 74%, 76% and 72%, respectively. Contradictorily, Queiroz and Boyd [22] tested a commercial probiotic in catfish (Ictalurus punctatus), noting a survival and net fish production significantly higher when the probiotic was applied. However, very little differences were significant for the determined water quality variables (ammonia, chemical oxygen demand, nitrate, soluble reactive phosphorus, and dissolved oxygen) between the treated and control ponds. Taoka et al. [51] studied effects of commercial probiotics formulated from mixed cultures of bacteria and yeast on survival of Japanese flounder Paralichthys olivaceus, and water quality in a closed recirculating system. The probiotics-treated groups showed significantly greater survival rate as compared to the control group at the end of rearing experiment (50 days of culture), and water quality parameters were significantly lower in probiotics diet groups (from 0.24 ± 0.22 to 0.12 ± 0.10 mg L^{-1} of NH4, from 0.15 ± 0.08 to 0.08 ± 0.08 mg L^{-1} of NO_2, and from 13.0 ± 3.9 to 10.2 ± 3.0 mg L^{-1} of PO_4). Meanwhile, Wang et al. showed that a commercial product made from Bacillus sp., Saccharomyces cerevisiae, Nitrosomonas sp., and Nitrobacter sp. had the ability to increase the beneficial bacterial microbiota of Penaeus vannamei shrimp, further reducing the concentrations of inorganic nitrogen from 3.74 to 1.79 mg L^{-1} and phosphate from 0.1105 to 0.0364 mg L^{-1} [107].

Stress Tolerance

Aquaculture practices demand intensive productions in shorter times, causing stress in crop species. For example, it has been reported that chronic stress in zebra fish, Danio rerio, induces a general depression on the synthesis of muscle protein [108]. As a result, it was sought to increase stress tolerance by using probiotics. One of the firsts formal reports on this field studied the supplementation of Lactobacillus delbrueckii ssp. delbrueckii in the diet of European sea bass (Dicentrarchus labrax), at time intervals of 25 to 59 days. In addition to evaluating the growth improvement, hormone cortisol was quantified in fish tissue as stress marker, since it is directly involved in the animal's response to stress. Cortisol levels obtained in the treated fishes were significantly lower than those in the control (3.6 ± 0.36 ng g^{-1} and 5.1 ± 0.47 ng g^{-1}, resp.) [49]. Another way to assess stress in fish involve subjecting them to heat shock, as in the case of Japanese flounder (Paralichthys olivaceus) grown in a recirculating system [72]. The stress tests were carried out until half the population died, thus calculated the mean lethal time (LT_{50}) in the absence and with addition of a commercial probiotic containing Bacillus subtilis, Lactobacillus acidophilus, Clostridium butyricum, and Saccharomyces cerevisiae. The group treated with probiotics showed greater tolerance in the stress test than the control group, the LT_{50} was 40 and 25 min, respectively.

Lactate and plasma glucose levels are considered appropriate indicators of stress as they increase as a secondary response during periods of stress to cover high energy requirements induced by this situation. Therefore, Varela et al. have conducted studies on the gilt-head bream (Sparus auratus); the glycogen and triglycerides reserves in the livers of the control group were significantly decreased in relation to concentrations obtained when the fish feed was supplemented with the probiotic Alteromonas sp. Strain Pdp 11 [50].

Castex et al. [52] evaluated the effect of Pediococcus acidilactici MA 18/5 in the antioxidative response of the shrimp Litopenaeus stylirostris to oxidative stress. The results showed high activities of antioxidant enzymes; superoxide dismutase (283.7 and 153.7 mol min^{-1} in control and treated groups, resp.) and catalase (3.08 and 1.34 mol min^{-1} for the control and probiotic-treated, resp.).

The results obtained so far raise the possibility of preparing the fish in advance with probiotic treatment, for conventional aquacultures practices that create stress in animals, such as transport, change in water temperature, and periodic manipulations [46].

Effect on Reproduction of Aquatic Species

According to Izquierdo et al. [109], breeding aquaculture species have high nutritional requirements, thus reproductive capacity depends on appropriate concentrations of lipids, proteins, fatty acids, vitamins C and E, and carotenoids. Furthermore, the relationship of these components influences reproduction in various processes such as fertility, fertilization, birth and development of larvae. At present, for most cultured fish species, there are commercially available "broodstock diets" that just are larger-sized diets. In practice, many fish hatcheries improve the nutrition of their broodstock by feeding them solely on fresh fish byproducts or in combination with commercial diets. The most common fresh organisms used to feed broodstock fishes include squid, cuttlefish, mussels, krill, and small crustaceans. The use of these unprocessed fish products often do not provide adequate levels of nutrients needed by broodstock fishes, increasing the risk of pathogens transmission to the parents and offspring, including parasites, bacteria, and viruses. Therefore, probiotics added to food or water were used in order to prevent infections and to explore their effect on reproduction.

The pioneer study on the effect of probiotic supplementation on reproductive performance of fish was carried out by Ghosh et al. [53], using a strain of B. subtilis isolated from intestine of Cirrhinus mrigala, incorporated at different concentrations to four species of ornamental fishes: Poecilia reticulata, P. sphenops, Xiphophorus helleri, and X. maculates, in a one-year experiment. The results showed that using B. subtilis concentrations of 10^6–10^8 cells g^{-1} of food, produced increases in the gonadosomatic index, fecundity, viability, and production of fry from the females of all four species. Furthermore, these authors proposed that complex B vitamins synthesized by the probiotic, especially thiamine (vitamin B1) and vitamin B12, contribute to reduce the number of dead or deformed alevins.

Abasali and Mohamad [55] carried out similar studies with X. helleri, using a commercial probiotic containing Lactobacillus acidophilus, L. casei, Enterococcus faecium, and Bifidobacterium thermophilum. The total production of alevin per female and the relative fecundity were evaluated. Their results showed significant differences between the control and probiotic-treated groups; in the first parameter 105 and 150 alevins were found on average, respectively. While in the second, 28 females were fecundated in the control and 41 for the probiotic treatment.

SAFETY CONSIDERATIONS OF PROBIOTIC

Traditionally, probiotics used in food industry have been deemed safe, in fact, no human risks have been determined, remaining as the best proof of its safety [110]. Theoretically, probiotics may be responsible for four types of side effects in susceptible individuals: systemic infections, deleterious metabolic activities, excessive immune stimulation, and gene transfer. However, no hard evidence has been found.

In practice, there are few reports of bacteremia in humans, where isolation of probiotic bacteria from infections seems to be the result of an opportunistic infection caused by skin lesions, cancer, chronic illness, or a druginduced abnormality. These conditions lead to a decreased intestinal barrier that promotes the passage of the bacteria through the mucosal epithelium. Subsequently, these microorganisms are transported to the mesenteric lymph nodes and other organs, leading to bacteremia that may progress to septicemia [111]. All reported cases of bacteremia occurred in patients with chronic illness or a weakened immune system [69, 112].

In the case of some lactic acid bacteria that are regarded as probiotics, the resistance to antibiotics can be linked to genes on chromosomes, plasmids, or transposons. However, there is insufficient information about the circumstances in which these genetic elements could be mobilized [113]. Moreover, it is recognized that some enterococci may possess virulence characteristics and have the ability to transfer antibiotic resistance elements, so it is recommended that no reference should be made to these organisms as probiotics for human consumption, unless the producer can demonstrate that this strain cannot acquire or transfer antibiotic resistance or induce infection [112].

Regarding safety of aquaculture products, in Asia and more recently in Latin America cultures of Penaeus monodon have been reported with bacterial white spot syndrome (BWSS), in farms with recurrent use of probiotics based on Bacillus subtilis. The spots are similar to those generated in the white spot viral syndrome (WSS), which is a deadly disease that spreads rapidly and causes mass mortalities in shrimp cultures [114, 115]. Wang et al. [116] demonstrated in 2000 that BWSS is a nonsystemic infection in P. monodon and lesions usually disappear after molting, under this condition cultures are still active and grow normally without significant mortality. However, it is of great concern because most farmers cannot distinguish BWSS from WSS. Farmers are advised in case of suspicion to send samples to the laboratory for confirmatory diagnosis.

Furthermore, since some aquaculture products are consumed raw or half cooked, has raised the question of whether residual probiotics may cause any infection in the final consumer. Shakibazadeh et al. [117] assessed the potential risk to humans caused by the use of probiotic Shewanella algae in shrimp farms. Studies were performed in mice that were given up to 10^{36} cfu to reach the LD_{50} value for S. algae, proving the safety of using probiotics in mice. Based on their results, these authors state that the use of S. algae is safe for the consumer of shrimps, such as for workers in the processing plants and farms.

Because there was no international consensus to ensure efficiency and safety of probiotics, FAO and WHO recognized the need to create guidelines for a systematic approach for the evaluation of probiotics in food, in order to substantiate their health claims. A working group with experts in the field was formed in order to recommend criteria and methodology for the evaluation of probiotics, based on scientific evidence [118]. As a result the "Guide for the Evaluation of Probiotics in Food" was presented, providing guidelines on the evaluation of health and nutrition properties of probiotics in food. The working group stated that no pathogenic or virulent properties were found in lactobacilli, bifidobacteria, or lactococci, although they acknowledged that under certain conditions, some strains of lactobacilli have been associated with rare cases of bacteremia. However, its incidence does not increase with raising the use of lactobacillus in probiotics. It was also mentioned that enterococci may possess virulence characteristics; therefore, it is not recommended as a probiotic for human consumption [113]. Although the guide is not focused on aquaculture products, it creates a precedent for conducting studies to evaluate the safety of probiotics in this area.

To date, the use of animal models including mice, rats, and fish has not revealed specific determinants of virulence or pathogenicity of the studied probiotic microorganisms, suggesting the overall safety of them [112]. However, it is important to continue research using three approaches: (i) analyzing the intrinsic properties of probiotic strains, (ii) studying its pharmacokinetics (survival, activity in the intestine, dose response, and recovery from mucosa), and (iii) understanding the interactions between the microorganism and the host [69].

CONCLUDING REMARKS

The current global food crisis and increasing production costs has put pressure on governments and the international community to ensure sufficient food supply for a growing population. Thus, aquaculture is presented as a way to meet the growing demand for fresh water food or seafood, and to meet current challenges relating to the ongoing globalization of trade, intensification and diversification of aquaculture, progress in technological innovations for food production, changes in ecological systems and human behavior, including a greater awareness to protect biodiversity, public health, and the environment. These challenges will lead to increased attention for improving aquaculture practices, and will become an important alternative to overexploitation and modification of aquatic ecosystems caused by capture fisheries. The use of probiotics can potentiate the benefits of this activity because, as presented in this paper, it offers viable alternatives for the generation of a higher-quality livestock product in terms of size, production time, and health.

In the near future, it is necessary to conduct studies relating to probiotics resistance to antibiotics, and the chances of transmission of genetic elements to other microorganisms in the fish GIT, and thus to humans when consuming the aquaculture product.

On the other hand, there is a need to strengthen studies of microbial ecology in aquaculture systems, correlating microbial communities (microorganisms on water and in the GIT of aquatic species) with animal growth and its relationship to the water quality. Nowadays, a variety of probiotic strains present in the GIT of aquatic animals and nitrifying bacteria from biofilters have been isolated and characterized using biochemical, morphological, and molecular techniques [68, 119–121]. The development of molecular techniques such as PCR, FISH (fluorescent in situ hybridization), DGGE (denaturing gradient gel electrophoresis), and generation of genomic libraries have started to unveil the diversity present in aquaculture systems. Currently, next-generation sequencing methodologies offer great potential for phylogenetic identification of probiotic microorganisms without using conventional cultivation techniques.

CONFLICT OF INTERESTS

The authors declare that there is no conflict of interests between authors and the commercial probiotics and trademarks mentioned in the paper. Mentioning of those products is related to the scientific literature cited in the text with the mere purpose of illustrating the wide range of probiotic applications.

REFERENCES

1. FAO, The State of World Fisheries and Aquaculture, 2012.
2. M. Timmons, J. Ebeling, F. Wheaton, S. Summerfelt, and B. Vinci, Recirculating Aquaculture Systems, Cayuga Aqua Ventures, 2nd edition, 2002.
3. C. E. Boyd and C. S. Tucker, Pond Aquaculture Water Quality Management, Kluwer, Norwell, Mass, USA, 1998.
4. D. Cressey, "Aquaculture: future fish," Nature, vol. 458, no. 7237, pp. 398–400, 2009.
5. P. G. Amaya and D. F. Castellano, Pesca, Acuicultura e Investigacion en M ′ ′exico, Edited by Cedrssa, 2006.
6. Y. B. Wang and Z. R. Xu, "Probiotics treatment as method of biocontrol in aquaculture," Feed Research, vol. 12, pp. 42–45, 2004.
7. M. G. Bondad, R. P. Subasinghe, J. R. Arthur et al., "Disease and health management in Asian aquaculture," Veterinary Parasitology, vol. 132, no. 3-4, pp. 249–272, 2005.
8. N. Kautsky, P. Ronnb ¨ ack, M. Tedengren, and M. Troell, ¨ "Ecosystem perspectives on management of disease in shrimp pond farming," Aquaculture, vol. 191, no. 1–3, pp. 145–161, 2000.
9. F. A. O. /OIE/WHO, "Antimicrobial use in aquaculture and antimicrobial resistance," Report of a Joint. Expert Consultation on Antimicrobial Use in Aquaculture And Antimicrobial Resistance, 2006.
10. K. Nomoto, "Prevention of infections by probiotics,"

Journal of Bioscience and Bioengineering, vol. 100, no. 6, pp. 583–592, 2005.

11. WHO, "Antimicrobial resistance. Fact sheet N◦ 194," 2012, http://www.who.int/mediacentre/factsheets/fs194/es/index .html.
12. C. S. Lewin, "Mechanisms of resistance development in aquatic microorganisms," in Chemothe-Rapy in Aquaculture: From Theory To Reality, C. Michel and D. Alderman, Eds., pp. 288–301, Office International des Epizooties, Paris, France, 1992.
13. J. L. Balcazar, I. D. Blas, I. Ruiz-Z, D. Cunningham, D. ´ Vendrell, and J. L. Muzquiz, "The role of probiotics in ´ aquaculture," Veterinary Microbiology, vol. 114, no. 3-4, pp. 173–186, 2006.
14. M. Kozasa, "Toyocerin (Bacillus toyoi) as growth promotor for animal feeding," Microbiologie Aliments Nutrition, vol. 4, no. 2, pp. 121–135, 1986.
15. R. S. Porubcan, "Reduction of ammonia nitrogen and nitrite in tanks of Penaeus monodon using floating biofilters containing processed diatomaceous earth media pre-inoculated with nitrifying bacteria," in Proceedings of the Program and Abastracts of the 22nd Annual Conference and Exposition, World Aquaculture Society, Puerto Rico, Spain, June 1991.
16. R. S. Porubcan, "Reduction in chemical oxygen demand and improvement in Penaeus monodon yield in ponds inoculated with aerobic Bacillus bacteria," in Proceedings of the Program and Abastracts of the 22nd Annual Conference and Exposition, World Aquaculture Society, Puerto Rico, Spain, June 1991.
17. G. B. Gomez, A. Roque, and J. F. Turnbull, "The use and ´ selection of probiotic bacteria for use in the culture of larval aquatic organisms," Aquaculture, vol. 191, no. 1–3, pp. 259– 270, 2000.
18. D. J. W. Moriarty, "Control of luminous Vibrio species in penaeid aquaculture ponds," Aquaculture, vol. 164, no. 1–4, pp. 351–358, 1998.
19. Y. B. Wang, J. R. Li, and J. Lin, "Probiotics in aquaculture: challenges and outlook," Aquaculture, vol. 281, no. 1–4, pp. 1–4, 2008.
20. A. Irianto and B. Austin, "Probiotics in aquaculture," Journal of Fish Diseases, vol. 25, no. 11, pp. 633–642, 2002.
21. S. Rengpipat, W. Phianphak, S. Piyatiratitivorakul, and P. Menasveta, "Effects of a probiotic bacterium on black tiger shrimp Penaeus monodon survival and growth," Aquaculture, vol. 167, no. 3-4, pp. 301–313, 1998.
22. J. F. Queiroz and C. E. Boyd, "Effects of a bacterial inoculum in channel catfish ponds," Journal of the World Aquaculture Society, vol. 29, no. 1, pp. 67–73, 1998.
23. A. Gildberg, H. Mikkelsen, E. Sandaker, and E. Ringø, "Probiotic effect of lactic acid bacteria in the feed on growth and survival of fry of Atlantic cod (Gadus morhua)," Hydrobiologia, vol. 352, no. 1–3, pp. 279–285, 1997.
24. P. A. Douillet and C. J. Langdon, "Use of a probiotic for the culture of larvae of the Pacific oyster (Crassostrea gigas Thunberg)," Aquaculture, vol. 119, no. 1, pp. 25–40, 1994.
25. F. J. Gatesoupe, "The use of probiotics in aquaculture," Aquaculture, vol. 180, no. 1-2, pp. 147–165, 1999.
26. A. R. S. Harzeveli, H. VanDuffel, P. Dhert, J. Swing, and P. Sorgeloos, "Use of a potential probiotic Lactococcus lactis AR21 strain for the enhancement of growth in the rotifer Brachionus plicatilis (Muller)," Aquaculture Research, vol. 29, no. 6, pp. 411–417, 1998.
27. S. Dharmaraj and K. Dhevendaran, "Evaluation of Streptomyces as a probiotic feed for the growth of ornamental fish Xiphophorus helleri," Food Technology and Biotechnology, vol. 48, no. 4, pp. 497–504, 2010.
28. L. H. H. Hernandez, T. C. Barrera, J. C. Mejia et al., "Effects of the commercial probiotic Lactobacillus casei on the growth, protein content of skin mucus and stress resistance of juveniles of the Porthole livebearer Poecilopsis gracilis (Poecilidae)," Aquaculture Nutrition, vol. 16, no. 4, pp. 407– 411, 2010.
29. M. Rahiman, J. Yousuf, T. Ambat, and M. Hatha, "Probiotic effect of Bacillus NL110 and Vibrio NE17 on the survival, growth performance and immune response of Macrobrachium rosenbergii (de Man)," Aquaculture Research, vol. 41, no. 9, pp. e120–e134, 2010.
30. Lin Sh, Y. Guan, L. Luo, and Y. Pan, "Effects of dietary chitosan oligosaccharides and Bacillus coagulans on growth, innate immunity and resistance of koi (Cyprinus carpio koi)," Aquaculture, vol. 342-343, pp. 36–41, 2012.
31. C. I. Chang and W. Y. Liu, "An evaluation of two probiotic bacterial strains, Enterococcus faecium SF68 and Bacillus toyoi, for reducing edwardsiellosis in cultured European eel, Anguilla anguilla L," Journal of Fish Diseases, vol. 25, no. 5, pp. 311–315, 2002.
32. S. Nikoskelainen, A. Ouwehand, S. Salminen, and G. Bylund, "Protection of rainbow trout (Oncorhynchus mykiss) from furunculosis by Lactobacillus rhamnosus," Aquaculture, vol. 198, no. 3-4, pp. 229–236, 2001.
33. A. Irianto and B. Austin, "Use of probiotics to control furunculosis in rainbow trout, Oncorhynchus mykiss (Walbaum)," Journal of Fish Diseases, vol. 25, no. 6, pp. 333–342, 2002.
34. L. Gram, J. Melchiorsen, B. Spanggaard, I. Huber, and T. F. Nielsen, "Inhibition of Vibrio anguillarum by Pseudomonas fluorescens AH2, a possible probiotic treatment of fish," Applied and Environmental Microbiology, vol. 65, no. 3, pp. 969–973, 1999.
35. L. Gram, T. Løvold, J. Nielsen, J. Melchiorsen, and B. Spanggaard, "In vitro antagonism of the probiont Pseudomonas fluorescens strain AH2 against Aeromonas salmonicida does not confer protection of salmon against furunculosis," Aquaculture, vol. 199, no. 1-2, pp. 1–11, 2001.
36. B. Spanggaard, I. Huber, J. Nielsen et al., "The probiotic potential against vibriosis of the indigenous microflora of rainbow trout," Environmental Microbiology, vol. 3, no. 12, pp. 755–765, 2001.
37. C. Ruiz-Ponte, J. F. Samain, J. L. Sanchez, and J. L. Nicolas, ´ "The benefit of a Roseobacter species on the

survival of scallop larvae," Marine Biotechnology, vol. 1, no. 1, pp. 52–59, 1999.

38. U. Scholz, D. G. Garcia, D. Riccque, S. L. E. Cruz, A. F. Vargas, and J. Latchford, "Enhancement of vibriosis resistance in juvenile Penaeus vannamei by supplementation of diets with different yeast products," Aquaculture, vol. 176, no. 3-4, pp. 271–283, 1999.

39. B. Austin, L. F. Stuckey, P. A. W. Robertson, I. Effendi, and D. R. W. Griffith, "A probiotic strain of Vibrio alginolyticus effective in reducing diseases caused by Aeromonas salmonicida, Vibrio anguillarum and Vibrio ordalii," Journal of Fish Diseases, vol. 18, no. 1, pp. 93–96, 1995.

40. B. Austin, E. Baudet, and M. Stobie, "Inhibition of bacterial fish pathogens by Tetraselmis suecica," Journal of Fish Diseases, vol. 90, pp. 389–392, 1992.

41. Y. Youping, M. Dongdong, L. Shijiang, and W. Zhongkang, "Effects on growth and digestive enzyme activities of the Hepialus gonggaensis larvae caused by introducing probiotics," World Journal of Microbiology and Biotechnology, vol. 27, no. 3, pp. 529–533, 2011.

42. A. M. Abdullah, R. Hashim, and P. M. Aliyu, "Evaluating the use of Lactobacillus acidophilus as a biocontrol agent against common pathogenic bacteria and the effects on the haematology parameters and histopathology in African catfish Clarias gariepinus juveniles," Aquaculture Research, vol. 42, no. 2, pp. 196–209, 2011.

43. S. D. Moreira, S. S. Medeiros, L. L. Pereira, L. A. Romano, W. Wasielesky, and B. E. Cupertino, "The use of probiotics during the nursery rearing or the pink shrimp Farfantepenaeus brasiliensis (Latreille, 1817) in a zero exchange system," Aquaculture Research, pp. 1–10, 2011.

44. S. Zhang, Y. Sing, M. Long, and Z. Wei, "Does dietary administration of Lactococcus lactis modulate the gut microbiota of grouper, Epinephelus coioides," Journal of the World Aquaculture Society, vol. 43, no. 2, pp. 198–207, 2012.

45. A. Dohail, M. Abdullah, H. Roshada, and M. Aliyu, "Effects of the probiotic, Lactobacillus acidophilus, on the growth performance, haematology parameters and immunoglobulin concentration in African Catfish (Clarias gariepinus, Burchell 1822) fingerling," Aquaculture Research, vol. 40, no. 14, pp. 1642–1652, 2009.

46. P. A. Tapia, R. P. D´ıaz, R. J. Leon et al., "Use or the probiotic ´ Shewanella putrefaciens Pdp11 on the culture of Senegalenses sole (Solea senegalensis, Kaup 1858) and gilthead seabream (Sparaus aurata L.)," Aquaculture International, vol. 21, pp. 1–15, 2012.

47. X. X. Zhou, Y. B. Wang, and W. F. Li, "Effect of probiotic on larvae shrimp (Penaeus vannamei) based on water quality, survival rate and digestive enzyme activities," Aquaculture, vol. 287, no. 3-4, pp. 349–353, 2009.

48. F. Shishehchian, F. M. Yusoff, and M. Shariff, "The effects of commercial bacterial products on macrobenthos community in shrimp culture ponds," Aquaculture International, vol. 9, no. 5, pp. 429–436, 2001.

49. O. Carnevali, L. de Vivo, R. Sulpizio et al., "Growth improvement by probiotic in European sea bass juveniles (Dicentrarchus labrax, L.), with particular attention to IGF- 1, myostatin and cortisol gene expression," Aquaculture, vol. 258, no. 1–4, pp. 430–438, 2006.

50. J. L. Varela, I. Ru´ız, L. Vargas et al., "Dietary administration of probiotic Pdp11 promotes growth and improves stress tolerance to high stocking density in gilthead seabream Sparus auratus," Aquaculture, vol. 309, no. 1–4, pp. 265–271, 2010.

51. Y. Taoka, H. Maeda, J. Y. Jo et al., "Growth, stress tolerance and non-specific immune response of Japanese flounder Paralichthys olivaceus to probiotics in a closed recirculating system," Fisheries Science, vol. 72, no. 2, pp. 310–321, 2006.

52. M. Castex, P. Lemaire, N. Wabete, and L. Chim, "Effect of dietary probiotic Pediococcus acidilactici on antioxidant defences and oxidative stress status of shrimp Litopenaeus stylirostris," Aquaculture, vol. 294, no. 3-4, pp. 306–313, 2009.

53. S. Ghosh, A. Sinha, and C. Sahu, "Effect of probiotic on reproductive performance in female livebearing ornamental fish," Aquaculture Research, vol. 38, no. 5, pp. 518–526, 2007.

54. G. Gioacchini, F. Maradonna, F. Lombardo, D. Bizzaro, I. Olivotto, and O. Carnevali, "Increase of fecundity by probiotic administration in zebrafish (Danio rerio)," Reproduction, vol. 140, no. 6, pp. 953–959, 2010.

55. H. Abasali and S. Mohamad, "Effect of dietary supplementation with probiotic on reproductive performance of female livebearing ornamental fish," Research Journal of Animal Sciences, vol. 4, no. 4, pp. 103–107, 2010.

56. J. Schrezenmeir and M. De Vrese, "Probiotics, prebiotics, and synbiotics—approaching a definition," American Journal of Clinical Nutrition, vol. 73, pp. 361S–364S, 2001.

57. E. Metchnikoff, "Lactic acid as inhibiting intestinal putrefaction," in The Prolongation of Life, Optimistic Studies, pp. 161– 183, Mitchell Heinemann, London, UK, 1907.

58. D. M. Lilly and R. H. Stillwell, "Probiotics: growthpromoting factors produced by microorganisms," Science, vol. 147, no. 3659, pp. 747–748, 1965.

59. G. S. Sperti, Probiotics, Avi Publishing, West Point, Miss, USA, 1971.

60. R. B. Parker, "Probiotics, the other half of the antibiotics story," Animal Nutrition Health, vol. 29, pp. 4–8, 1974.

61. R. Fuller, "Probiotics in man and animals," Journal of Applied Bacteriology, vol. 66, no. 5, pp. 365–378, 1989.

62. R. Havenaar and I. Huis, The Lactic Acid Bacteria in Health and Disease, Volume 1, Edited by B. J. B. Wood, Elsevier, New York, NY, USA, 1992.

63. F. Guarner and G. J. Schaafsma, "Probiotics," International Journal of Food Microbiology, vol. 39, no. 3, pp. 237–238, 1998.

64. D. Myers, "Probiotics," Journal of Exotic Pet Medicine, vol. 16, no. 3, pp. 195–197, 2007.

65. L. Verschuere, G. Rombaut, P. Sorgeloos, and W.

Verstraete, "Probiotic bacteria as biological control agents in aquaculture," Microbiology and Molecular Biology Reviews, vol. 64, no. 4, pp. 655–671, 2000.

66. L. Vazquez, R. Hern ´ andez, E. Sainz et al., "Cambio en la flora ´ intestinal de ratones por la administracion de bifidobacterias ´ y jugo de girasol," Veterinaria M´exico, vol. 27, no. 2, pp. 127– 131, 1996.

67. N. G. Vine, W. D. Leukes, and H. Kaiser, "Probiotics in marine larviculture," FEMS Microbiology Reviews, vol. 30, no. 3, pp. 404–427, 2006.

68. G. Burr, D. Gatlin, and S. Ricke, "Microbial ecology of the gastrointestinal tract of fish and the potential application of prebiotics and probiotics in Finnish aquaculture," Journal of the World Aquaculture Society, vol. 36, no. 4, pp. 425–435, 2005.

69. C. Soccol, L. Porto, M. Rigon et al., "The potential of probiotics: a review," Food Technology and Biotechnology, vol. 48, no. 4, pp. 413–434, 2010.

70. E. Haroun, A. Goda, and M. Kabir, "Effect of dietary probiotic Biogen supplementation as a growth promoter on growth performance and feed utilization of Nile tilapia Oreochromis niloticus (L.)," Aquaculture Research, vol. 37, no. 14, pp. 1473–1480, 2006.

71. Y. B. Wang, Z. R. Xu, and M. S. Xia, "The effectiveness of commercial probiotics in northern white shrimp Penaeus vannamei ponds," Fisheries Science, vol. 71, no. 5, pp. 1036– 1041, 2005.

72. Y. Taoka, H. Maeda, J. Y. Jo et al., "Use of live and dead probiotic cells in tilapia Oreochromis niloticus," Fisheries Science, vol. 72, no. 4, pp. 755–766, 2006.

73. C. P. Mart´ınez, L. Mayorga, T. Ponce et al., "Efecto de los fructooligosacaridos en la poblaci ´ on bacteriana fecal de un ´ neonato, crecida en cultivo por lote," Revista Mexicana De Ciencias Farmac´euticas, vol. 39, no. 1, pp. 32–37, 2008.

74. K. M. Tuohy, H. M. Probert, C. W. Smejkal, and G. R. Gibson, "Using probiotics and prebiotics to improve gut health," Drug Discovery Today, vol. 8, no. 15, pp. 692–700, 2003.

75. A. Y. Dakar, S. M. Shalaby, and I. P. Saoud, "Assessing the use of a dietary probiotic/prebiotic as an enhancer of spinefoot rabbitfish Siganus rivulatus survival and growth," Aquaculture Nutrition, vol. 13, no. 6, pp. 407–412, 2007.

76. Smart Microbials Inc, "Un universo de microbios inteligentes a su servicio. Technical details on the use of commercial probiotic product Aqua BOOSTER," 2012, http:// smartmicrobialsinc.com/aqua.php.

77. C. Lacroix and S. Yildirim, "Fermentation technologies for the production of probiotics with high viability and functionality," Current Opinion in Biotechnology, vol. 18, no. 2, pp. 176–183, 2007.

78. S. Rokka and P. Rantamaki, "Protecting probiotic bacteria by ¨ microencapsulation: challenges for industrial applications," European Food Research and Technology, vol. 231, no. 1, pp. 1–12, 2010.

79. P. Rosas-Ledesma, J. M. Leon-Rubio, F. J. Alarc ´ on, M. A. ´ Morinigo, and M. C. Balebona, "Calcium alginate capsules ˜ for oral administration of fish probiotic bacteria: aassessment of optimal conditions for encapsulation," Aquaculture Research, vol. 43, pp. 106–116, 2012.

80. M. A. Morinigo, V. S ˜ anchez, and T. F. Mart ´ ´ınez, "Encapsulation of a bacterial fish probiiotic in alginate beads: protective effect under in vitro simulations of fish gastric conditions," in Abstracts of the International Conference on Fish Diseases and Fish Immunology, pp. 6–9, Iceland University, Reykjavik, Iceland, 2008.

81. J. A. Muller, R. P. Ross, G. F. Fitzgeralk, and C. Stanton, "Manufacture of Probiotic Bacteria," Prebiotics and Probiotics Science and Technology, pp. 125–759, 2009.

82. R. C. Medina and E. B. Cordero, "Crecimiento y composicion´ bioqu´ımica de la diatomea Chaetoceros muelleri Lemmerman, mantenida en cultivo estatico con un medio comercial," ´ Ciencia y Mar, pp. 19–25, 2011.

83. G. Gomez, A. Roque, and B. Velasco, "Culture of´ Vibrio alginolyticus C7b, a potential probiotic bacterium, with the microalga Chaetoceros muelleri," Aquaculture, vol. 211, no. 1– 4, pp. 43–48, 2002.

84. M. Planas, J. Vazquez, J. Marqu ´ es, R. P ´ erez, M. Gonz ´ alez, and ´ M. Murado, "Enhancement of rotifer (Brachionus plicatilis) growth by using terrestrial lactic acid bacteria," Aquaculture, vol. 240, no. 1–4, pp. 313–329, 2004.

85. F. Lara, N. Olvera, M. Guzman, and M. L ´ opez, "Use of the ´ bacteria Streptococcus faecium and Lactobacillus acidophilus, and the yeast Saccharomyces cerevisiae as growth promoters in Nile tilapia (Oreochromis niloticus)," Aquaculture, vol. 216, no. 1–4, pp. 193–201, 2003.

86. S. Ghosh, A. Sinha, and C. Sahu, "Dietary probiotic supplementation on growth and health of live-bearing ornamental fishes," Aquaculture Nutrition, vol. 14, no. 4, pp. 289–299, 2008.

87. B. M. Macey and V. E. Coyne, "Improved growth rate and disease resistance in farmed Haliotis midae through probiotic treatment," Aquaculture, vol. 245, no. 1–4, pp. 249–261, 2005.

88. T. Nakano, Microorganism. En dietary supplements for the health and quality of cultured fish, CAB International, London, UK, 2007.

89. P. J. Ronson and R. Medina, ´ Probioticos en la Acuicultura ´ , Ciencia y Mar Notas, 2002.

90. P. A. W. Robertson, C. O'Dowd, C. Burrells, P. Williams, and B. Austin, "Use of Carnobacterium sp. as a probiotic for Atlantic salmon (Salmo salar L.) and rainbow trout (Oncorhynchus mykiss, Walbaum)," Aquaculture, vol. 185, no. 3-4, pp. 235–243, 2000.

91. A. Irianto and B. Austin, "Use of dead probiotic cells to control furunculosis in rainbow trout, Oncorhynchus mykiss (Walbaum)," Journal of Fish Diseases, vol. 26, no. 1, pp. 59– 62, 2003.

92. A. V. Ravi, K. S. Musthafa, G. Jegathammbal, K. Kathiresan, and S. K. Pandian, "Screening and evaluation of probiotics as a biocontrol agent against pathogenic Vibrios in marine aquaculture," Letters in Applied Microbiology, vol. 45, no. 2, pp. 219–223, 2007.

93. K. K. Vijayan, I. S. Bright Singh, N. S. Jayaprakash et al., "A brackishwater isolate of Pseudomonas PS-102, a potential antagonistic bacterium against pathogenic Vibrios in penaeid and non-penaeid rearing systems," Aquaculture, vol. 251, no. 2–4, pp. 192–200, 2006.
94. N. G. Vine, W. D. Leukes, and H. Kaiser, "In vitro growth characteristics of five candidate aquaculture probiotics and two fish pathogens grown in fish intestinal mucus," FEMS Microbiology Letters, vol. 231, no. 1, pp. 145–152, 2004.
95. N. G. Vine, W. D. Leukes, H. Kaiser, S. Daya, J. Baxter, and T. Hecht, "Competition for attachment of aquaculture candidate probiotic and pathogenic bacteria on fish intestinal mucus," Journal of Fish Diseases, vol. 27, no. 6, pp. 319–326, 2004.
96. R. G. Gomez, J. L. Balc ´ azar, and M. A. Shen, "Probiotics as ´ control agents in aquaculture," Journal of Ocean University of China, vol. 6, no. 1, pp. 76–79, 2007.
97. D. Tovar, J. Zambonino, C. Cahu, F. Gatesoupe, R. Vazquez, ´ and R. Lesel, "Effect of live yeast incorporation in compound diet on digestive enzyme activity in sea bass (Dicentrarchus labrax) larvae," Aquaculture, vol. 204, no. 1-2, pp. 113–123, 2002.
98. M. C. Hidalgo, A. Skalli, E. Abellan, M. Arizcun, and G. ´ Cardenete, "Dietary intake of probiotics and maslinic acid in juvenile dentex (Dentex dentex L.): Effects on growth performance, survival and liver proteolytic activities," Aquaculture Nutrition, vol. 12, no. 4, pp. 256–266, 2006.
99. D. L. Merrifield, G. Bradley, R. T. M. Baker, and S. J. Davies, "Probiotic applications for rainbow trout (Oncorhynchus mykiss Walbaum) II. Effects on growth performance, feed utilization, intestinal microbiota and related health criteria postantibiotic treatment," Aquaculture Nutrition, vol. 16, no. 5, pp. 496–503, 2010.
100. R. Tovar, D. Mazurais, J. Gatesoupe, P. Quazuquel, C. Cahu, and J. Zambonino, "Dietary probiotic live yeast modulates antioxidant enzyme activities and gene expression of sea bass (Dicentrarchus labrax) larvae," Aquaculture, vol. 300, no. 1–4, pp. 142–147, 2010.
101. Z. Heizhao, G. Zhixun, Y. Yingying, Z. Wenhui, and J. Zhuojia, "Effect of dietary probiotics on apparent digestibility coefficients of nutrients of white shrimp Litopenaeus vannamei Boone," Aquaculture Research, vol. 35, no. 15, pp. 1441–1447, 2004.
102. S. Ziaei, M. Habibi, G. Azari, D. Lovett, A. Mirvaghefi, and M. Shakouri, "The effect of Bacillus spp. bacteria used as probiotics on digestive enzyme activity, survival and growth in the Indian white shrimp Fenneropenaeus indicus," Aquaculture, vol. 252, no. 2–4, pp. 516–524, 2006.
103. D. J. W. Moriarty, "Probiotics and bioremediation in aquaculture," Asian Shrimp News, vol. 26, article 3, 1996.
104. V. M. Maillard, G. D. Boardman, J. E. Nyland, and D. D. Kuhn, "Water quality and sludge characterization at racewaysystem trout farms," Aquacultural Engineering, vol. 33, no. 4, pp. 271–284, 2005.
105. G. Rafiee and C. R. Saad, "Nutrient cycle and sludge production during different stages of red tilapia (Oreochromis sp.) growth in a recirculating aquaculture system," Aquaculture, vol. 244, no. 1–4, pp. 109–118, 2005.
106. R. Lalloo, S. Ramchuran, D. Ramduth, J. Gorgens, and N. ¨ Gardiner, "Isolation and selection of Bacillus spp. as potential biological agents for enhancement of water quality in culture of ornamental fish," Journal of Applied Microbiology, vol. 103, no. 5, pp. 1471–1479, 2007.
107. L. Jiqiu, T. Beiping, M. Kangsen et al., "Comparative study between probiotic bacterium Arthrobacter XE-7 and chloramphenicol on protection of Penaeus chinensis postlarvae from pathogenic Vibrios," Aquaculture, vol. 253, no. 1–4, pp. 140–147, 2006.
108. S. Vianello, L. Brazzoduro, V. Dalla, P. Belvedere, and L. Colombo, "Myostatin expression during development and chronic stress in zebrafish (Danio rerio)," Journal of Endocrinology, vol. 176, no. 1, pp. 47–59, 2003.
109. M. Izquierdo, H. Palacios, and A. Tacon, "Effect of broodstock nutrition on reproductive performance of fish," Aquaculture, vol. 197, no. 1–4, pp. 25–42, 2001.
110. M. Saxelin, H. Rautelin, S. Salminen, and P. H. Makela, "Safety of commercial products with viable Lactobacillus strains," Infectious Diseases in Clinical Practice, vol. 5, no. 5, pp. 331–335, 1996.
111. N. Ishibashi and S. Yamazaki, "Probiotics and safety," American Journal of Clinical Nutrition, vol. 73, no. 2, pp. 465S–470S, 2001.
112. S. J. Lahtinen, R. J. Boyle, A. Margolles, R. Fr´ıas, and M. Gueimonde, "Safety assessment of probiotics," Prebiotics and Probiotics Science and Technology, pp. 1193–1235, 2009.
113. FAO/WHO, Guidelines for the Evaluation of Probiotics in Food, 2006.
114. H. Y. Chou, X. Y. Huang, C. H. Wang, H. C. Chiang, and C. F. Lo, "Pathogenicity of a baculovirus infection causing white spot syndrome in cultured penaeid shrimp in Taiwan," Diseases of Aquatic Organisms, vol. 23, no. 3, pp. 165–173, 1995.
115. Y. G. Wang, M. D. Hassan, M. Shari ff, S. M. Zamri, and X. Chen, "Histopathology and cytopathology of white spot syndrome virus (WSSV) in cultured Penaeus monodon from peninsular Malaysia with emphasis on pathogenesis and the mechanism of white spot formation," Diseases of Aquatic Organisms, vol. 39, no. 1, pp. 1–11, 1999.
116. Y. G. Wang, K. L. Lee, M. Najiah, M. Shari ff, and M. D. Hassan, "A new bacterial white spot syndrome (BWSS) in cultured tiger shrimp Penaeus monodon and its comparison with white spot syndrome (WSS) caused by virus," Diseases of Aquatic Organisms, vol. 41, no. 1, pp. 9–18, 2000.
117. S. Shakibazadeh, C. R. Saad, A. Christianus, M. S. Kamarudin, K. Sijam, and P. Sinaian, "Assessment of possible human risk of probiotic application in shrimp farming," International Food Research Journal, vol. 18, pp. 433–437, 2011.

118. M. Pineiro and C. Stanton, "Probiotic bacteria: Legislative framework—requirements to evidence basis," Journal of Nutrition, vol. 137, pp. 850S–835S, 2007.

119. A. D. Schulze, O. A. Abayomi, S. A. Tattersall, and K. M. Miller, "Bacterial diversity in a marine hatchery: balance between pathogenic and potentially probiotic bacterial strains," Aquaculture, vol. 256, no. 1–4, pp. 50–73, 2006.

120. H. Sugita, H. Mizuki, and S. Itoi, "Diversity of siderophoreproducing bacteria isolated from the intestinal tracts of fish along the Japanese coast," Aquaculture Research, vol. 43, no. 4, pp. 481–488, 2012.

121. Y. H. Ling, S. Y. Shang, M. R. Long, and Y. J. Dan, "PCRDGGE analysis of the autochthonous gut microbiota of grouper Ephinephelus coioides following probiotic Bacillus clausii administration," Aquaculture Research, vol. 43, no. 4, pp. 489–497, 2012.

CHAPTER 3

PROBIOTICS APPLICATION IN AQUACULTURE

Ekundayo Taiye Michael, Sogbesan Olukayode Amos and Lauratu Tahir Hussaini

Keywords: Probiotics, Aquaculture, Disease outbreak, Non-antibiotic agent

INTRODUCTION

Aquaculture has grown tremendously during the last years becoming an economically important industry [1]. Today it is the fastest growing food-producing sector in the world with the greatest potential to meet the growing demand for aquatic food [2]. Although aquaculture activity in Nigeria started about 50 years ago [3], aquaculture production in Nigeria is currently about 40,000 metric tonnes contributing only 6% of domestic fish production [4]. Nigerians are high fish consumers and offer the largest market for fisheries production in Africa. Thus, Nigeria has become one of the largest fish importers in the developing world, importing about 600,000 metric tonnes annually [3].

With the increasing intensification and commercialization of aquaculture production, disease is a major problem in the fish farming industry [5]. However the utility of antimicrobial agents as a preventive measure has been questioned, given extensive documentation of the evaluation of antimicrobial resistance among pathogenic bacteria. On the other hand antibiotics inhibit or kill beneficial microbiota in the gastrointestinal (GI) ecosystem but it also made antibiotic residue accumulated in fish products to be harmful for human consumption [6]. Because of the health risks associated with the use of antibiotics in animal production, there is a growing awareness that antibiotics should be used with more care [7]. In view of the risk associated with the use of antibiotics, the development of non-antibiotic agents is one of the key factors for health management in aquaculture. According to Browdy [8], one of the most significant technologies that evolved in response to disease control problems is the use of probiotics. The term probiotic means life; it was derived from two Greek words 'pro' and 'bios' [9]. Probiotics are live microbes that can be used to improve the host intestinal microbial balance and growth performance. Development of probiotics in aquaculture management will reduce the use of antimicrobial drugs which were prophylactive alone and whose over dependence in recent times poses potential hazards to man who consume them [10]. This report summarizes the probiotics selection procedures, application and its significance in aquaculture.

Citation: Michael ET, Amos SO, Hussaini LT (2014) Probiotics Application in Aquaculture. Fish Aquac J 5:111. doi:10.4172/2150-3508.1000111.

Selection of Probiotics

The initial major purpose of using probiotics is to maintain or reestablish a favourable relationship between friendly and pathogenic microorganism that constitute the flora of intestinal or skin mucus of fish. A successful probiotic is expected to have a few specific properties in order to certify, a beneficial effect. And in order to produce probiotics for commercialization, the following steps are to be put into consideration as in Figure 1:

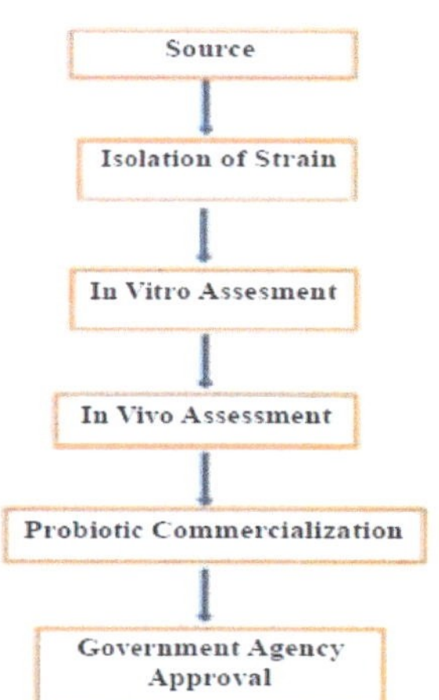

Source: Modified (Balcazar et al., 2006)

Figure 1: Flow chart for selection of probiotics

Figure 1: Flow chart for selection of probiotics.

A healthy source of microorganisms from a digestive tract of healthy aquatic animals must be selected.

The microorganisms with which the work is to be carried out are isolated and identified by means of selective culture.

A new culture with only the colonies of interest for conducting in vitro evaluations such as inhibition of pathogens; pathogenicity to target species; resistance conditions of host; among others are performed.

In case of the absence of restrictions on the use of the target species, experiments with in vivo supplementation, and small and large scale, are carried out to check if there are real benefits to the host.

Finally, the probiotic that presented significantly satisfactory result can be produced commercially and utilized.

Characteristics of Good Probiotics

Fuller [11] listed the following as features of good probiotic bacteria:

- It should be a strain, which is capable of exerting a beneficial effect on the host animal e.g. increased growth or resistance to disease.
- It should be non-pathogenic and non-toxic.
- It should be present as viable cells preferable in large numbers.
- It should be capable of surviving and metabolizing in the gut environment e.g. resistance to low pH and organic acid.
- It should be stable and capable of remaining viable for periods under storage and field conditions.

A probiotic agent with all these features has considerable advantage over antibacterial supplements such as antibiotics currently in use. They do not induce resistance to antibiotics which will compromise therapy. They are not toxic and therefore will not produce undesirable side effect when being fed and in the case of food animal will not produce toxic residues in the carcass. They may stimulate immunity whereas the immune status remains unaffected by antibiotics.

Constraints to Probiotics in Aquaculture

- Inability of strains to be produced in commercial quantities and consequent demonstration on a large scale.
- Difficulty in proving performance at the farm level.
- Inability of companies to conduct extensive research on how to make product specifically for aquaculture purposes.

Probiotics Significance in Aquaculture

There are some possible benefits linked to the administering of probiotics which have already been suggested as:

Improvement in Water Qualities

Nitrogenous compounds contamination such as ammonia, nitrite and nitrate in fish culture systems/ponds has been a serious concern. The susceptibility of cultured aquatic species to high concentration of these compounds is generally species-specific, but in high concentrations, these compounds may be extremely harmful and cause mass mortality in all cases. Ma et al. [12] reported the ability of *Lactobacillus spp*. JK-8 and JK-11 simultaneously removes nitrogen and pathogens from contaminated shrimp farms. In several other studies, water quality has been improved by the addition of probiotics especially Bacillus spp. [13,14]. The reason is that gram – positive Bacillus spp. according to Stanier et al. [15] are generally more efficient in converting organic matter back to CO_2 than gram – negative bacteria, which would convert a greater percentage of organic carbon to bacterial biomass or slime.

As Growth Promoters

It has been demonstrated experimentally that probiotics indeed may enhance the growth of fish. The ability of organisms to out-grow the pathogens in favour of host or to improve the growth of the host and yet no side effect on the host made it a probiotic bacteria.Yassir et al. [16] in attempt to use probiotic bacteria as growth promoter on tilapia (*Oreochromis niloticus*) identified that the highest growth performance was recorded with *Micrococcus luteus* a probiotic and the best feed conversion ratio was observed with the same organism. So *M. luteus* may be considered as a growth promoters in fish aquaculture. Lactic acid bacteria also had an effect as growth promoters on the growth rate in juvenile carp though not in Sea bass [17].

For Disease Prevention

Probiotics or their products for health benefits to the host have been found useful in aquaculture, terrestrial animals and in human disease control. These include microbial adjunct that prevent pathogens from proliferating in the intestinal tract, on the superficial surfaces and in culture environment of the culture species [13]. The effect of these beneficial organisms is achieved through optimizing the immune system of culture organism, increasing their resistance to disease, or producing inhibitory-substance that prevent the pathogenic organisms from establishing disease in the host.

Source of Nutrients and Enzymatic Contribution to Digestion

Some researchers have suggested that microorganisms have a beneficial effect in the digestive processes of aquatic animals. In fish, it has been reported that *Bacteroides* and *Clostridium sp.* have contributed to the host's nutrition, especially by supplying fatty acids and vitamins [18]. Some microorganisms such as *Agrobacterium sp., Pseudomonas sp., Brevi-bacterium sp., Microbacterium sp.,* and *Staphylococcus sp.*may contribute to nutritional processes in *Salvelinus alpinus L* [19].

Enhancement of the Immune Response

Among the numerous beneficial effects of probiotics, modulation of immune system is one of the most commonly purported benefits of probiotics. Fish larvae shrimps and other invertebrates have immune systems that are less well developed than adult stage and are dependent primarily on non-specific immune responses for their resistance to infection [13, 20] evaluated the ability of *Lactobacillus fermentum*

LbFF4 isolated from Nigerian fermented food ('fufu') and *L. plantarum* LbOGI from a beverage 'Ogi' to induce immunity in *Clarias gariepinus* (Burchell) against some selected fish bacterial pathogens.

Probiotics in Aquaculture Management

These organisms can be administered to the aquaculture management through feeding, injection or immersion of the probiotic bacteria [21].

Application in Feed

Probiotics are applied with the feed and a binder (egg or cod liver oil) and most commercial preparation contain either *Lactobacillus sp* or*Saccharomyces cerevisiae* [22]. According to FAO and WHO guidelines, probiotic organisms used in food must be capable of surviving passages through the gut i.e. they must have the ability to resist gastric juices and exposure to bile [23]. Furthermore they must be able to proliferate and colonize the digestive tract and they must be safe, effective and maintain their effectiveness and potency for the duration of the shelf life of the product [23].

Direct Application to Pond Water

The water probioics contain multiple strains of bacteria like *Bacillus acidophilus, B. subtilis B. lecheniformis, Nitrobacter sp, Aerobacter and Sacharomyces cerevisiae.* Application of probiotic through water of tanks and ponds may also have an effect on fish health by improving several qualities of water, since they modify the bacteria composition of the water and sediments [24,25].

Application through Injection

Application of probiotics by injection is a possibility. Austin et al., [26] suggested the possibility of freeze-drying the probiont like vaccine and applied either through bathing, or injection. Yassir et al. [16] has demonstrated the experimental administration of probiotic *Micrococcus luteus*to *Oreochromis niloticus* by injection through intra peritoneal route which had only 25% mortality as against 90% with *Pseudomonas* using the same route. According to Yassir et al. [16,27] the use of probiotics stimulate *Rainbow trout* immunity by stimulating phagocytes activity, complement mediated bacterial killing and immunoglobulin production [17].

CONCLUSION

Increased use of antibiotics has led to the high proportion of antibiotic-resistant bacteria which provide threat to fish and man through consumption of the infected fish. Inefficiencies in antibiotic treatment of fish illnesses lead to significant economic losses. But the use of probiotics in aquaculture has shown to have beneficial impact on fish health and thereby economic performance of fish farming. At the same time, the use of probiotics has also important environmental benefits. By reducing the risk of diseases, the necessity of medication and thereby the risk of residues left in the environment is reduced. Therefore the use of probiotics in fish feed should also be seen as an important step in aquaculture sustainability.

RECOMMENDATIONS

Fish farmers and other stakeholders in aquaculture management should make use of probiotics because of its colonization ability as preventive measures against over dependency on antibiotic therapy which is costly.

Fish farmers are also encouraged to incorporate probiotics in their feed formulations because of its importance in digestibility improvement.

Close network of aquaculture experts, fish nutritionists and microbiologists necessary to develop such aquatic foods.

REFERENCES

1. Subasinghe R, Soto D, Jia J (2009) Global aquaculture and its role in sustainable development. Reviews in Aquacul 1: 2-9.
2. FAO (2006). The State of Food Insecurity in the world. Food and Agriculture Organization of the United Nations, Rome, Italy: 235-240.
3. Olagunju FI, Adesiyan IO, Ezekiel AA (2007) Economic viability of catfish production in Oyo state, Nigeria. Journal of Human Ecology 21:121-124.
4. Adeogun OA, Ogunbadejo HK, Ayinla OA, Oresegun A (2007) Urban Aquaculture: Producer perceptions and practices in Lagos State Nigeria. Middle-East Journal of Scientific Research 2: 21-27.
5. Bondad-Reantaso MG, Subasinghe RP, Arthur JR, Ogawa K, Chinabut S et.al.,(2005). Disease and health management in Asian aquaculture. Veterinary Parasitol 132:249-272.
6. WHO(2006) Report of a joint FAO/OIE/WHO expert consultation on antimicrobial use in aquaculture and antimicrobial resistance: Seoul, Republic of Korea.
7. Defoirdt T,Sorgeloo P, Bossier P (2011) Alternatives to antibiotics for the control of bacterial disease in aquaculture. Current Opinion in Microbiology. 14: 251-258.
8. Browdy C (1998) Recent developments in penaeid brood stock and seed production technologies: improving the outlook for superior captive stocks. Aquaculture 164:3-21.
9. GismondoMR, L Drago, A Lombardi(1999) Review of probiotics available to modify gastrointestinal flora. International Journal of Antimicrobial Agents12: 287-292.
10. Salminen S, A Ouwehand Y Benno, YK Lee(1999) Probiotics: How should they be defined. Trends Food Science Technology 10: 107-110.
11. Fuller R(1989) A review: probiotics in man and animals. Journal of Applied Bacteriology 66:365–378.
12. Ma GW, Cho YS, Oh KH (2009)Removal of pathogenic bacteria and nitrogen byLactobacillus spp. JK-8- JK-11. Aquaculture 287: 266-270.
13. Verschuere, G Rombaut, P Sorgeloos,WVerstraete (2000)"Probiotic bacteria as biological control agents in aquaculture," Microbiology and Molecular Biology Reviews. 64 : 655-671.

14. Kolndadacha OD, Adikwu IA, Atiribom RY, Mohammed A, Musa YM, Ladu GB(2009) The role of probiotics in aquaculture in Nigeria – A Review: Continental Journal of Fisheries andAquatic Science 3:42 -49
15. Stanier RY, DoudoroffM, Adelberg EA(1963)The microbial world. Prentice- Hall Inc., Englewood Cliffs. N.J: 153-156.
16. Yassir, ALAdel, ME, Azze A (2002)Use of probiotic bacteria as growth promoters, antibacterial and the effect on physiological parameters of Oreochronusniloticus. Journal of Fish Diseases 25: 633-642.
17. Noh SH,Han K, Won TH, ChoiYJ(1994)Effect of antibiotics, enzymes, yeast culture and probiotics on the growth performance of Israeli carp. Korean Journal of Animal Science 36:480-486.
18. Sakata T (1990) Microflora in the digestive tract of fish and shellfish. Proceedings of the International Symposium on Microbiology in Poecilotherms: 171–176.
19. Ringø E, Strøm E, TabachekJA(1995) Intestinal microflora of salmonids: a review. Aquaculture Research 26: 773–789.
20. OgunsheAAO,Olabode PO (2009) Antimicrobial potentials of indigenous lactobacillus strains on gram–negative indicator bacterial species from Clariasgariepinus(Burchell) microbial inhibition of fish borne pathogens. African Journal of Microbiology Research 3:870-876.
21. Irianto A, Austin B(2002) Probiotics in aquaculture. Journal of Fish Disease 25:633-642.
22. AbidiR(2003) Use of probiotics in larval rearing of mew candidate species. AquacultureAsiavolume 8:15-16.
23. Senok AC, Ismeel AY,BottaGA(2005) Probiotics: facts and myths. Clinical Microbiology and Infection Diseases 11:958-960.
24. Ashraf, Ali (2000) Probiotics in fish farming–Evaluation of a candidate bacterial mixture. Vattenbruksinstitutionen :1-18.
25. VenkateswaraAR(2007) Bioremediation to restore the health of aquaculture. Pond Ecosystem. Hyderabad 500:1-12.
26. Austin B, Stuckey LF, Robertson PAW, Effendi I, Griffith DRW(1995) A probiotic strain of Vibrio alginolyticus effective in reducing diseases caused by Aeromonassalmonicida, Vibrio anguillarum and Vibrio ordalii. Journal of Fish Disease 18:93-96.
27. Nikoskelainen S, OuwehandAC, Bylund G, Salminen S, Lilius E (2003): Immune enhancement in rainbow trout (Oncorhynchusmykiss) by potential probiotic bacteria (Lactobacillus rhamnosus). Fish and Fish Immunology 15: 443-452.

CHAPTER 4

BACILLUS SUBTILIS A POTENTIAL PROBIOTIC BACTERIUM TO FORMULATE FUNCTIONAL FEEDS FOR AQUACULTURE

Jorge Olmos and J Paniagua-Michel

Keywords: *Bacillus subtilis*, Probiotics, Functional Feed, Vegetable ingredients

BACILLUS SUBTILIS PROPERTIES

Bacteria of the Bacillus genus are among the most widespread microorganisms in nature, they can be found in soil, water and air. Bacillus constitutes a diverse group of rod-shaped, Gram-positive bacteria, characterized by their ability to produce a robust spore [1-5]. The Bacillus subtilis genome is totally sequenced, leading to generation of a great amount of basic knowledge in this bacterium [6]. Additionally, developments of molecular and genetic methodologies are well established in B. subtilis [7-9]. B. subtilis is not harmful to mammals, including humans, and is commercially important as producer of a high and diverse amount of secondary metabolites like antibiotics, fine chemicals and enzymes, as well as heterologous proteins, antigens and vaccines [1,10-15]. B. subtilis grow efficiently with low-cost carbon and nitrogen sources, because its enzymes are very efficient breaking down a great variety of proteins, carbohydrates and lipids from animal and vegetable origin, into their constituent units [1,16-18]. The enzymes also degrade organic accumulated debris from shrimp/fish cultures inducing ponds bioremediation and consequently the prevention of viral and bacterial diseases [19-22]. On the other hand, the antimicrobial activity of Bacillus is greatly determined by their ability to produce antibiotics, principally from peptide origin [23]. There were identified 795 antibiotics from Bacillus species [11]. B. subtilis is the genus most productive species devoting 4-5% of its genome to their synthesis, producing 66 antibiotics [11]. Furthermore, Bacillus subtilis is Generally Recognized as Safe (GRAS) by the FDA, meaning this bacterium is not harmful to animals or humans (Figure 1). Taking into account beneficial properties of B. subtilis, this bacterium is a potential probiotic candidate to be considered in "Functional Feeds" of crustaceans and fishes [16-19,24].

Citation: Olmos J, Paniagua-Michel J (2014) Bacillus subtilis A Potential Probiotic Bacterium to Formulate Functional Feeds for Aquaculture. J Microb Biochem Technol 6: 361-365. doi:10.4172/1948-5948.1000169.

BACILLUS SUBTILIS PROBIOTIC CAPACITIES

B. subtilis is between the oldies species on earth, reason why animals and humans since the beginning of their existence have been in contact with this bacterium [25,26]. In this sense, recognition of B. subtilis for animal and human immune system is well established and a symbiotic relationship had been developed for a long period of time [27]. On the other hand, million years of evolution has created an amazing quorum sensing communication-recognition mechanism between beneficial and pathogenic bacteria [1,28,29]. However, animals and humans only sense pathogens when the disease had been developed, due equilibrium between beneficial and pathogenic bacteria had been broken because; environmental, nutritional and/or metabolic changes has occurred favoring pathogens proliferation. Synthetic antibiotics was the first option to control pathogens overgrowth in humans and animals, however the unregulated use of these compounds induced a multi-resistant mechanisms preferentially developed by the pathogenic bacteria [30]. Today, antibiotics utilization is well regulated in humans and also in animals including aquaculture. In this sense, the utilization of beneficial bacteria (probiotics) has emerge as an alternative due good results are been obtained in both; animals and humans [27].

The term "probiotics" was derived from the Greek word, meaning "for life" [31]. According to the currently adopted definition by FAO/ WHO, probiotics are: "Live microorganisms which when administered in adequate amounts confer a health benefit on the host [32]. Since 50 year ago exists a scientific and commercial interest in the use of beneficial bacteria for the prevention and treatment of diseases [27,33]. Lactobacillus and Bifidobacterium genera has been used almost exclusively for the competitive exclusion of pathogenic bacteria from the gut of humans, farmed animals and most recently aquaculture species [23,27,33,34]. However, the utilization of probiotic to application others than pathogenic bacterial exclusion are now been develop in several areas like; immune system stimulation, vaccine production, bioremediation and functional feed development [15-19,24,35]. In this sense, and based in our own experiences, we redefined probiotics as a living microbial supplement that: (a) positively affects host health status by modifying the host-associated microbial community and its immune system; (b) secrete a variety of enzymes to increases feed degradation-assimilation enhancing its nutritional values; (c) improves quality of environmental parameters by the bioremediation of waste products and (d) support extreme changing environmental and physical parameters (Figure 2).

However, to accomplish all these objectives the utilization of "Bacillus subtilis spore producing multifunctional probiotic bacterium" it is recommended, to support environmental, nutritional and metabolic changing conditions where probiotic will be involve and most importantly, to take advantage of them [17-19,24,35-37]. In this sense, B. subtilis spores have the capacity to resist extreme pH conditions, UV irradiation, high temperatures, solvents and long time periods of storage without refrigeration [1] In summary, B. subtilis has a) versatility of growth nutrients utilization, b) high level of enzymes production, c) secretion of antimicrobial compounds, d) spore producer, e) develops in aerobic and anaerobic conditions, and f) B. subtilis is Generally Recognized As Safe (GRAS) by the Food and Drug Administration (FDA) (Figure 1). In this sense, B. subtilis in "theory" could be considered as a perfect multifunctional probiotic bacterium for humans and animals [1,16-19,23,24,27,33,38,39].

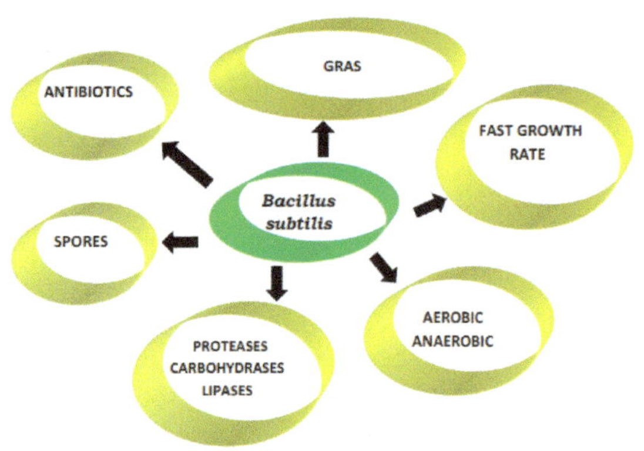

Figure 1: Bacillus subtilis intrinsic properties.

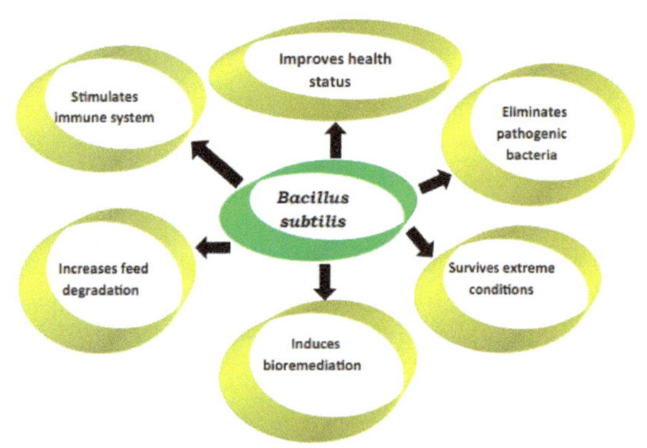

Figure 2: Bacillus subtilis probiotic capacities in humans and animals.

BACILLUS SUBTILIS RECOMBINANT PROBIOTIC DEVELOPMENT

The global sales of probiotic supplements were predicted to rise 48% from \$2.7 bn in 2011 to \$4 bn in 2016. In US, per capita spending on probiotic supplements is expected to nearly double by 2016 and overtakes Japan [34]. In this sense, the utilization of non-recombinant probiotic strains in humans, farmed animals and aquaculture systems is well established, principally for the competitive exclusion of pathogenic bacteria (Table 1) [27,33,40-42].

On the other hand, utilization of recombinant probiotic strains in humans and animals it is not accepted due Genetically Modified Organism (GMO) regulations [27,43]. B. subtilis strains with overproduction of antimicrobial peptides or enzymes already exist scientifically and commercially, but only their products are being sold not the strains [12,23,44,45]. Overproduction of antimicrobial peptides and/or enzymes by recombinant probiotic strains could induce health problems in animals and consequently in the consumers, as the unregulated utilization of synthetic antibiotics did. Human and animal market "supposedly" consumes non recombinant Bacillus probiotic strains, due health FDA regulation and restrictions that must be solved, before acceptance of GMO products commercialization.

However, diseases like the one induced by White Spot Syndrome Virus (WSSV) in shrimps, cannot be controlled at the moment by any non-recombinant bacterial probiotic strain. In this sense, production of vaccines using B. subtilis probiotic recombinant strains, are starting to be develops to prevent disease impacts on animals and humans [15,46-48]. Environmental contamination and utilization of strong stressful agents are inducing development of new bacterial and virus diseases. In this sense, recombinant probiotic utilization could be the only effective way to fight versus pathogen agent's proliferation. However, even when promising results have been obtained inducing immunization with recombinant B. subtilis probiotic strains, several FDA tests must be carried out in animal models before a massive food/feed inclusion and commercialization of recombinant probiotics can be done.

AQUACULTURE DEVELOPMENT RESTRICTIONS

The United Nations Food and Agriculture Organization (FAO) define aquaculture as the farming of aquatic organisms including mollusks, crustaceans, fish and plants. Marine capture fisheries and aquaculture, supplied the world with about 154 million tons of marine products in 2011 [49]. However, global population demanding for marine food products is increasing around the world, but fisheries capture production has leveled off, due principal fishing areas have reached their maximum potential and some have started to go down [24,49,50]. In this sense, aquaculture the multibillion dollar fastest growing industry of the world food economy, is increasing by more than 10% per year and currently accounts for more than 50% of all shrimp/ fish consumed [49]. For this reason, prices of fishmeal and oil utilized precisely to produce aquaculture feeds have increased considerably last years, making the industry unprofitable in many countries [24,50]. Worry about negative impacts of fishmeal and oil overproduction in ecology of global fisheries and environmental alterations have arisen [51,52]. Aquaculture represents the greatest potential to meet demands for aquatic food products; however, instead of helping to ease the crisis in wild fisheries, unsustainable aquaculture development may aggravate the problems and create new ones, damaging our important and already-stressed coastal areas [53]. In order to solve these aquaculture restrictions, the development of Functional Feeds using alternativeeconomical vegetable protein, carbohydrates, lipids and "Bacillus subtilis spore producing multifunctional probiotic bacterium" are recommended, to increase aquaculture profitability and environmental and ecological sustainability [16-19,24].

Table 1: Bacillus subtilis commercial products from humans and animals

BACILLUS PROBIOTIC FOR HUMAN USE			
Product	**Manufacturer**		**Comments**
Bio-Kult®	Probiotics International Ltd (Protexin), UK. http://www.bio-kult.com		*B. subtilis* is one component of 14 strains carried in this probiotic supplement (minimum 2x10^9 CFU/capsule).
Medilac-Vita®	Hanmi Pharmaceutical Co. Ltd., Seoul, Korea http://www.hanmipharm.com		*B. subtilis* in combination with *Streptococcus faecium* (at 10^8/g).
Primal Defense®	Garden of Life®, Florida, USA. http://www.gardenoflife.com		*B. subtilis* is one of the 12 bacteria species in this probiotic.
Biobaby®	Ildong Pharmaceutical Ho Chi Minh, Vietnam http://www.thuocbo.com.vn/Danh-muc-thuoc-bo/men-vi-sinh/biobaby.html		Granules for solution with *B. subtilis* 3 mg/g (3x10^6 CFU/g), in combination with *Lactobacillus sporogenes* and *Clostridium butyricum* (1x10^7 CFU/g).
Totalflora	Roux-Ocefa Laboratorios Buenos Aires, Argentina http://www.roux-ocefa.com		Each capsule contains: *Lactobacillus acidophilus, Bifidobacterium bifidum, Bacillus subtilis, Saccharomyces cerevisiae* and *Bacteriophages*.
BACILLUS PROBIOTIC FOR AQUACULTURE			
Brand	**Manufacturer**		**Comments**
Naturalle *Bacillus subtilis*	Wuhan Nature´s Favour Bioengineering Co., Ltd, Wuhan City, China http://www.wuhannature.com		*Bacillus subtilis* (2x10^{10} CFU/g).
Biozyme	Bio-Pharmachemie Joint-Venture Company, Ho Chi Minh, Vietnam http://www.biopharmachemie.com		*Bacillus subtilis* and *Saccaromyces cerevisiae*.
Fubon *Bacillus subtilis*	Angel Yeast Co., Ltd. Hubei, China http://www.angelyeast.com		*Bacillus subtilis* (≥ 20 billion CFU/g).
Bioron	American Pharma International, India http://www.americanpharmainternational.com		Each kg contains: *Bacillus subtilis* (4.5x10^8 CFU), *B. licheniformis, B. megaterium, Lactobacillus lactis, L. helveticus, Nitrosomonas sp. Nitrobacter sp. Saccharomyces cerevisiae* and *Aspergillus oryzae*.
Lactomin	American Pharma International, India http://www.americanpharmainternational.com		Each kg contains: *Bacillus subtilis* (45,000 million CFU), *B. licheniformis, Lactobacillus acidophilus, L. sporogenes* and *Saccharomyces cerevisiae*.
BACILLUS PROBIOTIC FOR VETERINARY USE			
Brand	**Animal**	**Manufacturer**	**Comments**
BioGrow®	Poultry, calves and swine	Provita Eurotech Ltd., Omagh, Northern Ireland, UK http://www.provita.co.uk	Listed as containing spores of *B. licheniformis* (1.6x10^9 CFU/g) and *B. subtilis* (1.6x10^9 CFU/g)
BioPlus® 2B	Piglets, chickens, turkeys for fattening	Christian Hansen Hoersholm, Denmark http://www.chr-hansen.com	Mixture (1/1) of *B. licheniformis* (DSM 5749) and *B. subtilis* (DSM 5750) at 1.6x10^9 CFU/g of each bacterium, EU approved
Biozyme	Poultry, swine, shrimp, fish	Bio-Pharmachemie Joint-Venture Company Ho Chi Minh, Vietnam http://www.biopharmachemie.com	Formulation: *B. subtilis* and *Saccharomyces cerevisiae*.
Proflora	Poultry	Zoetis, New Jersey, USA http://www.zoetisus.com	Proflora is a unique combination product with a Live DFM *B. subtilis* strain QST 713 plus Beta Mos®, a yeast-extract prebiotic.
Vime-Baciflor	Pigs, poultry	Vemedin Corporation, Can Tho, Vietnam http://www.vemedim.vn	*Bacillus subtilis*.

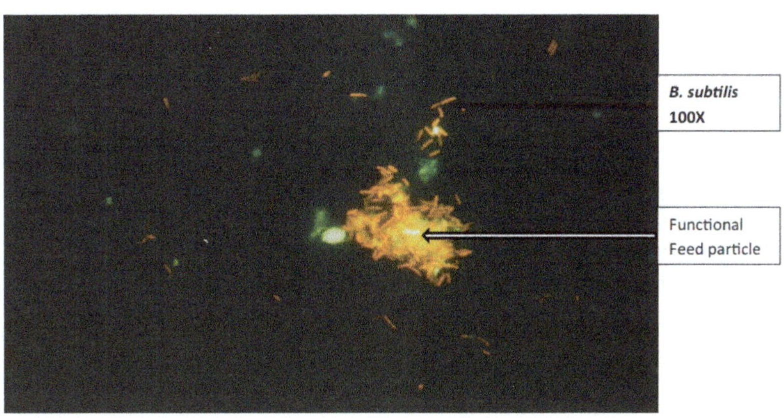

Figure 3: Functional Feed particle formulated with soybean meal, complex carbohydrates, vegetable oil and Bacillus subtilis multifunctional probiotic strain [20].

IMPORTANCE OF FUNCTIONAL FEED DEVELOPMENT IN AQUACULTURE

The rapid development of aquaculture from last years has increased feed ingredients demand and its prices, for this reason feeding represent 40-60% of the total production costs in shrimp/fish farming. In this sense, aquaculture industry development has become a great challenge for future generations; not only for feeding expenses but also from availability of fishmeal and oil ingredients. Reduced inclusion of fishmeal and oil in formulations is a tendency for almost a decade. It is estimated that these ingredients will be minimal or null in diets by the year 2020 [50]. Consequently, alternative feed formulation ingredients development is urgent and they must meet the animal nutritional requirements, be less expensive and sustainable with the environment [17,18,35,51,52,54-57]. Partial or complete fishmeal and oil substitution by alternative-economic vegetable protein, carbohydrates and lipids has been the goal to the field for more than 10 years, without satisfactory results due toxic effects and anti-nutritional nature of some vegetable molecules [58-63]. However, feeds evolution with inclusion of vegetable ingredients inducing; growth, health, environmental and economic benefits beyond traditional feeds, is a great opportunity to assure the future of a profitable and sustainable aquaculture [16-19,24]. Nevertheless, proteases, carbohydrases and lipases limited expression in shrimp/fish is the major obstacle to digestion and assimilation of vegetable ingredients, regulating its high level inclusion in feed formulations [58-63]. Thus, the utilization of specifically selected "Bacillus subtilis spore producing multifunctional probiotic bacterium" has emerged as a solution with huge applications in the aquaculture feeding industry (Figures 1 and 2). In this sense, excellent results have been obtained in Litopenaeus vannamei, Oreochrmis niloticus and Atractoscion nobilis; increasing digestion-assimilation of vegetable ingredients, improving Food Conversion Ratio (FCR) and inducing optimal growth on cultivated animals, when a B. subtilis probiotic strain have been added to feeds. Additionally, inhibition of pathogens proliferation and increased health status of animals was also improved. Finally, bioremediation of pond contaminated water was also achieved efficiently by B. subtilis probiotic bacteria used [16-19, 24]. These results open great opportunities to new Functional Feeds production for shrimp and fish aquaculture (Figure 3).

Complex carbohydrates degradation, proteases inhibitors inactivation and lipids assimilation, were not a problem to animals when "Bacillus subtilis spore producing multifunctional probiotic bacterium" was added to vegetable based-feed formulations [16-19, 24].

CONCLUSION

The world market for probiotics supplements has been growing for the last 50 years; today most dairy human products contain probiotic bacteria. Lactobacillus and Bifidobacterium are the most used genera principally for the exclusion of pathogens in humans and animals. However, species of these genera do not present multifunctional probiotic capacities as B. subtilis does [15-19,24]. In this sense, animal feed preparation processes usually require high temperatures and only spore producer bacteria are optimal to support these conditions. Other Bacillus than B. subtilis can also be used for feed preparation, however B. subtilis is one of the few recognized as safe by the FDA. B. subtilis enzymes support extreme environmental conditions increasing digestive human and animal capabilities, to assimilate ingredients from animal and plant origin [16-19,24]. The capacity of B. subtilis to grow aerobically and anaerobically allows this bacterium to survive and growth adequately inside and outside humans and animals, and in aquaculture water systems. All these properties give B. subtilis the potential capacity to become one of the "most perfect multifunctional probiotic bacteria" to formulate Functional Feeds for aquaculture. However, B. subtilis probiotic properties are strain-specific meaning bioprospection and selection methods needs to be well designed.

ACKNOWLEDGEMENT

The authors wish to thanks Rosalia Contreras and Leonel Ochoa for technical assistance. This work was done with support of CONACYT project CB-131115.

REFERENCES

1. Sonnenschein AL, Losick R, Hoch JA (1993) Bacillus subtilis and Others Gram-Positive Bacteria: Biochemistry, Physiology and Molecular Genetics. American Society for Microbiology, Washington, DC.
2. Garbeva P, van Veen JA, van Elsas JD (2003) Predominant Bacillus spp. in agricultural soil under different management regimes detected via PCR-DGGE. Microb Ecol 45: 302-316.
3. Nicholson WL (2004) Ubiquity, longevity, and ecological roles of Bacillus spores. In Bacterial Spore Formers: Probiotics and Emerging Applications, Ricca E, Henriques AO, Cutting SM (eds). Norflok, Horison Bioscience, pp 1-15.
4. Ivanova EP, Vysotskii MV, Svetashev VI, Nedashkovskaya OI, Gorshkova NM, et al. (1999) Characterization of Bacillus strains of marine origin. Int Microbiol 2: 267-271.
5. Andersson AM, Weiss N, Rainey F, Salkinoja-Salonen MS (1999) Dust-borne bacteria in animal sheds, schools and children's day care centres. J Appl Microbiol 86: 622-634.
6. Kunst F, Ogasawara N, Moszer I, Albertini AM, Alloni G, et al. (1997) The complete genome sequence of the gram-positive bacterium *Bacillus subtilis*. Nature 390: 249-256.
7. Harwood CR, Cutting SM (1990) Molecular Biological Methods for *Bacillus*. Wiley, Chichester, England, pp. 1-581.
8. Olmos J, Bolaños V, Causey S, Ferrari E, Bollvar F, et al. (1996) A functional Spo0A is required for maximal aprE expression in Bacillus subtilis. FEBS Lett 381: 29-31.
9. Olmos J, de Anda R, Ferrari E, Bolívar F, Valle F (1997) Effects of the sinR and degU32 (Hy) mutations on the regulation of the aprE gene in *Bacillus subtilis*. Mol Gen Genet 253: 562-567.
10. Bérdy J (2005) Bioactive microbial metabolites. J Antibiot (Tokyo) 58: 1-26.
11. Stein T (2005) Bacillus subtilis antibiotics: structures, syntheses and specific functions. Mol Microbiol 56: 845-857.
12. Ferrari E, Jarnagin AS, Schmidt BF (1993) Commercial production of extracellular enzymes. In: Sonenshein AL, Hoch JA, Losick R (Eds.), *Bacillus* subtilis and Other Gram-Positive Bacteria. American Society for Microbiology, Washington, DC, pp. 917-937.
13. Olmos-Soto J, Contreras-Flores R (2003) Genetic system constructed to overproduce and secrete proinsulin in Bacillus subtilis. Appl Microbiol Biotechnol 62: 369-373.
14. Westers L, Westers H, Quax WJ (2004) Bacillus subtilis as cell factory for pharmaceutical proteins: a

biotechnological approach to optimize the host organism. Biochim Biophys Acta 1694: 299-310.

15. Valdez A, Yepiz-Plascencia G, Ricca E, Olmos J (2014) First Litopenaeus vannamei WSSV 100% oral vaccination protection using CotC::Vp26 fusion protein displayed on Bacillus subtilis spores surface. J Appl Microbiol 117: 347-357.
16. Ochoa SJL, U.S. Patent 2012/0128827 A1, May 24, 2012
17. Ochoa-Solano J, Olmos-Soto J (2006) The functional property of *Bacillus* for shrimp feeds. Food Microbiol 23: 519-525.
18. Arellano CF, Olmos SJ (2002) Thermostable a-1,4- and a-1–6-glucosidase enzymes from Bacillus spp. isolated from a marine environment. World J Microb Biot 18: 791-795.
19. Olmos J, Ochoa L, Paniagua-Michel J, Contreras R (2011) Functional feed assessment on Litopenaeus vannamei using 100% fish meal replacement by soybean meal, high levels of complex carbohydrates and Bacillus probiotic strains. Mar Drugs 9: 1119-1132.
20. Rengpipat S, Phianphak W, Piyatiratitivorakul S, Menasvetac P (1998) Effects of a probiotic bacterium on black tiger shrimp Penaeus monodon survival and growth. Aquaculture 167: 301-313.
21. Guo JJ, Liu KF, Cheng SH, Chang CI, Lay JJ, et al. (2009) Selection of probiotic bacteria for use in shrimp larviculture. Aquac Res 40: 609-618.
22. Paniagua-Michel J, Franco-Rivera A, Cantera JJL, Stein LY (2005) Activity of nitrifying biofilms constructed on low-density polyester enhances bioremediation of a coastal wastewater effluent. World J Microb Biot 21: 1371-1377.
23. Sorokulova I (2013) Modern status and perspectives of Bacillus bacteria as probiotics. J Prob Health 1: 1-5.
24. Olmos SJ, Paniagua-Michel J, Lopez L, Ochoa S L (2014) Functional feeds in aquaculture. S.-K. Kim (Ed.), Handbook of Marine Biotechnology. NY: Springer, ISBN: 978-3-642-53970-1, 1800 pp.
25. Vreeland RH, Rosenzweig WD, Powers DW (2000) Isolation of a 250 million-year-old halotolerant bacterium from a primary salt crystal. Nature 407: 897-900.
26. Fajardo-Cavazos P, Nicholson WL (2006) Bacillus endospores isolated from granite: close molecular relationships to globally distributed Bacillus spp. from endolithic and extreme environments. Appl Environ Microbiol 72: 2856-2863.
27. Cutting SM (2011) Bacillus probiotics. Food Microbiol 28: 214-220.
28. Bassler BL, Wright M, Showalter RE, Silverman MR (1993) Intercellular signalling in Vibrio harveyi: sequence and function of genes regulating expression of luminescence. Mol Microbiol 9: 773-786.
29. Bassler BL, Greenberg EP, Stevens AM (1997) Cross-species induction of luminescence in the quorum-sensing bacterium Vibrio harveyi. J Bacteriol 179: 4043-4045.
30. Nikaido H (2009) Multidrug Resistance in Bacteria. Annu Rev Biochem 78: 119-146.
31. Reid G, Hass J, Sebulsdy MT, McCormick JD (2003) Potential use of probiotics in clinical practice. Clin Microbiol Rev 16: 658-672.
32. FAO/WHO (2006) Probiotic in foods: Health and nutritional properties and guidelines for evaluation. In FAO food and nutrition; 2005 pp 85, ISBN 92-5-105513-0.
33. Gatesoupe FJ (1999) The use of probiotics in aquaculture. Aquaculture 180: 147-165.
34. Naidu KSB, Adam JK, Govender P (2012) The use of probiotics and safety concerns: A review. Afr J Microbiol Res 6: 6871-6877.
35. Rurangwa E, Laranja JL, Van Houdt R, Delaedt Y, Geraylou Z, et al. (2009) Selected nondigestible carbohydrates and prebiotics support the growth of probiotic fish bacteria mono-cultures in vitro. J Appl Microbiol 106: 932-940.
36. Lara-Flores M (2011) The use of probiotic in aquaculture: an overview. Int Res J Microbiol 2: 471-478.
37. Shen WY, Fu LL, Li WF, Zhu YR (2010) Effect of dietary supplementation with Bacillus subtilis on the growth, performance, immune response and antioxidant activities of the shrimp. Aquac Res 41: 1691-1698.
38. Caruso A, Flamminio G, Folghera S, Peroni L, Foresti I, et al. (1993) Expression of activation markers on peripheral-blood lymphocytes following oral administration of Bacillus subtilis spores. Int J Immunopharmacol 15: 87-92.
39. Huang Q, Xu X, Mao YL, Huang Y, Rajput IR, et al. (2013) Effects of Bacillus subtilis B10 spores on viability and biological functions of murine macrophages. Anim Sci J 84: 247-252.
40. Balcázar JL, Rojas-Luna T, Cunningham DP (2007) Effect of the addition of four potential probiotic strains on the survival of pacific white shrimp (Litopenaeus vannamei) following immersion challenge with Vibrio parahaemolyticus. J Invertebr Pathol 96: 147-150.
41. Verschuere L, Rombaut G, Sorgeloos P, Verstraete W (2000) Probiotic bacteria as biological control agents in aquaculture. Microbiol Mol Biol Rev 64: 655-671.
42. Merrifield DL, Bradley G, Baker RTM, Davies SJ (2010) Probiotic applications for rainbow trout (Oncorhynchus mykiss Walbaum) II. Effects on growth performance, feed utilization, intestinal microbiota and related health criteria postantibiotic treatment. Aquac Nut 16; 496-503.
43. Sorokulova IB, Beliavskaia VA, Masycheva VA, Smirnov VV (1997) Recombinant probiotics: problems and prospects of their use for medicine and veterinary practice. Vestn Ross Akad Med Nauk 46-49.
44. Liu L, Liu Y, Shin HD, Chen RR, Wang NS, et al. (2013) Developing Bacillus spp. as a cell factory for production of microbial enzymes and industrially important biochemicals in the context of systems and synthetic biology. Appl Microbiol Biotechnol 97: 6113-6127.
45. Terpe K (2006) Overview of bacterial expression systems for heterologous protein production: from molecular and biochemical fundamentals to commercial systems. Appl Microbiol Biotechnol 72: 211-222.

46. Witteveldt J, Vlak JM, van Hulten MC (2006) Increased tolerance of Litopenaeus vannamei to white spot syndrome virus (WSSV) infection after oral application of the viral envelope protein VP28. Dis Aquat Organ 70: 167-170.

47. Jha RK, Xu ZR, Shen J, Bai SJ, Sun JY, et al. (2006) The efficacy of recombinant vaccines against white spot syndrome virus in Procambarus clarkii. Immunol Lett 105: 68-76.

48. Cutting SM, Hong HA, Baccigalupi L, Ricca E (2009) Oral vaccine delivery by recombinant spore probiotics. Int Rev Immunol 28: 487-505.

49. FAO (2012) The state of world fisheries and aquaculture 2011. Food and Agriculture Organization of United Nations. FAO, Viale delle Terme di Caracalla, 00153 Rome, Italy.

50. Tacon AGJ, Metian M (2008) Global overview on the use of fish meal and fish oil in industrially compounded aquafeeds: Trends and future prospects.

51. Aquaculture 285: 146-158.

52. Li X, Li J, Wang Y, Fu L, Fu Y, et al. (2011) Aquaculture Industry in China: Current State, Challenges, and Outlook. Rev Fish Sci 19: 187-200.

53. Zhanga S, Li G, Wua H, Liub X, Yaoa Y, et al. (2011) An integrated recirculating aquaculture system (RAS) for land-based fish farming: The effects on water quality and fish production. Aquacult Eng 45: 93-102.

54. Anh P, Kroeze C, Bush S, Mol A (2010) Water pollution by intensive brackish shrimp farming in south-east Vietnam: Causes and options for control. Agr Water Manage 97: 872-882.

55. Bautista-Teruel MN, Eusebio P, Welsh T (2003) Utilization of feed pea, Pisum sativum, meal as a protein source in practical diets for juvenile tiger shrimp, Penaeus monodon. Aquaculture 225: 121-131.

56. Bomba A, Nemcová R, Gancarcíková S, Herich R, Guba P, et al. (2002) Improvement of the probiotic effect of micro-organisms by their combination with maltodextrins, fructo-oligosaccharides and polyunsaturated fattyacids. Br J Nutr 88: 95-99.

57. Boyd CE, Tucker CS (1998) Pond Aquaculture Water Quality Management Kluwer Academic Publishers, USA.

58. Chen JC, Liu PC, Lei SC (1990) Toxicity of ammonia and nitrite to Penaeus monodon adolescents. Aquaculture 89: 127-137.

59. Colt JE, Armstrong DA (1981) Nitrogen toxicity to crustaceans, fish and mollusks. Proceedings of the bio-engineering symposium for fish culture. Fish culture section of the American fisheries society (FCS Publ. 1): 34-47.

60. Rosas C, Cuzon G, Taboada G (2001) Effect of dietary protein and energy levels on growth, oxygen consumption, haemolymph and digestive gland carbohydrates, nitrogen excretion and osmotic pressure of Litopenaeus vannamei (Boone) and L. setiferus (Linne) juveniles (Crustacea, Decapoda; Penaeidae). Aquac Res 32: 531-547.

61. Daranee S, Davis DA (2011) Pond production of Pacific white shrimp (Litopenaeus vannamei) fed high levels of soybean meal in various combinations. Aquaculture 319: 141-149.

62. Enes P, Panserat S, Kaushik S, Oliva-Telesa A (2008) Growth performance and metabolic utilization of diets with native and waxy maize starch by gilthead sea bream (Sparus aurata) juveniles. Aquaculture 274: 101-108.

63. German DP, Horn MH, Gawlicka A (2004) Digestive enzyme activities in herbivorous and carnivorous prickleback fishes (Teleostei: Stichaeidae): ontogenetic, dietary, and phylogenetic effects. Physiol Biochem Zool 77: 789-804.

CHAPTER 5

EVALUATION OF PROBIOTIC STRAIN BACILLUS SUBTILIS C-3102 AS A FEED SUPPLEMENT FOR KOI CARP (CYPRINUS CARPIO)

Suxu He, Wenshu Liu, Zhigang Zhou, Wei Mao, Pengfei Ren, Toshihiro Marubashi and Einar Ringø[3]

Keywords: Probiotic, Intestinal microbiota community, HSP70, Koi carp, Immune protection

INTRODUCTION

Probiotics in aquaculture have been reported to provide beneficial effects and now the use of probiotics is an important management tool in fish culture [1]. The positive effects of probiotic administration to fish growth and immune response are well-documented [2,3]. Probiotics may promote growth, non-specific immune responses, disease resistance and the survival rate of aquatic animals [4,5].

Bacillus subtilis is a Gram positive, spore forming bacterium, used commercially in probiotic products because its beneficial effects have been documented and the pathogenic potential is generally described as low or absent [6]; moreover, as a spore former, Bacillus preparations are resistant and have a long lasting shelf life and thus can be stored in a state of dehydration [7]. B. subtilis can promote survival and growth [8], by stimulating the immune system [9] and by controlling pathogenic bacteria [10].

The ornamental fish trade is an economically important and profitable area of fish culture. The prosperity of the ornamental fish industry has induced the indiscriminate use of antibiotics and chemotherapeutants for improved health and nutrition, which has led to the development of drug-resistant strains of pathogenic microorganisms [11]. Koi carp is a high-value ornamental fish species. The benefits of probiotics on koi carp health, growth, and intestinal bacteria are well documented [12-14] but there are few reports investigating the variation intestinal microbiota community of koi carp treated with B. subtilis.

The aim of the present study was to evaluate the efficacy of Calsporin®, a probiotic based on viable spores of Bacillus subtilis C-3102, on growth performance, intestinal microbiota and cytokine regulation of koi carp. We chose HSP70 (heat shock protein), TGF-β (transforming growth factor), IL-1β (interleukin-1 beta), TNF-α (tumour necrosis factor) and IL-10 (interleukin-10) genes in the liver, intestine and kidney to assess the effects of B. subtilis C-3102 on stress and immune responses. In addition, skin pigmentation, which is one of the most important quality criteria in dictating koi carp market value, was measured. Finally, koi carps were challenged by Aeromonas hydrophila with intraperitoneal injection to determine the immune protection.

METHODS

Bacteria Strain

Calsporin® (Bacillus subtilis C-3102, live spore $\geq 1\times10^{10}$ cfu/g) was supplied by Calpis Co., Ltd (Tokyo, Japan) and kept in a sterilized container prior to use.

Diets and Experimental Design

Sixteen tanks with eight replicate tanks per treatment were used. The formulation and chemical composition of the basal diet (which served as the control diet) is shown in Table 1. These ingredients were mixed, extruded and air-dried at room temperature. The experimental diet was prepared using the basal formulation with the supplementation of 1 g/ kg of Calsporin® (containing $> 1 \times 10^{10}$ cfu/g Bacillus subtilis C-3102 spores). The diets were presented to the fish as pellets. Feeds were stored in a cool dry place until used. Table 2 shows the experimental design. The dietary level of viable B. subtilis was enumerated on selective medium. Briefly, approximately 1 g feed was homogenized in 100 mL of deionized water containing 0.05%

Citation: He S, Liu W, Zhou Z, Mao W, Ren P, et al. (2011) Evaluation of probiotic strain Bacillus subtilis C-3102 as a feed supplement for koi carp (Cyprinus carpio). J Aquac Res Development S1:005. doi:10.4172/2155-9546.S1-005.

Tween 20. Serial dilutions were made and spread onto plates of B. subtilis-selective medium and incubated for 3 d at 24 °C. The selective medium consisted of 326 mL vegetable juice (V8, Campbell Soup Co., USA), 33 g NaCl, 0.8 g dextrose, 490 mL water, and 16 g agar, with the pH adjusted to 5.2 using NaOH [15]. Forty-five mg of cyclohexamide and 22.5 mg of polymixin B were added to each liter of medium to inhibit fungi and Gram-negative bacteria, respectively. Populations were reported as cfu/g feed. Juvenile koi carp were transported from a private hatchery (Beijing, China) to the Feed Research Institute, Chinese Academy of Agricultural Sciences. All fish were fed the basal diet for 2 weeks to acclimate to the experimental conditions. After the acclimation, 384 fish (initial body weight 4.48± 0.02 g) were randomly assigned to sixteen 100-L tanks (eight replicate groups per dietary treatment) at a density of 24 fish per tank. The tanks were part of a recirculation system, and 10% of the water was exchanged daily. During the experiment, the temperature ranged from 23 to 26 °C, dissolved oxygen (DO) ranged from 5.6 to 7.8 mg/L, pH was 7.82±0.05 and the total ammonium and nitrite were kept below 0.1 and 0.05, respectively.

Fish were fed twice daily at 9:00 and 15:00 at 1 - 2% body weight per day. After feeding, uneaten food from each tank was removed by siphoning to calculate feed utilization.

Morphometric Analysis and Survival Rate

At the beginning and end of the 35 day trial, fish were batch weighed. At the same time, the total numbers of remaining juveniles were counted and survival rate was calculated. In addition, the color index was measured by the SalmoFan™ Lineal based on the surface erythema.

The weight gain (%) was calculated as:

$$\frac{\text{final weight (g) - initial weight (g)}}{\text{initial weight (g)}} \times 100\%$$

The feed conversion ratio (FCR) was calculated as:

$$\frac{\text{total feed consumption (total feed casting - total food residue) (g)}}{\text{total final weight (g) - total initial weight (g) + total morality weight (g)}} \times 100\%$$

Table 1: Diet formulation and chemical composition of the basal diet

Ingredient	Percentage (%)
Fish meal	47.0
Soybean meal	24.0
Wheat flour	24.0
Soybean oil	2.0
$Ca(H_2PO_4)_2$	2.2
Vitamin C phosphate ester	0.1
Choline chloride	0.3
Vitamin mix	0.2
Mineral mix	0.2
Chemical composition	
Crude protein	42.0
Crude lipid	7.3

Table 2: The experimental treatments

Treatment	Diet	Calsporin® (g/ kg feed)
Control	Basal	0
Calsporin®	Basal + Calsporin®	1

Intestinal Microbial Analyses

Three fish per tank were sampled at day 0, 10, 20, 35 for intestinal microbial analysis using the method described by Zhou et al. [16]. Intestines of the koi carp were aseptically taken out, opened, and gently agitated thrice for 1 min in phosphate-buffered saline (PBS: 130 mM NaCl, 10 mM NaH2 PO4 , pH 7.2) to remove the contents. Genomic DNA was obtained using the extraction method described by Yu and Morrison [17]. The target sequence (V3 region of gene ssr) was amplified by PCR using primers 338f (5'-ACTCCTACGGGAGGCAGCAG-3') and 519r (5'-ATTACCGCGGCTGCTGG-3') for DGGE. DGGE gels were stained with ethidium bromide solution (5 µg/mL; 20 min), washed with deionized water, and viewed by UV transillumination. Computerassisted comparison of DGGE banding patterns was performed with BIO-1 D++ gel analysis software (VilberLourmat, Torcy, France). Cluster analysis was based on the unweighted pair group method using the arithmetic mean algorithm (UPGMA). The excised bands were reamplified and purified, then sequenced.

Gene Expression Analyses

Three fish per tank were sampled for liver, kidney and intestinal gene expression analysis at day 0, 10, 20 and 35, respectively. To reduce individual variation of gene expression, the sampled fish organs from each tank were pooled and homogenized using a glass homogenizer. Total RNA was extracted using a TRIzon Reagent RNA kit according to the manufacturer's instructions (Promega, Germany), thereafter the quality was analyzed on a 1.2% agarose gel. RNA was dissolved in 50µL RNase-Free water and stored at −80°C until use. The reverse transcription polymerase chain reaction (RT-PCR) was performed by using Rever Tra Ace-α-RT-PCR kit (TOYOBO, Shanghai, China) according to the manufacturer's instructions.

The RT-PCR primers were cautiously designed using Primer premier 5.0 software based on the available sequences in GenBank (Table 3), the primers were designed on the following principles: 1) the PCR product should be between 200 and 500 bp, 2) the GC content within 40-60%, and 3) the annealing temperature is between 58-60°C. The qPCRs were performed with the SYBR Green Premixus Ex Taq TMII (TaKaRa, Beijing, China) in an iQ5 multicolor real time PCR Detection system (Bio-Rad). The total volume of PCR was 20 µL and consisted of: 10 µL SYBR Green Premix Ex TaqII (2×), 1 µL primer of each, 2 µL cDNA and 6 µL ddH2O. The cycling conditions were as follows: 95°C 3min and then 40 cycles of 95°C 20, 55°C 20s and 72°C 20s. All real-time PCRs were performed at least in triplicate. Data analysis was conducted using the 2-ΔΔCT method [18], and 40S gene was chosen as internal standard.

Assessment of the Protective Effect of Probiotic on Aeromonas Hydrophila Infection

Aeromonas hydrophila NJ-1 was obtained from a natural outbreak in one of the culture ponds at the facility and donated by Professor Yongjie Liu (Nanjing Agricultural University, Nanjing, China). After the 35 day feeding experiment, carp were infected by A. hydrophila NJ-1 by an intraperitoneal injection of 50 µL at a dose of 107 cfu/mL. The Protective effects were evaluated by recording the accumulated mortalities over 7 days.

Statistical Analysis

Results are expressed as the mean ± S.D. Difference between groups were determined using a paired t-test with the statistical software package SPSS 17.0. Significant differences were accepted at $P < 0.05$.

In this study, a change in the expression of mRNA level by more or less than 50%, relative to baseline levels (i.e. day 0) is regarded as significant difference.

RESULTS

Growth Performance

The koi carp survival rate during the nutrition trial was 100% after 35 days. Data on growth performance and feed utilization including final body weight (FBW), relative weight gain (WG), feed conversion ratio (FCR) and survival rate (SR) are shown in Table 4. The mean final weight of the probiotic group was significantly higher ($P < 0.05$) than that of the control group. The relative WG of the probiotic group was significantly higher ($P < 0.05$) than that of the control group, moreover, the probiotic group presented lower FCR than that of control group.

The body surface colour index of koi carp is presented in Table 5. The mean value of the colour index of the probiotic fed fish was not significantly different to that of the control fed fish.

Intestinal Microbial Analyses

The 16S rDNA-V3 PCR-DGGE fingerprints of the intestinal autochthonous bacterial communities of the experimental koi carp are showed in Figure 1. Thirty two different bands from the DGGE gel were excised and the eluted DNA subjected to PCR and successful sequencing. The closest relative (obtained from BLAST search) of bacteria excised from the 16S V3 PCR-DGGE are presented in Table 6.

Obvious variation of the bacterial communities members were observed, especially at day 20 and 35. The visual band numbers for the initial fish, B. subtilis C-3102 treatment (T) and the control group (CK) are as follows: initial (20), CK10 (18), T10 (18), CK20 (12), T20 (10), CK35 (11) and T35 (11). The potential pathogenic bacteria such as Vibrio cholerae (band 4), Aeromonas sp. (band 24), Aeromonas jandaei (band 30) were not present at day 10; the abundance of Plesiomonas shigelloides was reduced at day 10 in the probiotic group, and then disappeared at days 20 and 35; Aeromonas sp. (band 21, 24), Aeromonas veronii (band 29), Aeromonas jandaei (band 30) were not detected in any groups at day 20. But another pathogenic bacteria, Vibrio cholerae (band 12), was detected from day 20 in all groups. In addition, B. subtilis C-3102 wasn't detected in either treatment at any time point. However, the concentration of B. subtilis C-3102 in the feed was 9.3x106 cfu/g at the end of the experiment in the probiotic diet (B. subtilis C-3102 was absent in the control diet).

Table 3: Primer sequences for QPCR

Gene name	Primer sequences	Product size (bp)	Genebank number
IL-10	F:GCACCATTACTCGATGAAAACG R:CTGGCGAACTCAAAGGGATT	236	JF957369
TNF-α	F: GCCATAGGAATCAGAGTAGCG R: GACCAGGCTTTCACTTCAGG	183	JF957372
HSP70	F: CACAATCACCAACGATAAGGG R: TTGGCAGACACCTTTTCACGC	114	JF957366
TGF-β	F: CCTGGGCTGGAAGTGGATAC R:GTAAAAGATGGGCAGTGGGTC	190	JF957371
40S	F:GTTGTACTTGCGGATGTAATGC R:TGGCTGTTGAAGGAAGTGGC	233	JF957364

Table 4: Effects of dietary Bacillus subtilis C-3102 on the growth performance, diet conversion and survival rate of koi carp

	IBW (g)	FBW (g)	WG (%)	FCR	SR (%)
Control	4.49±0.01	11.20±0.21	149.61±4.69	1.41±0.06	100.0
Calsporin®	4.47±0.02	11.73±0.38**	162.13±8.22**	1.31±0.04**	100.0

** $p < 0.01$.

The similarity coefficients of the treatments are shown in Figure 2. The similarity coefficients between the dietary B. subtilis C-3102 treatment and the control group gradually increased. In particular at day 35, the similarity coefficients exceeded 0.90. These results indicated that the effect of dietary B. subtilis C-3102 on intestinal microbiota of koi carp decreased with time/feeding duration in this study.

mRNA Analysis by Real-time PCR

The expression of the IL-1β, TNF-α, HSP70, TGF-β, IL-10 genes were evaluated at day 0, 10 ,20, and 35 in the liver, intestine and kidney to investigate the effects of B. subtilis C-3102 on immunity and stress at the molecular level.

In the liver (Figure 3a), dietary B. subtilis C-3102 caused a small, but significant increase in TNF-α and TGF-β mRNA levels at days 35 relative to day 0, also in HSP70 mRNA level at days 10 and 35, and in IL-10 mRNA level at days 20. However, IL-1β mRNA production was significantly down-regulated at day 10.

In the intestine (Figure 3b), a significant increase was detected in IL-1β, TNF-α and TGF-β gene expression at day 20 and 35. IL-10 gene expression was significantly increased at day10, 20 and 35. But IL-1β mRNA production was significant down-regulated at day 10. No significant effect was observed to HSP70 gene expression.

In the kidney (Figure 3c), IL-1β gene expression level was significantly reduced at day 35, and TGF was significantly downregulated at day 10 compared to initial levels. No significant effects were detected for TNF-α or HSP70 gene expression.

Immune Protection against Disease

The result of survival rate 7 days after the Aeromonas hydrophila challenge is displayed in Figure 4. No significant difference ($P > 0.05$) was detected between the CK and treated trials. This result suggests that Bacillus subtilis C-3102 could not enhance the immune protection to a great enough extent to protect against the A. hydrophila in the present challenge model.

DISCUSSION

Probiotics have been applied as dietary additives or water additives to improve growth performance and immune response of experimental fish. With the development of the ornamental fish trade, more investigations of the effects of probiotics on ornamental fish have been conducted [19-22]. Spores of Bacillus strains are thermo-stabile which makes them easy to incorporate into compound diets. In this particular instance, diets supplemented with B. subtilis C-3102 spores resulted in growth performances and feed utilization of koi carp significantly better than that of the koi carp fed the control basal diets (Table 4). Thus, indicating growth improving properties of the probiotic Calsporin. Similar results were also observed in studies on gilthead seabream (Sparus aurata L.) [19,23], tilapia (Oreochromis niloticus) [24] and large yellow croaker (Larimichthys crocea) [9]. These studies show that the Bacillus sp. can improve fish growth performance, immune response and digestive enzyme activity. In the present study, the administration of the probiotic had no effect on survival (Table 4) of koi carp or skin pigmentation (Table 5).

Table 5: Effect of dietary Bacillus subtilis C-3102 on the body surface color of koi carp$

	Color index
Control	27.8±1.2
Calsporin®	26.2±2.0

$Color index was measured by the SalmoFan™ Lineal based on the surface erythema

In the present study, the DGGE profile (Figure 1) and sequencing results (Table 6) showed that the intestinal bacteria varied with culturing time. Species such as Vibrio cholerae and Aeromonas spp. only appeared at the beginning experiment; some Vibrio, Plesiomonas, Cetobacterium, Escherichia and Aeromonas species were detected only on day 10 and some uncultured and Acinetobacter bacteria appeared only at day 20.

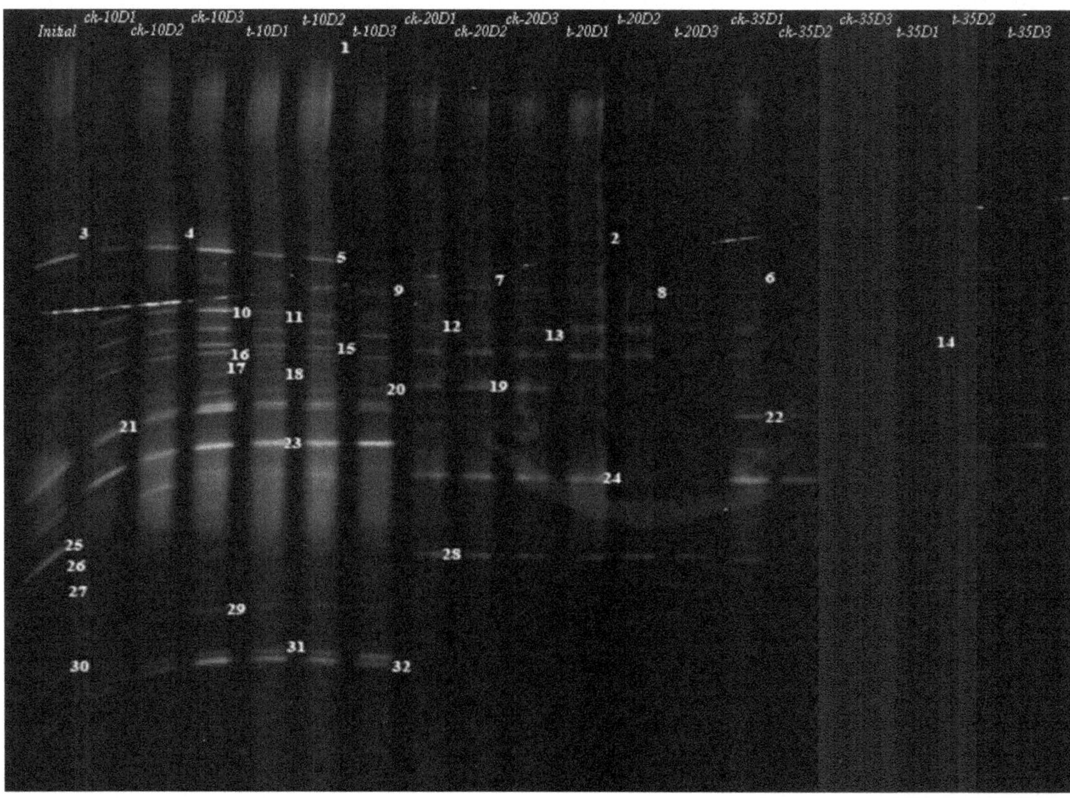

Figure 1: Effects of dietary Bacillus subtilis C-3102 on the intestinal autochthonous microbiota of koi carp by 16S rDNA v3 Denaturing Gradient Gel Electrophoresis (DGGE). Initial: the initial sample; ck-10D1-3: the triplicates of the control at the end of 10 feeding days; t-10D1-3: the triplicates of the Bacillus subtilis C-3102 treatment at the end of 10 feeding days; ck-20D1-3: the triplicates of the control at the end of 20 feeding days; t-20D1-3: the triplicates of the Bacillus subtilis C-3102 treatment at the end of 20 feeding days; ck-35D1-3: the triplicates of the control at the end of 35 feeding days; t-35D1-3: the triplicates of the Bacillus subtilis C-3102 treatment at the end of 35 feeding days. 1-32: the band number.

Furthermore, the number of autochthonous strains was dramatically decreased on day 20 and 35 compared to the initial or day 10 levels. The intestinal bacterial flora of hybrid tilapia cultured in earthen ponds has been reported to vary with seasonal changes: Al-Harbi & Uddin [25] reported that Pseudomonas spp. were found only in winter while other bacteria (e.g. Photobacterium damselae, Pasteurella spp., Cellulomomus sp. and Bacillus sp.) were present in some other seasons of the year. The observation of hybrid tilapia intestinal autochthonous bacterial variations over feeding period has also been reported [26]. In our study, Aeromonas veronii appeared at the initial and day 10 of control group, which were inhibited by dietary B. subtilis C-3102 (as indicated by their absence in the DGGE profiles of the probiotic group at all time points). A. veronii is a pathogenic bacteria which can cause fish mortalities [27]. Previous studies have also reported that dietary probiotics can reduce levels of potential pathogenic bacteria in the GI tract of fish [24,28]. Therefore, microbial manipulation constitutes a viable tool to reduce or eliminate the incidence of opportunistic pathogens [29]. The probiotics B. subtilis C-3102 was not observed as one of the dominant (i.e. was not detected) species in the GI tract of the probiotic fed fish at any timepoint. This similar phenomenon has also been observed in a previous study which showed that dietary Bacillus toyoi could not be detected in the intestine of European eels (Anguilla anguilla L.) [30]. The different effects of probiotics on GI microbiota may be attributed to the different mechanisms of action of probiotics. Mechanisms of microbiota modulation by probiotics include: (1) competition for attachment sites and space, (2) competition for nutrients, (3) production of antimicrobial substances, and (4) enhancement of the immune response against pathogenic microorganisms [1]. But the exact mechanisms and the interactions between the host and microbe, microbe and microbe, are still unknown.

Table 6: Representative of bacteria or clones and their relative abundance (RA, %) isolated from the gut wall of Koi carp fed Bacillus subtilis C-3102

Band No.	Closest relative (obtained from BLAST search)	Identity (%)	Relative abundance (RA, %)						
			Initial	ck-10D	t-10D	ck-20D	t-20D	ck-35D	t-35D
1	Uncultured *Ferrimonas* sp. (HQ914736.1)	99	0	0.0	0.9±0.5	1.1±0.5	2.1±1.4	0.9±0.7	1.5±0.6
2	Uncultured Mollicutes bacterium (HM749044.1)	100	0	0.0	0.0	0.0	1.6±2.3	0.0	0.0
3	Uncultured bacterium (EU468590.1)	96	3.2	0.0	0.0	0.0	0.0	0.0	0.0
4	Vibrio cholerae (GQ359963.1)	100	0	0.8±0.5	0.6±0.2	0.0	0.0	0.0	0.0
5	*Plesiomonas shigelloides* (HM007572.1)	100	20.6	10.9±2. 5	2.9±0.3	0.0	0.0	0.0	0.0
6	Uncultured bacterium (EU468590.1)	96	0	0.0	0.0	0.0	0.00	4.90±0.93	3.4±1.5
7	*Acinetobacter* sp. (HM582679.1)	100	0	0.0	0.0	0.4±0.3	0.6±0.9	0.0	0.0
8	Uncultured bacterium (EU468590.1)	96	0	0.0	0.0	4.2±1.1	4.0±1.8	4.4±1.5	2.3±1.0
9	*Cetobacterium somerae* (AB353124.1)	100	1.4	5.0±2.1	2.3±0.6	0.0	0.0	0.0	0.0
10	Uncultured bacterium (EU469094.1)	94	2.3	9.9±3.2	1.0±0.4	1.1±1.0	0.0	0.0	0.0
11	*Cetobacterium* sp. (HM778168.1)	100	0.3	0.4±0.1	0.3±0.05	0.0	0.0	0.0	0.0
12	*Vibrio cholerae* (GQ359963.1)	100	0	0.0	0.0	6.4±1.4	9.8±4.1	4.3±2.3	4.9±2.6
13	Uncultured bacterium (HM749031.1)	99	0	0.0	0.0	0.97±0.42	0.0	0.0	3.2±0.4
14	Uncultured bacterium (HQ905034.1)	100	1	10.6±5.8	3.5±0.9	12.2±2.5	0.0	2.3±1.8	6.0±2.7
15	Uncultured bacterium (HM749031.1)	100	1.4	5.3±2.3	3. 7±2.4	8.6±1.5	15.0±3.4	8.4±1.0	0.0
16	*Vibrio* sp. (AB182229.1)	100	1.2	4.5±1.6	1.6±0.2	0.0	0.0	3.4±2.3	7.4±1.5
17	Uncultured bacterium (HM778780.1)	98	0	5.4±3.7	1.5±0.2	0.0	0.0	0.0	0.0
18	*Cetobacterium somerae* (AB353124.1)	100	0.9	4.3±3.1	1.8±0.3	0.0	0.0	0.0	0.0
19	Uncultured bacterium (AB179538.1)	96	0	0.0	0.0	10.9±1.4	0.0	0.0	0.0
20	*Escherichia fergusonii* (HQ259962.1)	100	1.2	2.4±0.4	3.7±0.9	0.0	0.0	0.0	0.0
21	Aeromonas sp. (HQ841071.1)	100	0.5	2.0±1.3	3.0±1.0	0.0	0.0	0.0	0.0
22	*Uncultured bacterium* (GU531301.1)	100	3.9	9.2±4.4	13.1±2.6	5.67±0.88	9.7±5.4	18.8±3.3	11.9±5.0
23	Uncultured Mollicutes bacterium (HM749044.1)	100	29.8	24.5±3.7	33.8±4.5	0.0	14.9±2.4	9.8±2.0	21.9±3.1
24	*Aeromonas* sp. (HQ841071.1)	100	1.3	2.0±0.5	2.0±0.6	30.6±2.1	21.8±10.3	27.2±1.0	20.1±1.8
25	Uncultured *Exiguobacterium* sp. (HM749032.1)	100	15.7	0.0	0.0	0.0	0.0	0.0	0.0
26	Acinetobacter sp. (HQ684847.1)	100	1.4	0.0	0.0	0.0	0.0	0.0	0.0
27	Uncultured Mollicutes bacterium (HM749044.1)	98	2.1	0.0	0.0	0.0	0.0	0.0	0.0
28	*Acinetobacter* sp. (FR749840.1)	95	0	0.0	0.0	18.2±1.6	21.2±7.3	16.3±4.1	18.3±2.4
29	*Aeromonas veronii* (HQ832864.1)	99	0.6	0.4±0.5	0.0	0.0	0.0	0.0	0.0
30	*Aeromonas jandaei* (HQ683971.1)	100	0.9	0.0	0.0	0.0	0.0	0.0	0.0
31	*Acinetobacter* sp. (HM582678.1)	100	0	4.2±0.5	10.6±0.9	0.0	0.0	0.0	0.0
32	Bacterium DC21(2011) (HQ178949.1)	100	13.3	6.0±5.1	14.2±2.5	0.0	0.0	0.0	0.0

Fish growth is strictly related to environmental conditions, where improper captive rearing conditions induce stress and may negatively affect fish welfare. HSP genes, in particular HSP70, play essential roles in stress tolerance. A lower level of HSP70 gene expression was reported in Gilthead sea bream larvae fed on live prey enriched with probiotics (composed of three Bacillus strains) at both day 47 and 75 post hatch which indicated that a positive role of this mix of Bacillus reducing the severity of cellular stress [24]. In the present study, the HSP70 gene expression level was up-regulated in the liver at day 10 but reduced thereafter; no effects were exhibited in intestine and kidney of koi carp treated with the probiotic.

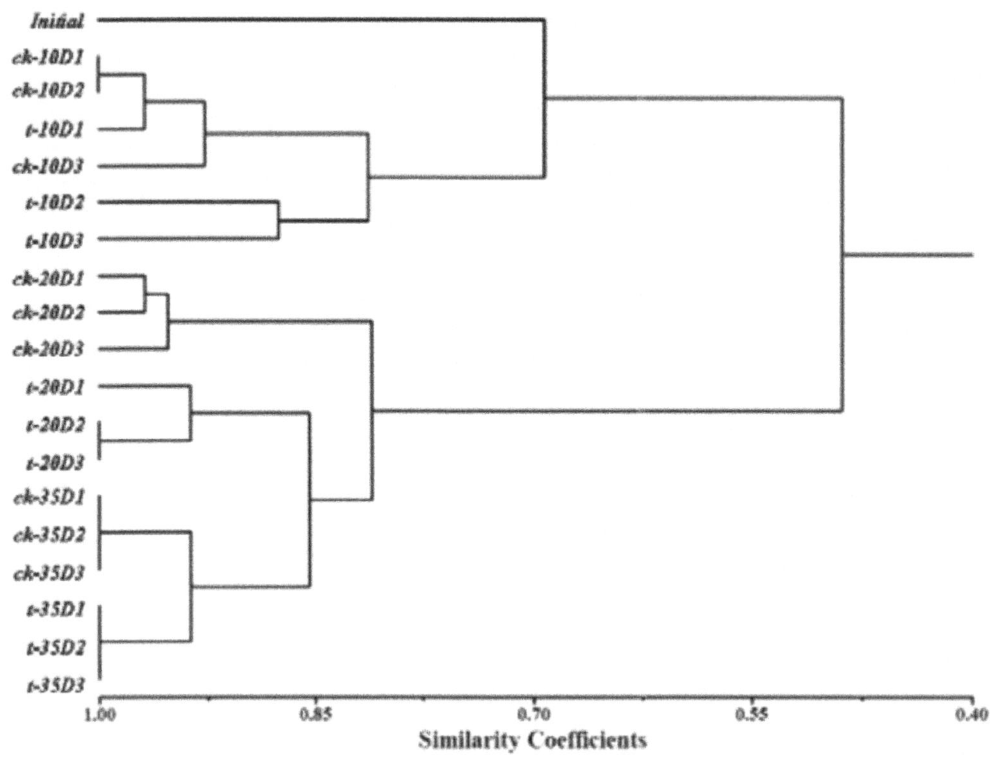

Figure 2: Cluster analysis of the intestinal autochthonous microbiota of koi carp fed Bacillus subtilis C-3102 based on 16S rDNA v3 Denaturing Gradient Gel Electrophoresis (DGGE). Initial: the initial sample; ck-10D1-3: the triplicates of the control at the end of 10 feeding days; t-10D1-3: the triplicates of the Bacillus subtilis C-3102 treatment at the end of 10 feeding days; ck-20D1-3: the triplicates of the control at the end of 20 feeding days; t-20D1-3: the triplicates of the Bacillus subtilis C-3102 treatment at the end of 20 feeding days; ck-35D1-3: the triplicates of the control at the end of 35 feeding days; t-35D1-3: the triplicates of the Bacillus subtilis C-3102 treatment at the end of 35 feeding day.

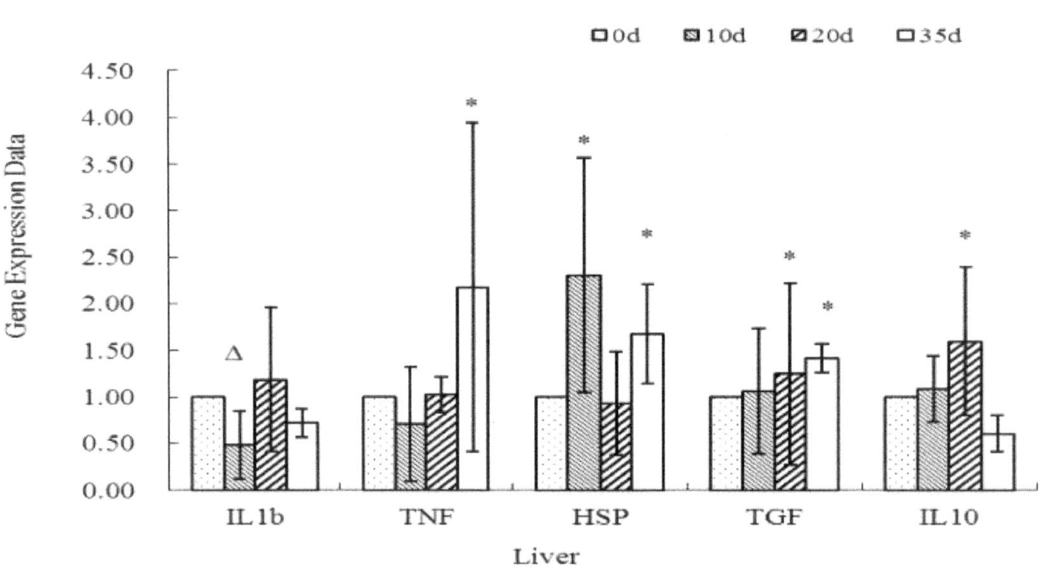

(a)

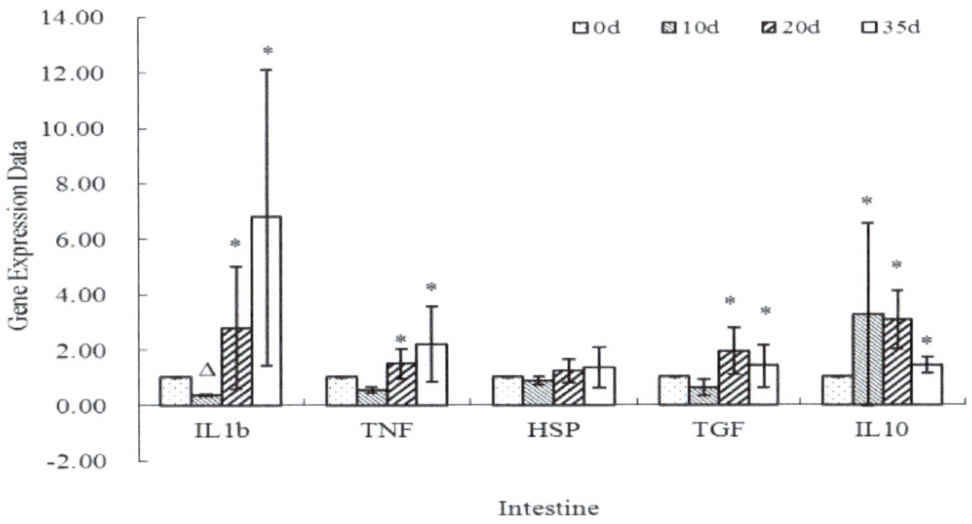

(b)

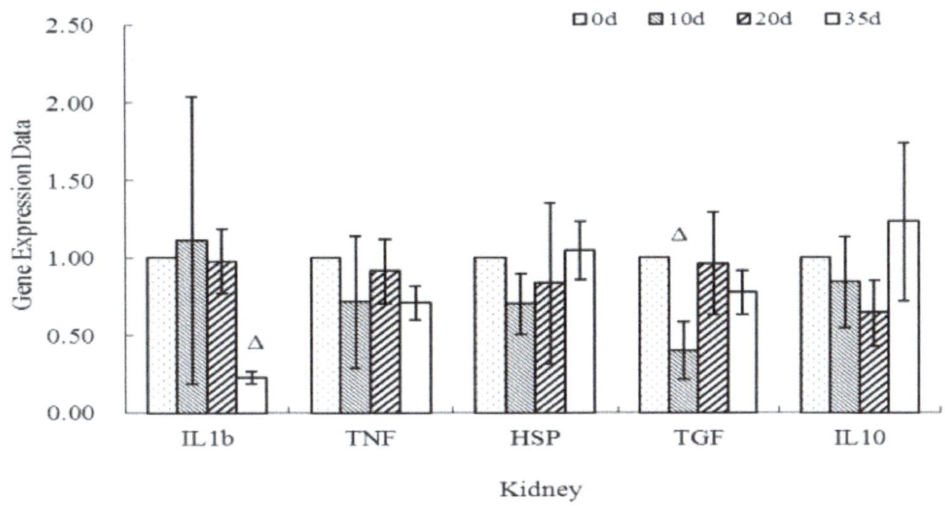

(c)

Figure 3: Effects of dietary Bacillus subtilis C-3102 on the immune responses of koi carp. (a) liver; (b) intestine; (c) kidney.

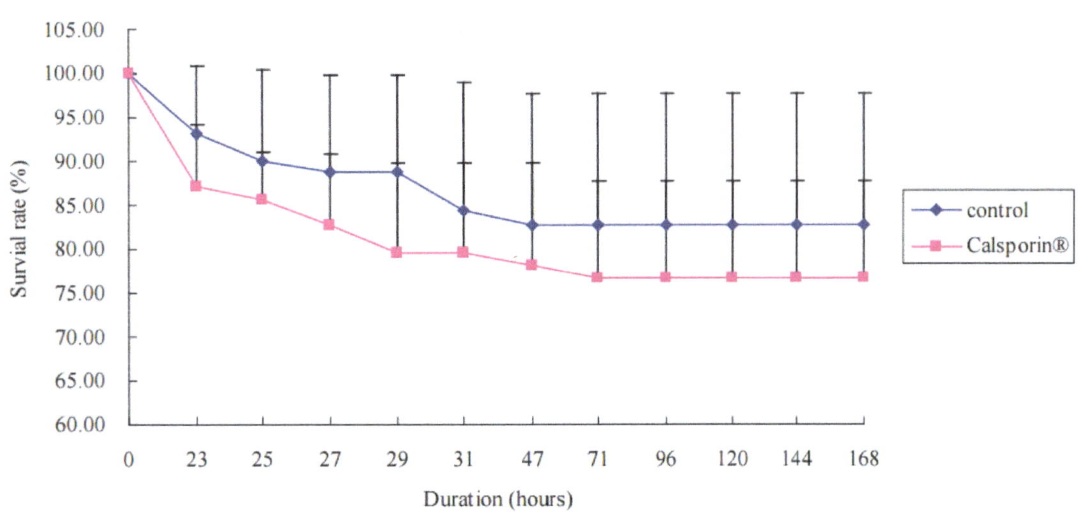

Figure 4: Effect of dietary Bacillus subtilis C-3102 on the survival rate of koi carp infected by Aeromonas hydrophila NJ-1.

TGF-β, TNF-α, IL-1β (all pro-inflammatory) and IL-10 (antiinflammatory) are inflammatory cytokines that play an important role in non-specific immune responses. TGF-β is a pleiotropic cytokine involved in tissue remodelling, wound repair, development and haematopoiesis, and is expressed in a wide range of cells and tissues. It acts to down-regulate the expression of cytokines and cytokineinduced effects. TNF-α is produced as a transmembrane protein or glycoprotein that is elevated in the extracellular C-terminal domain to release a biologically active mature peptide. Interleukin-1β is a major player in immune response of fish as in mammals. It's a key mediator in response to microbial invasion and tissue injury and can stimulate immune responses by activating lymphocytes or by inducing the release of other cytokines capable of triggering macrophages, NK cells and lymphocytes. Macrophages are the primary source of IL-1β although it is produced by a wide variety of other cell types [31]. As an autoregulatory mediator, IL-10 has important regulatory effects on immunological and inflammatory responses because of its capacity to inhibit the production of proinflammatory cytokines by monocytes [32]. Dietary Bacillus subtilis provided to rainbow trout have previously been reported to have no effects on the expression of the cytokine genes in the spleen and kidney [33], but the cytokine genes were significantly increased in rainbow trout head-kidney leucocytes when incubated in vitro with Carnobacterium maltaromaticum B26 and C. divergens B33 [34]. In the present study, a trend of elevated cytokine gene expression levels was observed in the intestine and liver but were unaffected in the kidney. The reason for the regulation of cytokine genes in fish organ induced by probiotic is not clear. Even though the probiotic was not detected as a prominent member of the gut microbiota from the DGGE analysis, we hypothesize that cell wall components or spores, which were not lysed effectively prior to DNA analysis, were present in the gut which may have stimulated the lymphoid cells in the mucosa to induce changes in the expression of TGF-β, TNF-α and IL-1β genes. In order to modulate immune response, IL-10 expression was significantly increased ($P < 0.05$) in the immune organs (liver and intestine). According to the previous studies and our results, Bacillus probiotics can cause different effects on cytokine gene expression levels in different organs of fish. Gatesoupe [35] concluded that probiotics for aquaculture should be antagonistic to pathogens and increase resistance of the host to pathogens. However, in our study, B. subtilis C-3102 did not show any protection against A. hydrophila, and the reason may be due to the too high dose of A. hydrophila used.

CONCLUSION

The results showed that dietary supplementation of B. subtilis C-3102 could provide beneficial effects on koi growth performance, strongly affect the intestinal microbiota at the initial stages (which diminished with time), and elevate the expression of some immune related genes. However, B. subtilis C-3102 did not improve the disease resistance of koi challenged with A. hydrophila. Further work is required to determine the function of the cytokine expressions related to host protection against A. hydrophila infection and disease recovery.

REFERENCES

1. Balcázar JL, de Blas I, Ruiz-Zarzuela I, Cunningham D, Vendrell D, et al. (2006) The role of probiotics in aquaculture. Vet Microbiol 114: 173-186.
2. Nikoselainen S, Ouwenhand AC, Bylund G, Salminen S, Lilius EM (2003) Immune enhancement in rainbow trout (oncorhynchus mykiss) by potential probiotic bacteria (Lactobacillus rhamnosus). Fish Shellfish Immunol 15: 443-452.
3. Kumar R, Mukherjee SC, Prasad KP, Pal AK (2006) Evaluation of bacillus subtilis as a probiotic to Indian major carp Labeo rohita (Ham.). Aquacult Res 37: 1215-1221.
4. Wang YB, Xu ZR (2006) Effect of probiotics for common carp (Cyprinus carpio) based on growth performance and digestive enzyme activities. Anim Feed Sci Technol 127: 283-292.
5. Venkat HK, Sahu NP, Jain KK (2004) Effect of feeding Lactobacillusbased probiotics on the gut microflora, growth and survival of postlarvae of Macrobrachium rosenbergii (de Man). Aquac Res 35: 501-507.
6. Kim Dh, Austin B (2006) Innate immune responses in rainbow trou (Oncorhynchus mykiss, Walbaum) induced by probiotics. Fish Shellfish Immunol 21: 513-524.
7. Boer AS, Diderichsen B (1991) On the safety of Bacillus subtilis and B.amyloliquefaciens: A review. Appl Microbio Biotechnol 36: 1-4.
8. Wang Y, Li J, Lin J (2008) Probiotics in aquaculture: challenges and outlook. Aquaculture 281: 1-4.
9. Ai QH, Xu HG, Mai KS, Xu W, Wang J, et al. (2011) Effects of dietary supplementation of Bacillus subtilis and fructooligosaccharide on growth performance, survival, nonspecific immune response and disease resistance of juvenile large yellow croaker, Larimichthys crocea. Aquaculture 317: 155-161.
10. Vaseeharan B, Ramasamy P (2003) Control of pathogenic Vibrio spp. by Bacillus subtili BT23, a possible probiotic treatment for black tiger shrimp Penaeus monodon. Lett Appl Microbiol 36: 83-87.
11. Amabile-Cuevas CF, Cárdenas-Garcia M, Ludgar M (1995) Antibiotic resistance. Anim Sci 83: 320-329.
12. EL-Nobi GA, Reda RM, Hassanin ME, Elmowalid G (2009) The impact of applications of probiotic "Bacillus subtilis" on health and growth of koi carp "Cyprinus carpio". Proceeding of the 2nd Global Fisheries and Aquaculture Research Conference.
13. Dhanaraj M, Haniffa MA, Singh SVA, Arockiaraj AJ, Ramakrishanan CM, et al. (2010) Effect of Probiotics on Growth Performance of Koi Carp (Cyprinus carpio). J Appl Aquacult 22: 202-209.
14. Chandra R, Rajan MR (2009) Probiotic effect of intestinal bacteria of Koi carp Cyprinus carpio var. koi. J Pure Appl Microbiol 3: 363-366.
15. Turner JT, Backman PA, (1991) Factors relating to peanut yield increases after seed treatment with Bacillus subtilis. Plant Disease 75: 347–353. Full Text via CrossRefTurner and Backman, 1991.

16. Zhou ZG, He SX, Liu YC, Shi P, Huang GX, et al. (2009) The Effects of Dietary Yeast Culture or Short-chain Fructo-oligosaccharides on the Intestinal Autochthonous Bacterial Communities in Juvenile Hybrid Tilapia, Oreochromis niloticus♀× Oreochromis aureus♂. J World Aquacult Soc 40: 450-459.

17. Yu Z, Morrison M (2004) Improved extraction of PCR-quality community DNA from digesta and fecal samples. Biotechniques 36: 808-812.

18. Livak KJ, Schmittgen TD (2001) Analysis of Relative Gene Expression Data Using Real-Time Quantitative PCR and the 2-ΔΔCT Method. Methods 25: 402- 408.

19. Salinas I, Cuesta A, Esteban MA, Meseguer J (2005) Dietary administration of Lactobacillus delbrüeckii and Bacillus subtilis, single or combined, on gilthead seabream cellular innate immune responses. Fish Shellfish Immunol 19: 67-77.

20. Ghosh S, Sinha A, Sahu C (2007) Dietary probiotic supplementation on growth and health of live-bearing ornamental fishes. Aquacult Nutr 14: 289-299.

21. Ghosh S, Sinha A, Sahu C (2007) Effect of probiotic on reproductive performance in female live-bearing ornamental fish. Aquacult Res 38: 518-526.

22. Avella MA, Olivotto I, Silvi S, Place AR, Carnevali O (2010) Effect of dietary probiotics on clownfish: a molecular approach to define how lactic acid bacteria modulate development in a marine fish. Am J Physiol Regul Integr Comp Physiol 298: 359-371.

23. Avella MA, Gioacchini G, Decamp O, Makridis P, Bracciatelli, et al. (2010) Application of multi-species of Bacillus in sea bream larviculture. Aquaculture 305: 12-19.

24. Aly SM, Ahmed YAG, Ghareeb AAZ, Mohamed MF (2008) Studies on Bacillus subtilis and Lactobacillus acidophilus, as potential probiotics, on the immune response and resistance of Tilapia nilotica (Oreochromis niloticus) to challenge infections. Fish Shellfish Immunol 25: 128-136.

25. Al-Harbi AH, Uddin NM (2004) Seasonal variation in the intestinal bacterial flora of hybrid tilapia (Oreochromis niloticus × Oreochromis aureus) cultured in earthen ponds in Saudi Arabia. Aquaculture 229: 37-44.

26. He SX, Zhou ZG, Liu YC, Shi PJ, Yao B, et al. (2009) Effects of dietary Saccharomyces cerevisiae fermentation product (DVAQUA®) on Growth performance, intestinal autochthonous bacterial community and non-specific immunity of hybrid tilapia (Oreochromisniloticus ♀×O. aureus ♂) cultured in cages. Aquaculture 294: 99-107.

27. Rahman M, Colque-Navarro P, Kühn I, Huys G, Swings J, et al. (2002) Identification and characterization of pathogenic Aeromonas veronii biovar sobria associated with epizootic ulcerative syndrome in fish in bangladesh. Appl Environ Microb 68: 650-665.

28. Avella MA, Olivotto I, Silvi S, Ribecco C, Cresci A, et al. (2011) Use of Enterococcus faecium to improve common sole (Solea solea) larviculture. Aquaculture 315: 384-393.

29. Balcázar JL (2002) Use of probiotics in aquaculture: general aspects. In: de Blas, I. (Ed.), Memorias del Primer Congreso Iberoamericano Virtual de Acuicultura, Zaragoza, Spain, pp. 877-881.

30. Chang CI, Liu WY (2002) An evaluation of two probiotic bacterial strains, Enterococcus faecium SF68 and Bacillus toyoi, for reducing edwardiellosis in cultured European eel, Anguilla anguilla L. J Fish Dis 25: 311-315.

31. Secombes CJ, Wang T, Hong S, Peddie S, Crampe M, et al. (2001) Cytokines and innate immunity of fish. Dev Comp Immunol 25: 713-23.

32. de Waal Malefyt R, Abrams J, Bennett B, Figdor CG, de Vries JE (1991) Interleukin 10(IL-10) inhibits cytokine synthesis by human monocytes: an autoregulatory role of IL-10 produced by monocytes. J Exp Med 174: 1209-1220.

33. Panigrahi A, Kirona V, Satoha S, Hirono I, Kobayashi T, et al. (2007) Immune modulation and expression of cytokine genes in rainbow trout Oncorhynchus mykiss upon probiotic feeding. Dev Comp Immunol 31: 372-382.

34. Kim Dh, Austin B (2006) Cytokine expression in leucocytes and gut cells of rainbow trout, Oncorhynchus mykiss Walbaum, induced by probiotics. Vet Immunol Immunopathol 114: 297-304.

35. Gatesoupe FJ (1999) The use of probiotics in aquaculture. Aquaculture 180: 147-165.

CHAPTER 6

STREPTOMYCES BACTERIA AS POTENTIAL PROBIOTICS IN AQUACULTURE

Loh Teng-Hern Tan, Kok-Gan Chan, Learn-Han Lee and Bey-Hing Goh

Keywords: *Streptomyces*, probiotic, aquaculture, fish pathogens, antibiotic resistance

INTRODUCTION

Statistics have revealed that the global aquaculture production continue to increase rapidly without the sign of reaching its peak. Meanwhile, the production from global capture fisheries has stabilized around 90 million tons since the mid-nineties (Mathieson, 2012). According to the United Nations Food and Agriculture Organization report (FAO, 2014), the global aquaculture production achieved another all-time high of 90.4 million tons including the 66.6 million tons of food fish and 23.8 million tons of aquatic algae in 2012 in response to the rising domestic and international seafood demand. Currently, it has been reported that food fish provides an average of one-fifth of total animal protein intake for the world population estimated at 7.3 billion people (Moffitt and Cajas-Cano, 2014). However, major disease outbreaks have been reported within the aquaculture sector in many part of the world due to the increased fish stocking density, overcrowding and lack of sanitary management with the rapid growth of aquaculture. The rapid spread of infections have led to global estimate of disease losses ranges about a quarter billion US$ annually (Bondad-Reantaso et al., 2005). For instance, the viral infections (white-spot syndromes, yellow head disease and taura syndrome) in shrimp industry has cost billions of dollars worldwide (Flegel, 2012; Lightner et al., 2012). Also, bacterial pathogens such as Vibrio sp. (Vibrio harveyi, V. parahaemolyticus, V. campbellii) caused luminous vibriosis in shrimp farms resulted in 50–100% mortality and vibrio infections in human (Shruti, 2012; Letchumanan et al., 2014; Wang et al., 2015).

Ever since the discovery of penicillin by Fleming in 1928 (Fleming, 1944), antibiotics have played unparalleled roles in disease prevention and treatment for human and animal health and welfare. In addition to the use in human medicine, antibiotics are widely utilized in food animals and aquaculture either as prophylactic or for growth enhancement (Marshall and Levy, 2011). Therefore, antibiotics are extensively used to ensure the development of the intensive and large-scale aquaculture industry. However, the uncontrolled and indiscriminate use of antibiotics has given rise to the emergence of antibiotic resistant bacteria in the aquaculture (Huang et al., 2015; Letchumanan et al., 2015a,b,c) and aquaculture ponds also have been evidenced as reservoirs for antibiotic resistance genes (Tomova et al., 2015; Xiong et al., 2015). These antibiotic resistance genes can be acquired by human and animal pathogens via horizontal gene transfer (Tomova et al., 2015), hence leading to difficulty in the treatment of infectious diseases. Moreover, the recent evidences of residual antibiotics in the cultured organisms could pose a potential health risk to human consumers (Chen et al., 2015; Pereira et al., 2015; Pham et al., 2015). In order to overcome the continuous emergence of antibiotic resistance pathogens due to abuse of antibiotics in aquaculture, an alternative to antibiotics is urgently needed for disease prevention and treatment and also improvement of quality and sustainability of aquaculture production. Extensive reviews have done indicating that probiotics could be a promising alternative for antibiotics in aquaculture, demonstrating beneficial effects to host by combating diseases, improving growth and also stimulating immune responses of host toward infections (NewajFyzul et al., 2014; Hai, 2015). Therefore, the aim of this review is to provide an insight on the use of the genus Streptomyces bacteria as an alternative to antibiotics, being a probiotic in controlling diseases and improving the health and quality of aquaculture production. Furthermore, this review also discusses the prospects and limitations of Streptomyces species as a probiotic in aquaculture.

Citation: Tan LT-H, Chan K-G, Lee L-H and Goh B-H (2016) Streptomyces Bacteria as Potential Probiotics in Aquaculture. Front. Microbiol. 7:79. doi: 10.3389/fmicb.2016.00079.

PROBIOTICS

The term 'probiotic' was initially defined as 'organisms and substances which contribute to intestinal microbial balance' (Parker, 1974). It was then revised as 'live microbial feed supplement which beneficially affects the host animal by improving its intestinal microbial balance' (Fuller, 1989). Meanwhile, in the case of aquatic animals which have much closer interactions with the external environment as compared to the terrestrial organisms, the external environment and feeding have substantial impacts on the microbial status of the aquatic animals. Hence, Verschuere et al. (2000) suggested that a probiotic for aquatic environments should be known as a live microbial adjunct exhibiting beneficial effect on the host by modulating the host-associated or ambient microbial community. Lately, probiotic was described as live, dead or component of a microbial cell that exerts beneficial effect on host by improving disease resistance, growth performance, feed utilization and health status, through the achievement of microbial balance in both host and ambient environments (Hai, 2015). Literatures have showed the possible mode of action of probiotics in aquaculture include (i) growth promoter, (ii) production of inhibitory compounds, (iii) improvement in nutrient digestion, (iv) water quality improvement, (v) enhancement of immune response, and (vi) competition for nutrient (Defoirdt et al., 2007; Martínez Cruz et al., 2012). In order to achieve a probiotic status, the microbes have to fulfill a number of criteria in term of their biosafety and functionality. The desirable characteristics of a potential probiotic include; (i) not harmful toward the host; (ii) ability to survive during transport to the active site; (iii) capability of colonizing and proliferating within the host; (iv) no virulence genes or antibiotic resistance genes (Hai, 2015). The common microorganisms used as probiotics are Lactobacillus acidophilus, Lactobacillus casei, Bacillus sp., Bifidobacterium bifidum, Lactococcus lactis and also the yeast, Saccharomyces cerevisiae (Ouwehand et al., 2002; Salamoura et al., 2014). However, less attention has been put on the use of Actinobacteria as probiotics in aquaculture despite being widely known as prolific producer for secondary metabolites, particularly the genus (Butler, 2008). The genus Streptomyces demonstrated promising results as probiotics (Das et al., 2010; Augustine et al., 2015). This review aim to discuss the prospects of using Streptomyces as a probiotic candidate in aquaculture. Table 1 summarizes all the features and mechanism of actions of the probiotic effects evidenced in the genus Streptomyces.

STREPTOMYCES SP. AS PROBIOTICS IN AQUACULTURE

The genus Streptomyces (phylum: Actinobacteria) are Grampositive, high G + C (70%) genome content, soil-living bacteria with characterized branching filamentous morphology. Streptomyces sp. has been widely recognized as industrially important microorganism due to its potential in producing diverse range of secondary metabolites (Lee et al., 2014b; Ser et al., 2015a,b; Tan et al., 2015) including antibiotics (Lee et al., 2014a), antitumor agents, antiparasitic, immunosuppressive agents, and enzymes (Manivasagan et al., 2013). The production of a variety of wide-spectrum chemical compounds as demonstrated by Streptomyces has the advantage of producing potential antagonistic and antimicrobial compounds that can be valuable as probiotics in aquaculture. The ability of producing antagonistic compounds may help the probiotics to compete for nutrients and attachment sites in the host. For instance, the production of bacteriocins (Desriac et al., 2010), siderophores (Lalloo et al., 2010), enzymes (protease, amylase, lipase; Augustine et al., 2015), hydrogen peroxide (Sugita et al., 2007) and organic acids (Sugita et al., 1997) have been documented from the probiotics used in aquaculture. You et al. (2005) reported a Streptomyces sp. with siderophores producing activities and suggested that the use of this Streptomyces sp. can influence the growth of pathogenic Vibrio sp. by competition for iron in the aquatic environment. Siderophores are ferric ion-specific chelating agents with low molecular masses that are usually produced by microorganisms and plants under Felimiting conditions (Ahmed and Holmstrom, 2014).

Table 1 The probiotic effects demonstrated by Streptomyces bacteria through different mechanism of actions

Features/Mechanism of actions	Probiotic *Streptomyces* bacteria	Outcomes	References
Antagonistic compounds production • Siderophore production	*Streptomyces cinerogriseus* A03 and A05 *Streptomyces griseorubroviolaceus* A26 and A42 *Streptomyces lavendulae* A41 *Streptomyces roseosporus* A45 *Streptomyces griseofuscus* B15	• All the strains positive for siderophore production, detected using CAS-agar • Displayed antagonistic activity toward *Vibrio* species tested (*V. harveyi*, *V. nereis*, *V. fluvialis*, *V. alginolyticus*, *V. parahaemolyticus*, *V. vulnificus* and *V. anguillarum*) ranging from <10 mm to >30 mm inhibition zones • Suggested the ability of the siderophore-producing *Streptomyces* strains controlled the *Vibrio* pathogens by competing for iron in the marine environment	You et al., 2005
• Anti-biofilm and anti-quorum sensing activity	*Streptomyces albus* A66	• Attenuated the biofilm formation of *V. harveyi* with inhibition rate of 99.3% at 2.5% (v/v) • Dispersed the mature biofilm of *V. harveyi* with degradation rate of 75.6% at 2.5% (v/v) • Suggested the anti-biofilm activity demonstrated by *Streptomyces* A66 through the degradation of the quorum-sensing factor *N*-AHSL (*N*-acylated homoserine lactone)	You et al., 2007
• Anti-virulence activity	*Streptomyces* sp. K01-0509	• Produced guadinomine B, a type III secretion system inhibitor of Gram-negative bacteria, including *Vibrio* sp., with IC_{50} at 14 nM	Iwatsuki et al., 2008
• Anti-viral activity	*Streptomyces* sp. AJ8	• Administrated intramuscularly ethyl acetate extract of the secondary metabolite reduced the white spot syndrome virus load significantly (85%) in the *Fenneropenaeus indicus* after third day of injection	Jenifer et al., 2015
Exoenzyme secretion	*Streptomyces* CLS-28 *Streptomyces* CLS-39 *Streptomyces* CLS-45	• All strains showed good proteolytic activity and variable amylolytic and lipolytic activities • Suggested to facilitate the feed utilization and digestion of the host, resulting in increased weight of *Penaeus monodon* when incorporated in the feed	Das et al., 2010

Growth enhancing effect	*Streptomyces fradiae* and *Streptomyces* sp.	• Improved growth of post-larval shrimp *P. monodon* and ornamental fish, *Xiphophorus helleri* • Produced growth-promoting hormone, indoleacetic acid which enhanced growth of *X. helleri*	Dharmaraj and Dhevendaran, 2010; Aftabuddin et al., 2013
Low pH tolerance and intestinal enzymes resistance	*Streptomyces* sp. JD9	• Showed excellent viability at pH 2 • Displayed resistance to pepsin at 3 mg/mL, bile at 0.3% and pancreatin at 1 mg/mL • Demonstrated good survivability in gastrointestinal conditions	Latha et al., 2015
Water quality amelioration	*Streptomyces fradiae* *Streptomyces* sp. *Streptomyces* CLS-28	• Reduced the ammonia level in the water • Increased the total heterotrophic bacterial populations in the water which helped to accelerate the decomposition of waste materials	Das et al., 2006, 2010; Aftabuddin et al., 2013
Single cell protein	*Streptomyces* sp.	• Used as a protein source for host, increased food conversion rate and food conversion efficiency, enhanced growth performance	Dharmaraj and Dhevendaran, 2010; Suguna, 2012; Selvakumar et al., 2013
In vivo protection/challenge experiment	*Streptomyces* CLS-28 *Streptomyces* CLS-39 *Streptomyces* CLS-45	Protection of *Artemia* against *V. harveyi* • *V. harveyi* at 10^6 CFU/mL killed all *Artemia nauplii* in 72 h • Addition of *Streptomyces* strains [at 1% (v/v)] increased the survival of *Artemia nauplii* by 67% and adults by 61% after 72 h exposure to *V. harveyi* at 10^6 CFU/mL Protection of *P. monodon* against *V. harveyi* • *V. harveyi* at 10^7 CFU/mL killed 55% of *P. monodon* after 5 days exposure • *Streptomyces* CLS-28 incorporated in the feed (after feeding for 15 days) increased the survival of *P. monodon* by 67% compared to control (without *Streptomyces*) in 5 days exposure	Das et al., 2010

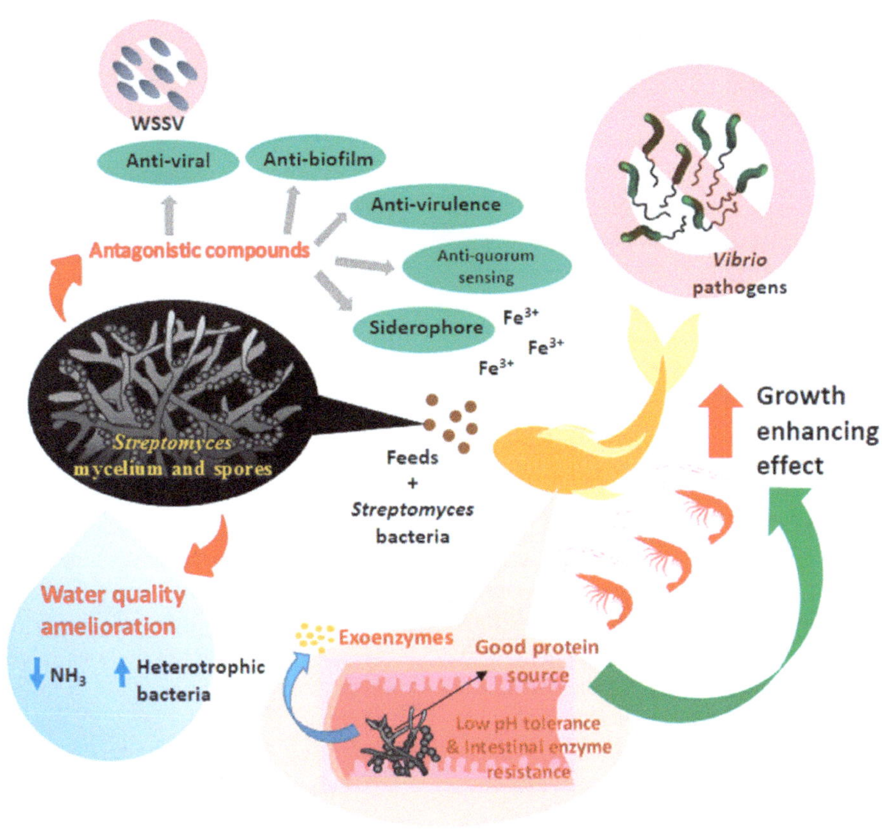

Figure 1 The mechanism of action of probiotic effects of Streptomyces bacteria and their applications in aquaculture. The Streptomyces as probiotic in aquaculture could protect the livestock from pathogens through the production of various antagonistic compounds (e.g., anti-biofilm, anti-quorum sensing and anti-virulence) against Vibrio pathogens. Streptomyces probiotics attenuate the growth of pathogens by producing siderophores which reduce the bioavailability of iron (Fe3+) for the pathogens in the aquatic environment. Anti-viral compounds are also produced by Streptomyces probiotic to prevent viral infection caused by white spot syndrome virus (WSSV) in aquaculture. Besides, the Streptomyces probiotics also play a role in ameliorating the water quality of aquaculture. Streptomyces probiotics help to regulate the microflora, especially to increase the protein mineralizing and ammonifying bacterial populations which accelerate the decomposition process of wastes materials and also ameliorate the water quality by reducing the ammonia level (NH_3). The consumption of feeds incorporated with the low pH tolerance and intestinal enzymes resistance Streptomyces probiotics could enhance the growth performance of the livestock by providing good protein sources. Streptomyces probiotics exhibit the ability to secrete hydrolytic exoenzymes which improve the amylolytic and proteolytic activity in the digestive tract of the livestock for more efficient use of the feed; eventually contribute to better growth performance of the livestock.

Probiotics with the capability of producing siderophores are believed to outcompete the pathogens by limiting the bioavailability of iron and resulting in growth attenuation of the pathogens as iron is essential for growth as well as biofilm formation (Weinberg, 2004). In addition, Streptomyces was also evidenced in the production of inhibitory compounds and metabolites involved in the attenuation of biofilm formation, anti-quorum sensing activity (You et al., 2007) and anti-virulence activity in Vibrio sp. (Iwatsuki et al., 2008). Besides displaying inhibitory effect on bacterial pathogens in aquaculture, Streptomyces also has been report to exhibit anti-viral activity, specifically against the white-spot syndrome virus (WSSV; Jenifer et al., 2015).

Streptomyces is primarily saprophytic, living in diverse soil habitats with the development of branching hyphal filaments under conducive environment (Flardh and Buttner, 2009). This unique growth adaptation allows Streptomyces in colonization of the solid substrates by adhering and penetrating to gain access on insoluble organic materials in the soil (Flardh and Buttner, 2009). Different hydrolytic enzymes such as amylase, protease and lipase can be produced by Streptomyces to break down the insoluble organic materials to provide nutrients for the formation of densely packed substrate mycelium which is reused to fuel the reproductive phase of aerial growth in producing chains of spores (Chater et al., 2010). These unique physiological adaptations of Streptomyces are believed to make them as potential probiotics such as the secretion of exoenzymes which may be helpful in facilitating the feed utilization and digestion once they colonize the host intestine in aquaculture. Das et al. (2010) demonstrated that the feed incorporated with Streptomyces increased the weight of Penaeus monodon shrimp, suggesting that these Streptomyces sp. secreted hydrolytic exoenzymes to improve the amylolytic and proteolytic activity in the shrimp digestive tract for more efficient use of the feed. The feed supplemented with Streptomyces fradiae isolated from mangrove sediment was also shown to enhance the growth of the post-larval P. monodon (Aftabuddin et al., 2013). Besides showing good growth promoting effects in shrimp, all the feeds supplemented with Streptomyces was also shown to improve growth performance of the ornamental fish, Xiphophorus helleri (red swordtail fish) after 50 days of feeding trial when compared to control without the Streptomyces sp. (Dharmaraj and Dhevendaran, 2010). Furthermore, the similar study also showed that the production of growth-promoting hormone, indoleacetic acid by the Streptomyces sp. could be contributed to the better growth rate as demonstrated by Xiphophorus helleri fed with Streptomyces supplemented feeds (Dharmaraj and Dhevendaran, 2010).

The formation of enzymatic digestion, sonic vibration and desiccation-resistant spores demonstrated by Streptomyces are also some of the attractive features for this genus of bacteria to resist the harsh environment conditions (McBride and Ensign, 1987), thereby allowing them to retain longer shelf life in the aquaculture ponds before being taken up or to resist the low pH in the gastrointestinal tracts of the animals. However, it should be noted that Streptomyces spore is only resistant to moderately high temperature (McBride and Ensign, 1987) as compared to the highly heat resistant endospores of Bacillus sp. which is compositionally and physiologically different from the Streptomyces spore. Nevertheless, Latha et al. (2015) reported that the Streptomyces sp. isolated from fecal sample of chicken showed excellent viability at pH 2, exhibited strong pepsin resistance (at 3 mg/mL) as well as the resistance toward both bile (at 0.3%) and pancreatin (at 1 mg/mL), suggesting that strains from the animal internal cavities would be better in adapting and colonizing the gastrointestinal of the animals. This is also demonstrated by Das et al. (2010) which isolated Streptomyces sp. from the sediment of the shrimp culture system able to reach the digestive system of the shrimp, hence allow easier establishment and growth of the probionts in the host. These findings indicate that the spore-forming capacity of Streptomyces with high acidity and bile acids tolerance makes them a more practical alternative than those bacteria with non-spore forming capability and further ascertain the potential of Streptomyces as probiotic in aquaculture (Das et al., 2010).

The in vivo challenge experiment conducted further proved that Streptomyces should be spotlighted as probiotics in aquaculture (Das et al., 2010). This study successfully demonstrated the protection effect of Streptomyces on both juvenile and adult Artemia (15 days old) from Vibrio pathogens. The study showed that the Streptomyces at 1% concentration (v/v) resulted in higher survival rates than the untreated control group of Artemia after challenged with V. harveyi or V. proteolyticus at 106 CFU/mL. The protective response shown by the study suggests that Streptomyces could be administrated to target organisms through bioencapsulation in Artemia as a vector for supplementing the beneficial Streptomyces probiotics in aquaculture. Bioencapsulation of probiotics in live food such as Artemia and rotifers was demonstrated to be more effective in delivery of the probiotics to the digestive tract of the target aquaculture organisms by previous studies (Gatesoupe, 2002; Suzer et al., 2008). The study also further evaluated the efficacy of the Streptomyces in protecting the shrimp P. monodon from the Vibrio pathogens. The feed supplemented with Streptomyces sp. CLS-28 for 15 days was found to be exerting protection effect on shrimp P. monodon against the 12 h challenge of V. harveyi (LD50 at 106.5 CFU/mL; Das et al., 2010). A more recent study reported a marine S. rubrolavendulae M56 (accession number KJ403746) was shown to exhibit antagonistic activity against all four Vibrio sp. including V. harveyi, V. alginolyticus, V. parahaemolyticus and V. fluvialis in an in vitro co-culture experiment (Augustine et al., 2015). In order to confirm the in vitro findings, Augustine et al. (2015) demonstrated that the biogranules S. rubrolavendulae M56 resulted in lower percentage of mortality of P. monodon post-larvae with the reduction of viable Vibrio sp. in the culture system after 28 days.

The build-up of ammonia and nitrite level is a major water quality problem which has considerable effects on the health status of the aquaculture livestock due the accumulation of metabolic waste of cultured organisms and the decomposition of the residual feed. The probiotic Streptomyces was also found to regulate the microflora of the aquaculture water besides controlling the pathogenic microorganisms and resulted in a better pond conditions. Literature showed that the application of probiotic product did not adversely affect the microflora of aquaculture in turn increased the protein mineralizing and ammonifying bacterial population which help to accelerate the decomposition process of the accumulated wastes materials (Devaraja et al., 2002). Several studies also demonstrated similar results indicating the reduction of ammonia level and increased in the total heterotrophic bacteria in the ponds/tanks treated with the probiotic Streptomyces as compared to the control ponds/tanks (Das et al., 2006, 2010; Aftabuddin et al., 2013). These findings suggested that Streptomyces could be applied as probiotics which ameliorate the water quality of aquaculture indirectly improve the growth performance and yield of the cultured organisms.

Traditionally, fish meal has been an indispensable ingredient in commercial aquaculture feeds due to its high protein content with excellent amino acid profile and is highly digestible (Gatlin et al., 2007). However, current feed formulations have shifted to other alternative protein source due to the high cost and limited availability of fish meal. Microbial single cell protein of Streptomyces is one of the alternative sources of protein and has been utilized and evaluated for better food conversion efficiency and growth for fish (Suguna, 2012; Selvakumar et al., 2013) and shrimp (Manju and Dhevendaran, 1997). Dharmaraj and Dhevendaran (2010) suggested that the use of Streptomyces not only showing beneficial effects as probiotic in aquaculture, the incorporation of Streptomyces in the feed is also a cost effective approach as the probiotic bacteria replaced around 30–40% of the fish meal used in a feed. The study demonstrated that Streptomyces can be a cheaper alternative protein source in the aquaculture feed (Dharmaraj and Dhevendaran, 2010).

LIMITATIONS OF STREPTOMYCES AS PROBIOTIC IN AQUACULTURE

Geosmin and MIB (2-methylisoborneol) are two common semivolatile terpenoid compounds that exhibit earthy/musty taste and odor produced by Streptomyces have been known to reduce the palatability of the cultured livestock and negative impact for aquaculture industries (Auffret et al., 2011). These off-flavor compounds are known to be absorbed and bioaccumulated in the gills, skin and flesh of fish up to 200- to 400-folds as compared to the ambient concentration, resulting in lower commercial value of the fish (Howgate, 2004). Many efforts have been shown in literatures for the removal of these earthy odor compounds from water involving the use of powdered activated carbon, ozonation and biofiltration (Elhadi et al., 2004). Among these technologies, ozonation is suggested to effective in this case with the use of Streptomyces as the probiotics in aquaculture. Ozone has been known to remove odorants such as geosmin and MIB from water via oxidation (Gonçalves and Gagnon, 2011). A study has demonstrated that the combined effect of ozonation (at 0.3 mg O3/L ROC) and probiotic diets (Bacillus sp. S11) was able to protect shrimp P. monodon from Vibrio challenge test without harming shrimp and the probiotic bacteria in the internal system of shrimp (Meunpol et al., 2003).

Furthermore, the risk of lateral gene transfer of antibiotic resistance genes could be an argument against the use of Streptomyces as probiotic in aquaculture. Despite that, there are increasing reports on the antibiotic resistance developed by most of the commonly used probiotics such as Lactobacillus sp. (Sharma et al., 2015), Bifidobacterium sp. and Bacillus sp. (Gueimonde et al., 2013). Furthermore, studies also reported that the antibiotic resistance phenotypes displayed by the probiotic Streptomyces strains were generally conferred by their intrinsic resistance properties (Das et al., 2010; Latha et al., 2015). Hence, systematic screening for potential antibiotic resistance gene determinants in potential probiotics genome has to be conducted to assess the potential risks and mobility. Furthermore, curative strategies can be valuable tool to remove the genetic element that harbor antibiotic resistance from the relevant probiotic strains (Morelli and Campominosi, 2002; Rosander et al., 2008). For instance, Rosander et al. (2008) successfully demonstrated the protoplast formation curing method able to remove two resistant plasmids from the parent Lactobacillus reuteri (ATCC 55730) and without affecting the probiotic properties of the strain. All in all, Streptomyces can be one of the interesting probiotics to be further exploited as an alternative to antibiotics in maintaining a sustainable aquaculture.

CONCLUSION AND FUTURE WORK

To date, the number of study employs Streptomyces as probiotics in aquaculture is still limited although promising results have been represented by previous studies. A schematic figure is also illustrated to show the mechanism of action of the probiotic effects demonstrated by the Streptomyces in aquaculture (Figure 1). In order for Streptomyces being included among the commonly used biological control agents in aquaculture, further extensive trials are still required to establish the probiotic nature of Streptomyces in disease prevention and growth enhancement of aquaculture animals. Furthermore, a better understanding is needed on the exact mode of action of Streptomyces involved in probiotic effects. Hence, further research could focus more on molecular techniques to elucidate the possible underlying mechanism portrayed by Streptomyces probiotic in aquaculture settings.

AUTHOR CONTRIBUTIONS

All authors listed, have made substantial, direct and intellectual contribution to the work, and approved it for publication.

ACKNOWLEDGMENTS

This work was supported by the Monash University Malaysia ECR Grant (5140077-000-00), MOSTI eScience Funds (02-02- 10-SF0215 and 06-02-10-SF0300), University of Malaya for High Impact Research Grant (UM-MOHE HIR Nature Microbiome Grant No. H-50001-A000027 and No. A000001-50001).

REFERENCES

1. Aftabuddin, S., Kashem, M. A., Kader, M. A., Sikder, M. N. A., and Hakim, M. A. (2013). Use of Streptomyces fradiae and Bacillus megaterium as probiotics in the experimental culture of tiger shrimp Penaeus monodon (Crustacea, Penaeidae). AACL Bioflux 6, 253–267.
2. Ahmed, E., and Holmstrom, S. J. (2014). Siderophores in environmental research: roles and applications. Microb. Biotechnol. 7, 196–208. doi: 10.1111/1751- 7915.12117
3. Auffret, M., Pilote, A., Proulx, E., Proulx, D., Vandenberg, G., and Villemur, R. (2011). Establishment of a real-time PCR method for quantification of geosmin-producing *Streptomyces* spp. in recirculating aquaculture systems. *Water Res.* 45, 6753–6762. doi: 10.1016/j.watres.2011.10.020

4. Augustine, D., Jacob, J. C., and Philip, R. (2015). Exclusion of *Vibrio* spp. by an antagonistic marine actinomycete *Streptomyces rubrolavendulae* M56. *Aquac. Res.* 1–10. doi: 10.1111/are.12746

5. Bondad-Reantaso, M. G., Subasinghe, R. P., Arthur, J. R., Ogawa, K., Chinabut, S., Adlard, R., et al. (2005). Disease and health management in Asian aquaculture. *Vet. Parasitol.* 132, 249–272. doi: 10.1016/j.vetpar.2005.07.005

6. Butler, M. S. (2008). Natural products to drugs: natural product-derived compounds in clinical trials. *Nat. Prod. Rep.* 25, 475–516. doi: 10.1039/b51 4294f

7. Chater, K. F., Biro, S., Lee, K. J., Palmer, T., and Schrempf, H. (2010). The complex extracellular biology of *Streptomyces*. *FEMS Microbiol. Rev.* 34, 171–198. doi: 10.1111/j.1574-6976.2009.00206.x

8. Chen, H., Liu, S., Xu, X. R., Liu, S. S., Zhou, G. J., Sun, K. F., et al. (2015). Antibiotics in typical marine aquaculture farms surrounding Hailing Island, South China: occurrence, bioaccumulation and human dietary exposure. *Mar. Pollut. Bull.* 90, 181–187. doi: 10.1016/j.marpolbul.2014. 10.053

9. Das, S., Lyla, P., and Ajmal Khan, S. (2006). Application of *Streptomyces* as a probiotic in the laboratory culture of *Penaeus monodon* (Fabricius). *Isr. J. Aquac.* 58, 198–204.

10. Das, S., Ward, L. R., and Burke, C. (2010). Screening of marine *Streptomyces* spp. for potential use as probiotics in aquaculture. *Aquaculture* 305, 32–41. doi: 10.1016/j.aquaculture.2010.04.001

11. Defoirdt, T., Boon, N., Sorgeloos, P., Verstraete, W., and Bossier, P. (2007). Alternatives to antibiotics to control bacterial infections: luminescent vibriosis in aquaculture as an example. *Trends Biotechnol.* 25, 472–479. doi: 10.1016/j.tibtech.2007.08.001

12. Desriac, F., Defer, D., Bourgougnon, N., Brillet, B., Le Chevalier, P., and Fleury, Y. (2010). Bacteriocin as weapons in the marine animal-associated bacteria warfare: inventory and potential applications as an aquaculture probiotic. *Mar. Drugs* 8, 1153–1177. doi: 10.3390/md8041153

13. Devaraja, T. N., Yusoff, F. M., and Shariff, M. (2002). Changes in bacterial populations and shrimp production in ponds treated with commercial microbial products. *Aquaculture* 206, 245–256. doi: 10.1016/s0044-8486(01)00721-9

14. Dharmaraj, S., and Dhevendaran, K. (2010). Evaluation of *Streptomyces* as a probiotic feed for the growth of ornamental fish *Xiphophorus helleri*. *Food Technol. Biotechnol.* 48, 497–504.

15. Elhadi, S. L., Huck, P. M., and Slawson, R. M. (2004). Removal of geosmin and 2-methylisoborneol by biological filtration. *Water Sci. Technol.* 49, 273–280.

16. FAO (2014). *The State of World Fisheries and Aquaculture 2014.* Rome: Food and Agriculture Organization.

17. Flardh, K., and Buttner, M. J. (2009). *Streptomyces* morphogenetics: dissecting differentiation in a filamentous bacterium. *Nat. Rev. Microbiol.* 7, 36–49. doi: 10.1038/nrmicro1968

18. Flegel, T. W. (2012). Historic emergence, impact and current status of shrimp pathogens in Asia. *J. Invertebr. Pathol.* 110, 166–173. doi: 10.1016/j.jip.2012.03.004 Fleming, A. (1944). The discovery of penicillin. *Br. Med. Bull.* 2, 4–5.

19. Fuller, R. (1989). Probiotics in man and animals: a review. *J. Appl. Bacteriol.* 66, 365–378. doi: 10.1111/j.1365-2672.1989.tb05105.x

20. Gatesoupe, F.-J. (2002). Probiotic and formaldehyde treatments of *Artemia nauplii* as food for larval pollack, *Pollachius pollachius*. *Aquaculture* 212, 347–360. doi: 10.1016/s0044-8486(02)00138-2

21. Gatlin, D. M., Barrows, F. T., Brown, P., Dabrowski, K., Gaylord, T. G., Hardy, R. W., et al. (2007). Expanding the utilization of sustainable plant products in aquafeeds: a review. *Aquac. Res.* 38, 551–579. doi: 10.1111/j.1365-2109.2007.01704.x

22. Gonçalves, A. A., and Gagnon, G. A. (2011). Ozone application in recirculating aquaculture system: an overview. *Ozone Sci. Eng.* 33, 345–367. doi: 10.1080/01919512.2011.604595

23. Gueimonde, M., Sanchez, B., de Los Reyes-Gavilán, C.G., and Margolles, A. (2013). Antibiotic resistance in probiotic bacteria. *Front. Microbiol.* 4:202. doi: 10.3389/fmicb.2013.00202

24. Hai, N. (2015). The use of probiotics in aquaculture. *J. Appl. Microbiol.* 119, 917–935. doi: 10.1111/jam.12886

25. Howgate, P. (2004). Tainting of farmed fish by geosmin and 2-methyl-iso-borneol: a review of sensory aspects and of uptake/depuration. *Aquaculture* 234, 155–181. doi: 10.1016/j.aquaculture.2003.09.032

26. Huang, Y., Zhang, L., Tiu, L., and Wang, H. H. (2015). Characterization of antibiotic resistance in commensal bacteria from an aquaculture ecosystem. *Front. Microbiol.* 6:914. doi: 10.3389/fmicb.2015.00914

27. Iwatsuki, M., Uchida, R., Yoshijima, H., Ui, H., Shiomi, K., Kim, Y.-P., et al. (2008). Guadinomines, type III secretion system inhibitors, produced by *Streptomyces* sp. K01-0509. *J. Antibiot. (Tokyo)* 61, 230–236. doi: 10.1038/ja.2008.33

28. Jenifer, J. S., Donio, M. B., Michaelbabu, M., Vincent, S. G., and Citarasu, T. (2015). Haloalkaliphilic *Streptomyces* spp. AJ8 isolated from solar salt works and its' pharmacological potential. *AMB Express* 5, 143. doi: 10.1186/s13568-015-0143-2

29. Lalloo, R., Moonsamy, G., Ramchuran, S., Görgens, J., and Gardiner, N. (2010). Competitive exclusion as a mode of action of a novel *Bacillus cereus* aquaculture biological agent. *Lett. Appl. Microbiol.* 50, 563–570. doi: 10.1111/j.1472-765x.2010.02829.x

30. Latha, S., Vinothini, G., John Dickson, D.C., and Dhanasekaran, D. (2015). In vitro probiotic profile based selection of indigenous actinobacterial probiont *Streptomyces* sp. JD9 for enhanced broiler production. *J. Biosci. Bioeng.* 121, 124–131. doi: 10.1016/j.jbiosc.2015.04.019

31. Lee, L.-H., Zainal, N., Azman, A.-S., Eng, S.-K., Ab Mutalib, N.-S., Yin, W.-F., et al. (2014a). *Streptomyces pluripotens* sp. nov., a bacteriocin-producing streptomycete that inhibits meticillin-resistant *Staphylococcus aureus*. *Int. J. Syst. Evol. Microbiol.* 64, 3297–3306. doi: 10.1099/ijs.0.065045-0

32. Lee, L.-H., Zainal, N., Azman, A. S., Eng, S. K., Goh, B.-H., Yin, W.-F., et al. (2014b). Diversity and antimicrobial activities of actinobacteria isolated from tropical mangrove sediments in Malaysia. *ScientificWorldJournal* 2014, 698178. doi: 10.1155/2014/698178

33. Letchumanan, V., Chan, K.-G., and Lee, L.-H. (2014). *Vibrio parahaemolyticus*: a review on the pathogenesis, prevalence, and advance molecular identification techniques. *Front. Microbiol.* 5:705. doi: 10.3389/fmicb.2014.00705

34. Letchumanan, V., Chan, K.-G., and Lee, L.-H. (2015a). An insight of traditional plasmid curing in *Vibrio* species. *Front. Microbiol.* 6:735. doi: 10.3389/fmicb.2015.00735

35. Letchumanan, V., Pusparajah, P., Tan, L. T.-H., Yin, W.-F., Lee, L.-H., and Chan, K.-G. (2015b). Occurrence and antibiotic resistance of *Vibrio parahaemolyticus* from shellfish in Selangor, Malaysia. *Front. Microbiol.* 6:1417. doi: 10.3389/fmicb.2015.01417

36. Letchumanan, V., Yin, W.-F., Lee, L.-H., and Chan, K.-G. (2015c). Prevalence and antimicrobial susceptibility of *Vibrio parahaemolyticus* isolated from retail shrimps in Malaysia. *Front. Microbiol.* 6:33. doi: 10.3389/fmicb.2015.00033

37. Lightner, D. V., Redman, R. M., Pantoja, C. R., Tang, K. F., Noble, B. L., Schofield, P., et al. (2012). Historic emergence, impact and current status of shrimp pathogens in the Americas. *J. Invertebr. Pathol.* 110, 174–183. doi: 10.1016/j.jip.2012.03.006

38. Manivasagan, P., Venkatesan, J., Sivakumar, K., and Kim, S. K. (2013). Marine actinobacterial metabolites: current status and future perspectives. *Microbiol. Res.* 168, 311–332. doi: 10.1016/j.micres.2013.02.002

39. Manju, K., and Dhevendaran, K. (1997). Effect of bacteria and actinomycetes as single cell protein feed on growth of juveniles of *Macrobrachium idella* (Hilgendorf). *Indian J. Exp. Biol.* 35, 53–55.

40. Marshall, B. M., and Levy, S. B. (2011). Food animals and antimicrobials: impacts on human health. *Clin. Microbiol. Rev.* 24, 718–733. doi: 10.1128/CMR.00002-11

41. Martínez Cruz, P., Ibáñez, A. L., Monroy Hermosillo, O. A., and Ramírez Saad, H. C. (2012). Use of probiotics in aquaculture. *ISRN Microbiol.* 2012, 916845. doi: 10.5402/2012/916845

42. Mathieson, A. (2012). *The State of World Fisheries and Aquaculture*. Rome: FAO Fisheries and Aquaculture Department.

43. McBride, M. J., and Ensign, J. C. (1987). Effects of intracellular trehalose content on *Streptomyces griseus* spores. *J. Bacteriol.* 169, 4995–5001.

44. Meunpol, O., Lopinyosiri, K., and Menasveta, P. (2003). The effects of ozone and probiotics on the survival of black tiger shrimp (*Penaeus monodon*). *Aquaculture* 220, 437–448. doi: 10.1016/s0044-8486(02)00586-0

45. Moffitt, C. M., and Cajas-Cano, L. (2014). Blue growth: the 2014 FAO state of world fisheries and aquaculture. *Fisheries* 39, 552–553. doi: 10.1080/03632415.2014.966265

46. Morelli, L., and Campominosi, E. (2002). Genetic stability of *Lactobacillus paracasei* subsp. paracasei F19. *Microb. Ecol. Health Dis.* 14, 14–16. doi: 10.3402/mehd.v14i1.8206

47. Newaj-Fyzul, A., Al-Harbi, A., and Austin, B. (2014). Review: developments in the use of probiotics for disease control in aquaculture. *Aquaculture* 431, 1–11. doi: 10.1016/j.aquaculture.2013.08.026

48. Ouwehand, A., Salminen, S., and Isolauri, E. (2002). "Probiotics: an overview of beneficial effects," in *Lactic Acid Bacteria: Genetics, Metabolism and Applications*, eds R. J. Siezen, J. Kok, T. Abee, and G. Schasfsma (Dordrecht: Springer Netherlands), 279–289. doi: 10.1007/978-94-017-2029-8_18

49. Parker, R. (1974). Probiotics, the other half of the antibiotic story. *Anim. Nutr. Health* 29, 4–8.

50. Pereira, A. M., Silva, L. J., Meisel, L. M., and Pena, A. (2015). Fluoroquinolones and tetracycline antibiotics in a portuguese aquaculture system and aquatic surroundings: occurrence and environmental impact. *J. Toxicol. Environ. Health A* 78, 959–975. doi: 10.1080/15287394.2015.1036185

51. Pham, D. K., Chu, J., Do, N. T., Brose, F., Degand, G., Delahaut, P., et al. (2015). Monitoring antibiotic use and residue in freshwater aquaculture for domestic use in Vietnam. *Ecohealth* 12, 480–489. doi: 10.1007/s10393-01 4-1006-z

52. Rosander, A., Connolly, E., and Roos, S. (2008). Removal of antibiotic resistance gene-carrying plasmids from *Lactobacillus reuteri* ATCC 55730 and characterization of the resulting daughter strain, *L. reuteri* DSM 17938. *Appl. Environ. Microbiol.* 74, 6032–6040. doi: 10.1128/AEM.00991-08

53. Salamoura, C., Kontogianni, A., Katsipi, D., Kandylis, P., and Varzakas, T. (2014). Probiotic fermented milks made of cow's milk, goat's milk and their mixture. *J. Biotechnol.* 185S, S125. doi: 10.1016/j.jbiotec.2014.07.262

54. Selvakumar, D., Jyothi, P., and Dhevendaran, K. (2013). Application of *Streptomyces* as a single cell protein to the juvenile fish *Xiphophorus maculatus*. *World J. Fish Mar. Sci.* 5, 582–586. doi: 10.5829/idosi.wjfms.2013.05.06. 74154

55. Ser, H.-L., Palanisamy, U. D., Yin, W.-F., Malek, A., Nurestri, S., Chan, K.-G., et al. (2015a). Presence of antioxidative agent, Pyrrolo [1, 2-a] pyrazine-1, 4-dione, hexahydro-in newly isolated *Streptomyces mangrovisoli* sp. nov. *Front. Microbiol.* 6:854. doi: 10.3389/fmicb.2015.00854

56. Ser, H.-L., Ab Mutalib, N. S., Yin, W.-F., Chan, K.-G., Goh, B.-H., and Lee, L.-H. (2015b). Evaluation of antioxidative and cytotoxic activities of *Streptomyces pluripotens* MUSC 137 isolated from mangrove soil in Malaysia. *Front. Microbiol.* 6:1398. doi: 10.3389/fmicb.2015.01398

57. Sharma, P., Tomar, S. K., Sangwan, V., Goswami, P., and Singh, R. (2015). Antibiotic resistance of *Lactobacillus* sp. isolated from commercial probiotic preparations. *J. Food Saf.* 36, 38–51. doi: 10.1111/jfs.12211

58. Shruti, C. (2012). *Vibrio* related diseases in aquaculture and development of rapid and accurate identification methods. *J. Marine Sci. Res. Dev.* S1, 002. doi: 10.4172/2155-9910.s1-002

59. Sugita, H., Matsuo, N., Hirose, Y., Iwato, M., and Deguchi, Y. (1997). *Vibrio* sp. strain NM 10, isolated from the intestine of a Japanese coastal fish, has an inhibitory effect against *Pasteurella piscicida. Appl. Environ. Microbiol.* 63, 4986–4989.

60. Sugita, H., Ohta, K., Kuruma, A., and Sagesaka, T. (2007). An antibacterial effect of *Lactococcus lactis* isolated from the intestinal tract of the Amur catfish, Silurus asotus Linnaeus. *Aquac. Res.* 38, 1002–1004. doi: 10.1111/j.1365-2109.2007.01765.x

61. Suguna, S. (2012). Production of probiotics from *Streptomyces* sp. associated with fresh water fish and its growth evaluation on *Xiphorous helleri*. *Int. J. Pharm. Biol. Sci. Arch.* 3, 601–603.

62. Suzer, C., Çoban, D., Kamaci, H. O., Saka, Ş., Firat, K., Otgucuoglu,˘ Ö, et al. (2008). *Lactobacillus* spp. bacteria as probiotics in gilthead sea bream (*Sparus aurata*, L.) larvae: effects on growth performance and digestive enzyme activities. *Aquaculture* 280, 140–145. doi: 10.1016/j.aquaculture.2008.04.020

63. Tan, L. T.-H., Ser, H.-L., Yin, W.-F., Chan, K.-G., Lee, L.-H., and Goh, B.-H. (2015). Investigation of antioxidative and anticancer potentials of *Streptomyces* sp. MUM256 isolated from Malaysia mangrove soil. *Front. Microbiol.* 6:1316. doi: 10.3389/fmicb.2015.01316

64. Tomova, A., Ivanova, L., Buschmann, A. H., Rioseco, M. L., Kalsi, R. K., Godfrey, H. P., et al. (2015). Antimicrobial resistance genes in marine bacteria and human uropathogenic *Escherichia coli* from a region of intensive aquaculture. *Environ. Microbiol. Rep.* 7, 803–809. doi: 10.1111/1758-2229. 12327

65. Verschuere, L., Rombaut, G., Sorgeloos, P., and Verstraete, W. (2000). Probiotic bacteria as biological control agents in aquaculture. *Microbiol. Mol. Biol. Rev.* 64, 655–671. doi: 10.1128/mmbr.64.4.655-671.2000

66. Wang, L., Chen, Y., Huang, H., Huang, Z., Chen, H., and Shao, Z. (2015). Isolation and identification of *Vibrio campbellii* as a bacterial pathogen for luminous vibriosis of *Litopenaeus vannamei*. *Aquac. Res.* 46, 395–404. doi: 10.1111/are.12191

67. Weinberg, E. D. (2004). Suppression of bacterial biofilm formation by iron limitation. *Med. Hypotheses* 63, 863–865. doi: 10.1016/j.mehy.2004.04.010

68. Xiong, W., Sun, Y., Zhang, T., Ding, X., Li, Y., Wang, M., et al. (2015). Antibiotics, antibiotic resistance genes, and bacterial community composition in fresh water aquaculture environment in china. *Microb. Ecol.* 70, 425–432. doi: 10.1007/s00248-015-0583-x

69. You, J., Cao, L., Liu, G., Zhou, S., Tan, H., and Lin, Y. (2005). Isolation and characterization of actinomycetes antagonistic to pathogenic *Vibrio* spp. from nearshore marine sediments. *World J. Microbiol. Biotechnol.* 21, 679–682. doi: 10.1007/s11274-004-3851-3

70. You, J., Xue, X., Cao, L., Lu, X., Wang, J., Zhang, L., et al. (2007). Inhibition of *Vibrio* biofilm formation by a marine actinomycete strain A66. *Appl. Microbiol. Biotechnol.* 76, 1137–1144. doi: 10.1007/s00253-007-1074-x

CHAPTER 7 YEASTS AS PROBIOTICS IN FISH AQUACULTURE

Paola Navarrete and Dariel Tovar-Ramírez

INTRODUCTION

According to the UN Food and Agriculture Organization (FAO), "aquaculture, probably the fastest growing food-producing sector, now accounts for nearly 50 percent of the world's food fish" [1]. However, production is hampered by unpredictable mortalities that may be due to the negative interactions between fish and pathogenic bacteria. Intensive fish farming has resulted in a problematic growth in bacterial diseases, prompting the necessary and intensive use of antimicrobials for their treatment.

Because of the rapid expansion of aquaculture, a limited supply of fishmeal has the potential to impede the future growth of this industry. Consequently, much effort has been given to studying other protein and oil sources, but finding a suitable alternative has proved to be challenging. Among the alternatives, plant-based formulations are the least expensive, and many such formulations have a suitable protein profile and long-term availability. Oilseeds, in particular soybean and grain products, have great potential as alternative sources of fish feed. Soybeans are rich in protein and represent the most commonly used plant protein source on the world market. Soybean meal (SBM) has already become an important protein source in fish feed. However, the inclusion of some vegetable proteins, such as SBM, in the diets of fish at levels of >20% may induce intestinal disorders including pathomorphological changes in the distal intestinal epithelium accompanied by diarrhea [2, 3], sometimes caused by the antinutritional factors that are present in SBM. The addition of probiotics (acid lactic bacteria) to starter diets appeared to improve SBM utilization in first feeding rainbow trout [4].

Citation: Paola Navarrete and Dariel Tovar-Ramírez, Yeasts as Probiotics in Fish Aquaculture, http://dx.doi.org/10.5772/57196.

In this context, two of the major challenges in fish aquaculture facilities are 1) the control of diseases, especially during the earliest life stages, and 2) the improvement of nutrition by optimizing food utilization, especially for new fish species.

It is well recognized that the bacterial microbiota of fish is beneficial to the host and affects important biological processes, including nutrient processing and absorption, the development of the mucosal immune system, and angiogenesis, as was demonstrate in gnotobiotic mice. In larval gnotobiotic zebrafish studies, was shown that the microbiota also influences enterocyte morphology and epithelial renewal, host-transcriptional responses to the micro- biota regarding epithelial proliferation, nutrient and xenobiotic metabolism, and immune responses [5].

Yeast have been identified as part of the normal microbiota of both wild and farmed fish, and their role in fish health and nutrition has been addressed in the literature, as yeast have been used either alive to feed live food organisms or after processing as a feed ingredient after demonstrating an artificial colonization of the intestinal host.

Even when accounting for less than 1% of the total microbial isolates in the host, yeast can represent a major physiological contribution beyond what has been observed for probiotic bacteria; in fact, cell volumes from yeast may be larger than those of bacteria by a hundredfold [6]. In contrast to bacteria, yeast cells utilize a wide spectrum of simple and more complex organic compounds. This phenomenon results from the extensive metabolic potential of yeast, which is reflected by the production of diverse enzymes. Polyamines secreted by yeasts are also involved in the maturation of the digestive tract of fish larvae. Furthermore, some yeast species and their components, such as β-glucans and mannoproteins, can stimulate the immune and antioxidant systems of the host. Understanding the participation of yeast microbiota in fish health and nutrition may improve both the sanitary conditions and the production performance of fish.

The aim of this chapter is to describe the current knowledge regarding the use of yeasts as probiotics in aquaculture systems. The chapter will include a recent review on the presence and diversity of yeast in marine and aquaculture systems, focusing on the yeast diversity found in the fish gut microbiota. The chapter will also include basic information on the molecular methods used for yeast identification. Finally, the chapter will emphasize topics related to the essential role of probiotic yeasts used in disease control and nutritional improvements in aquaculture, with a special focus on the beneficial effects of yeast β-glucans.

YEAST IDENTIFIED IN THE MARINE AND OTHER AQUATIC ENVIRONMENTS

Yeasts are unicellular eukaryotic microorganisms that are taxonomically placed within the phyla Ascomycota and Basidiomycota within the Kingdom Fungi [7]. Ascomycete yeasts comprise approximately 1,000 phylogenetically diverse species that have recently been assigned to 14 different lineages on the basis of multigene sequence analysis [8]. The other species of yeasts are classified as the basidiomycetes [9].

Table 1. Yeast identified in the marine and other aquatic environment

Marine samples	Identified yeast	Identification method	Reference
Northern Biscayne Bay	*Candida tropicalis* *Candida guilliermondii* *C. parapsilosis* *Rhodotorula rubra*	identification of cultivated yeasts by the methods described by Lodder and Kreger-van Rij [23] and Wickerham [24].	[25]
Marine grass flats (Soldier Key)	*Rhodotorula pilimanne* *R. rubra* *Cryptococcus*	identification of cultivated yeasts by the methods described by Lodder and Kreger-van Rij [23] and Wickerham [24].	[25]
Gulf Stream 15 miles east of the coast of South Florida.	*R. graminis* *R. glutinis*	identification of cultivated yeasts by the methods described by Lodder and Kreger-van Rij [23] and Wickerham [24].	[25]
Marine vegetation	*Cryptococcus albidus*	identification of cultivated yeasts by the methods described by Lodder and Kreger-van Rij [23] and Wickerham [24].	[25]
Suwannee Florida estuary (water)	*Candida guilliermondii* *Candida krusei* *Candida valida* *Cryptococcus laurentii var. laurentii* *Cryptococcus laurentii var. flavescens* *Hansenula saturnus var. saturnus* *Hansenula spp.* *Rhodotorula marina* *Rhodotorula minuta var. minuta* *Rhodotorula rubra* *Rhodotorula spp.* *Torulopsis candida* *Trichosporon cutaneum*	identification of cultivated yeasts by the methods described by Lodder [26].	[27]
Suwannee Florida estuary (sediment)	*Brettanomyces intermedius* *Candida boidinii* *Candida diversa* *Candida glaebosa* *Candida ingens* *Candida krusei* *Candida lambica* *Candida maritima* *Candida melibiosica* *Candida silvae* *Candida solani* *Candida valida* *Candida spp.* *Cryptococcus dimennae* *Cryptococcus laurentii var. laurentii* *Cryptococcus laurentii var. flavescens* *Debaryomyces cantarellii* *Debaryomyces phaffii* *Hansenula beijerinckii* *Hansenula saturnus var. saturnus* *Kluyveromyces polysporous* *Leucosporidium capsuligenum* *Pichia membranaefaciens* *Pichia ohmeri* *Rhodotorula glutinis* *Rhodotorula graminis* *Rhodotorula lactosa* *Rhodotorula marina* *Rhodotorula rubra* *Rhodotorula spp.* *Saccharomyces spp.* *Torulopsis candida* *Torulopsis inconspicua* *Torulopsis mogii* *Torulopsis spp.* *Trichosporon aculeatum* *Trichosporon cutaneum* *Trichosporon penicillatum*	identification of cultivated yeasts by the methods described by Lodder [26].	[27]
Western coast of Baja California Sur, Mexico	*Sporobolomyces roseus* *Sporobolomyces puniceus* *Sporobolomyces hosaticus*	morphological and biochemical identification	[28]
Suruga and Sagami Bay (sediments, crab and *Calyptogena*)	*Kluyveromyces nonfermentans*	18S rDNA, 5.8S rDNA and ITS sequencing	[29]
Northwest Pacific Ocean (benthic animals)	*R. aurantiaca* *R. glutinis* *R. minuta* *R. mucilaginosa* *Sporobolomyces salmonicolor* *S. shibatanus*	5.8S-ITS rDNA sequencing of cultivated yeasts	[30]
Northwest Pacific Ocean (sediments)	*R. glutinis* *R. mucilaginosa*	5.8S-ITS rDNA sequencing of cultivated yeasts	[30]
Sagami bay (deep-sea tubeworm)	*R. lamellibrachii*	sequencing of ITS, 5.8S rDNA, and D1/D2 of the 26S rDNA	[31]
Sagami Bay and Iheya Ridge (deep-sea tubeworm)	*R. bentica*	sequencing of ITS, 5.8S rDNA, and D1/D2 of the 26S rDNA	[32]
Sagami Bay (deep-sea clam)	*R. calyptogenae*	sequencing of ITS, 5.8S rDNA, and D1/D2 of the 26S rDNA	[32]
Deep-sea hydrothermal systems of the Mid-Atlantic Rift	*C. atlantica* *C. atmosphaerica* *C. lodderae* *C. parapsilosis* *D. hansenii* *P. guilliermondii* *Rhodosporidium babjevae* *R. diobovatum* *R. kratochvilovae* *R. sphaerocarpum*	26SrRNA gene sequencing of cultured yeasts	[33]

	R. toruloides *Rh. Mucilaginosa* *Rh. minuta* *S. dacryoides*		
Northwest Pacific ocean	*R. pacifica*	sequencing of ITS, 5.8S rDNA, and D1/D2 of the 26S rDNA	[34]
11 deep-sea samples	*Pichia fermentans* *Saccharomyces cerevisiae* *Debaryomyces hansenii*	Cloning and sequencing	[35]
Japan Trench (deep-sea sediments)	*Dipodascus tetrasporeus*	Sequencing of 18S rDNA, ITS1, 5.8S rDNA, ITS2 and D1/D2 of the 26S rDNA	[36]
Coastal waters of northeastern Taiwan	*Candida tropicalis* *Pichia anomala* *Issatchenkia orientalis* *C. glabrata* *Saccharomyces yakushimaensis* *Kodamaea ohmeri* *Hanseniaspora uvarum* *Kazachstania jiainicus* *Torulaspora delbrueckii*	5.8S-ITS rDNA sequencing of cultivated yeasts	[37]
Deep-sea shrimps Deep-sea mussels	*R. mucilaginosa*	26SrRNA gene sequencing of cultured yeasts	[14]
R. exoculata exuviae in decomposition on smoker rocks, *B. azoricus* and a sponge	*Rhodosporidium diobovatum* *Sporobolomyces roseus*	26SrRNA gene sequencing of cultured yeasts	[14]
carbonate colonization module	*Cryptococcus uzbekistanensis*	26SrRNA gene sequencing of cultured yeasts	[14]
B. azoricus mussel	*Leucosporidium scottii*	26SrRNA gene sequencing of cultured yeasts	[14]
R. exoculata, *M. fortunata*, a deep-sea coral and the gills of the gastropod *Ifremeria nautilei*	*Debaryomyces hansenii*	26SrRNA gene sequencing of cultured yeasts	[14]
R. exoculata exuviae in decomposition *B. azoricus*	*Candida atlantica*	26SrRNA gene sequencing of cultured yeasts	[14]
Deep-sea sponge	*Pichia guilliermondii* *Candida viswanathii*	26SrRNA gene sequencing of cultured yeasts	[14]
Deep-sea coral	*Candida sp.*	26SrRNA gene sequencing of cultured yeasts	[14]
B. azoricus	*Phaeotheca triangularis* *Hortaea werneckii*	26SrRNA gene sequencing of cultured yeasts	[14]
Arabian sea (200 m depth)	*Candida* *Lipomyces* *Yarrowia* *Rhodotorula* *Pichia*	morphological and biochemical identification	[38]
Arabian sea (500 m depth)	*Candida* *Yarrowia* *Lipomyces* *Rhodotorula* *Debaryomyces* *Pichia* *Wingea* *Dekkera*	morphological and biochemical identification	[38]
Arabian sea (1000 m depth)	*Lypomyces* *Candida* *Wingea* *Dekkera* *Rhodotorula*	morphological and biochemical identification	[38]
Bay of Bengal (200 m depth)	*Wingea* *Candida* *Cryptococcus* *Rhodotorula* *Bullera* *Lipomyces* *Oosporidium* *Dekkera*	morphological and biochemical identification	[38]
Bay of Bengal (500 m depth)	*Candida* *Rhodotorula* *Cryptococcus* *Yarrowia* *Pichia* *Bullera* *Wingea* *Dekkera* *Oosporidium*	morphological and biochemical identification	[38]
Bay of Bengal (1000 m depth)	*Candida* *Wingea* *Rhodotorula* *Bullera* *Lipomyces* *Trichosporon*	morphological and biochemical identification	[38]

Some of the general characteristics and ecological properties of each phylum include the following: 1) the cell wall polysaccharide composition is dominated by chitin in the basidiomycetes and by β-glucans in the ascomycetes; 2) the guanine + cytosine (G + C) composition of the nuclear DNA tends to be higher than 50% in basidiomycetes and lower than 50% in ascomycetes; 3) ascomycetes yeasts are generally more fermentative, more copiotrophic (but at the same time nutritionally specialized), more fragrant, and mostly hyaline, while basidiomycete yeasts more often form mucoid colonies, display intense carotenoid pigments and tend to use a broader range of carbon compounds more efficiently at lower concentrations [7]; and 4) ascomycete yeasts are often found in specialized niches involving interactions with plants and insects or other invertebrate animals that they rely upon for dispersal, while basidiomycete yeasts seem to be adapted to the colonization of nutrient-poor solid surfaces and may not rely to the same extent on animal vectors for their dispersal [10].

Yeasts are widely distributed in several natural environments such as soil, freshwater, and seawater. Their numbers and species distributions are dependent on the concentrations and types of available organic materials. Nearshore environments are usually inhabited by 10 to 1000 of yeast cell/L of water, whereas low organic surface to deep sea oceanic regions contain 10 or fewer cells/L. Marine yeasts are divided into "obligate" and "facultative" groups. When yeast are able to grow on a marine substrate and are frequently collected from the marine environment, they are called "obligate marine" yeasts; in contrast, "facultative marine" yeasts can also be recovered from terrestrial habitats. Marine yeasts participate in several ecological processes in the sea, especially in estuarine and nearshore environments, such as the decom- position of plant substrates [11], nutrient recycling [12], and the biodegradation of oil/ recalcitrant compounds [13]; they are also part of the microbiota of marine and aquaculture animals [6, 14]. This functional diversity is due, in part, to the fact that yeasts have extraordi- nary metabolic potential. This potential is available for exploitation [15-20], but notably, the vast majority of this potential has yet to be discovered. Several yeast compounds have significant biological value as reagents, cell proteins, vitamins, pigments, and enzymes. Different yeast species have been identified in several marine locations (Table 1). For excellent reviews on marine yeasts, see [21, 22]. The ascomycete yeasts Debaryomyces hansenii, and Candida spp. are typical ubiquitous species found in oceanic, and other aquatic environments. Basidiomycete yeasts often account for the majority of the total yeast population found in oligotrophic oceanic water. Among the basidiomycete yeasts, some species of Cryptococcus, Rhodotorula, and Sporobolomyces are widespread across various oceanic regions [22].

YEAST AS PART OF THE GUT MICROBIOTA OF FISH

Most of the literature on the yeast microbiota of fish is based on the identification of cultivable yeast (Table 2). Yeast have been isolated from the gills, skin, mouths, feces and guts of different fish species. The occurrence of yeast in the fish gut is variable and can fluctuate from nondetectable levels to 10^7 CFU/g of intestinal content [6]. Both ascomycete and basidiomycete yeasts have been isolated from fish intestines (Table 2): among ascomycetes, Saccharomyceta- ceae (which include Candida, Pichia, Saccharomyces, and Debaryomyces) is likely the most important family, while basidiomycetes, include the genera Rhodotorula, Cryptococcus, Sporobolomyces, and Trichosporon [6]. The yeasts Metschnikowia zobelii, Kloeckera apiculata, and Debaryomyces sp. dominate in some marine fish (Tachurus symmetricus and Atherinopis affinis littoralis) [39], and in these fish species, the yeast concentration was significantly higher inside the fish than in the surrounding sea water, suggesting that the yeast may grow inside the fish intestine [39]. The ascomycetes Debaryomyces hansenii, Candida sp., and Saccharomyces cerevi- siae, the basidiomycete Leucosporidium sp., and Rhodotorula have been frequently isolated as the dominant yeast found in the rainbow trout intestine [6]. Yeast can also be isolated from the waters of fishponds with different abundance and diversity depending on the season of the year. Fishponds from the Záhorie Lowlands in Slovakia, sampled in summer, harbor the most heterogeneous yeast species, with Aureobasidium, Sporobolomyces, Candida, and Cryptococcus as the most frequently isolated species [40]. In autumn, the yeast numbers were higher than in summer, with Candida, Hyphopichia burtonii, Aureobasidium pullulans, Hansenula anomala and Cryptococcus laurentii being most frequently identified [40].

It has been reported that yeasts isolated from the intestine of rainbow trout may adhere to and grow in intestinal mucus [41]. Some yeast cells can colonize the intestine of fish after dietary introduction [42], and this ability to colonize may be related to cell surface hydrophobicity [43] and the ability of the strains to grow on mucus [41, 44]. Some experiments have shown that high levels of yeast intestinal colonization can be achieved when a pure culture of yeast is inoculated into fish. Rainbow trout and turbot were inoculated with Rhodotorula glutinis and D. hansenii HF1, and up to 3.8×10^4 , 3.1×10^6 , and 2.3×10^9 viable yeast cells/g of intestine or feces were recovered in three separate colonization experiments [45]. It is important to note that the majority of the studies published until 2007, on the yeast species identified from the aquaculture fish gut, were published by the Gatesoupe Lab [6]. Later, studies focused more on the probiotic effects of different yeast strains in aquaculture fish (as described below), and less on the actual yeast species isolated from the fish gut.

Table 2. Yeast identified in the fish intestine

Location	Fish intestine	Identified yeast	Identification method	Reference
Estuarine and coastal areas (Biscayne Bay, Florida)	*Haemulon* *Carana* *Anisotremus*	*T. cutaneum* *C. parapsilosis* *C. guilliermondi* *C. tropicalis* *R. rubra* *R. pilimanae* *H. anomala* *D. kloeckeri* *II. valbyensis*	identification of cultivated yeast by the methods described by Lodder and Kreger-van Rij [23] and Wickerham [24].	[25]
Tropical island-(Bimini, The Bahamas, 1960)	*Haemulon* *Stenotomus* *Ocyurus* *Anisotremus* *Lachnolainus*	*R. minuta* *C. parapsolosis* *R. glutinis*	identification of cultivated yeast by the methods described by Lodder and Kreger-van Rij [23] and Wickerham [24].	[25]
Tropical island-(Bimini, The Bahamas, 1961)	*llaemulon* *Lutjanus* *Sphyraena* *Seriola* *Balistes* *Malacanthus* *Halichoeres* *Holocentrus* *Carana*	*C. tropicalis* *R. pilimanae* *Torulopsisspp.* *C. parapsilosis*	identification of cultivated yeast by the methods described by Lodder and Kreger-van Rij [23] and Wickerham [24].	[25]

	Anisotremus			
La Jolla coast	*Atherinopis affinis littoralis* *Trachurus symmetricus*	*Metschnikowia zobellii* *Kloeckera apiculata* *Debaryomyces sp.*	identification of cultivated yeast by the methods described by Lodder and Kreger-van Rij [23], Wickerham [24], and van Uden and Farinha [46].	[39]
Clyde estuary North Sea	*Herring* *Haddock* *Whiting* *Skate* *Halibut* *Flounder* *Lemon sole*	*Candida* *Cryptococcus* *Debaryomyces* *Rhodotorula* *Torulopsis* *Trichosporon pullulans*	identification of cultivated yeast by the methods described by Lodder and Kreger-van Rij [23] and Kreger-van Rij [47].	[48]
Sweden farm	*Salmo gairdneri*	*S. cerevisiae* *D. hansenii* *Cryptococcus* *Leucosporidium* *Rhodotorula rubra* *R. glutinis*	Identified by the CBS Yeast Collection	[45]
Swedish west coast	*P. platessa* *P. flesus*	*Rhodotorula*	Identified by the CBS Yeast Collection	[45]
Experimental fish farm at Sizun (France)	*Oncorhynchus mykiss*	*Debaryomyces hansenii* *D. hansenii var. fabryi* *Trichosporon* *Rhodotorula mucilaginosa*	ITS 1, 5.8S rRNA gene, ITS 2, and partial sequencing of 26S rRNA gene	[49]
Experimental fish farm at Sizun (France)	*Oncorhynchus mykiss*	*Debaryomyces hansenii*	ITS 1, 5.8S rRNA gene, ITS 2, and partial sequencing of 26S rRNA gene	[42]
Cabras (Oristano, Sardinia)	*Mugil auratus* *M. chelo* *M. capito* *M. saliens* *M. cephalus*	*Candida* *Metschnikowia* *Sporidiobolus* *Clavispora* *Sporobolomyces*	morphological, physiological and biochemical tests of the isolated yeast strains according to [50]	[51]

DEBARYOMYCES HANSENII, AN UBIQUITOUS YEAST FREQUENTLY ASSOCIATED WITH FISH AND THE MARINE ENVIRONMENT

Debaryomyces hansenii is a halotolerant, non-pathogenic ubiquitous yeast capable of growing in a variety of environments, such as the marine environment and the fish gut. D. hansenii has been described as one of the most frequently isolated yeast associated with fish (Table 2). This species is prevalent in seawater, which may explain its high incidence in the fish gut. One study reported the presence of sequences affiliated with D. hansenii in hydrothermal sediments [35]. A major biotechnological advantage of D. hansenii over Saccharomyces cerevisiae is that D. hansenii is osmotolerant and can grow in media containing up to 4 M NaCl, whereas the growth of S. cerevisiae is restricted to media containing less than 1.7 M NaCl. D. hansenii has been extensively studied because of its significant enzymatic potential [52]. For example, several enzymes of biological and biotechnological interest have been identified and characterized in this yeast, including inulinase [53], protease [54], superoxide dismutase (SOD) [55], lipase [56], catalase [57] and α-galactosidase [58-62]. Additionally, β-glucosidase from D. pseudopolymor- phus [63] and phytase from D. castellii [64] has also been identified. The ability of D. hansenii to synthesize α-galactosidase has been useful in the treatment of soybean products to reduce raffinose oligosaccharides [59], which are recognized as anti-nutritional factors for mammals and fish. Interestingly, D. hansenii SOD has been proposed as a therapeutic anti-inflammatory agent in animal models (Wistar rats) [55]. Because of these characteristics, D. hansenii is one of the yeast species that has been selected for complete sequencing [65]. The beneficial effects of this yeast species in cultured fish are described below.

METHODS TO ANALYZE THE YEAST MICROBIOTA

In the past, the identification of yeast species was a tedious and labor-intensive process that was generally based on the morphological and physiological properties of the isolated yeasts. To resolve these problems, the identification of cultivated yeasts is now based on DNA sequence analysis, which is a faster and more accurate process. Sequence-based approaches to the study of yeast biodiversity have resulted in a two-fold increase in the number of described species over the past decade, and a 100-fold increase is predicted in the coming decades. A previous work [8] has studied the phylogenetics of the Saccharomycetales by performing DNA sequence analysis based on five loci: 1) the nuclear small subunit (SSU) ribosomal RNA gene, 2) the D1/D2 region of the nuclear large subunit (LSU) 26S rDNA, 3) the elongation factor 1α gene (EF-1α), and 4) the largest and 5) the second largest subunits of the RNA polymerase II gene (RPB1 and RPB2). Based on the availability of the sequence data in the GenBank and AFTOL databases, the LSU rDNA genes were found to be more reliable for yeast identification. Sequencing the 400-650 bp D1/D2 region or a wider LSU region is extremely useful and distinguishes yeast rapidly at a near-species level. The D1/D2 LSU rDNA region has been sequenced for almost all known yeast, both as an identification tool and as a means for estimating phylogenetic relationships among the Saccharomycetales. A previously study [66] published the sequences of the fungal primers ITS1 and ITS4, which amplify the internal transcribed spacer (ITS) ITS1-5.8S-ITS2 region that has also been used for yeast identification. Some specialized databases are available on the web in order to help with yeast identification. The Centraal bureau voor Schimmelcultures (CBS) database aids BLAST analysis by allowing pairwise identification of LSU and SSU rDNA, ITS [67] and miscellaneous sequences [68].

It is generally accepted that in every ecosystem, there are cultivable organisms as well as viable organisms that cannot be cultivated in the laboratory. Less than 1% of the microbial species from the marine environment can be cultivated [69]. Similarly, only approximately 1% of yeast species has been described thus far [70]. In the last decades, several molecular methods have been developed to study natural samples. These methods allow for the identification of microorganisms without isolation and for the determination of the phylogenetic affiliation of community members, thus revealing the enormous extent of microbial diversity. Methods based on the amplification of fragments coding for the 16S rRNA gene have emerged as a powerful tool for studying the bacterial diversity. Denaturing or temperature gradient gel electrophoresis (PCR-DGGE/TTGE) techniques have been introduced into molecular microbial ecology to determine the genetic diversity of the bacterial communities found in the fish gut [71-76]. These techniques were also applied to characterize the dominant active bacteria in the intestine of different rainbow trout families using RNA that was extracted directly from the samples [77]. One important limitation of PCR-based methods, however, is low sensitivity, which can identify approximately 1% of the total number of species [78]. On the other hand, the use of rRNA gene fingerprinting requires sequencing of the cloned bands to identify the community's members accurately. Sequencing is necessary because the amplicons from different species may migrate to the same positions, or one species may give multiples bands because of multiple gene copies with intragene differences [74, 79]. The use of molecular approaches to study yeast communities has been scarce and generally limited to the study of food matrices. Yeast communities have been studied using PCR-DGGE, and amplification of a portion of the 26S rRNA gene of yeast [80-85], while PCR-TTGE has been applied to establish phylogenetic relationships of species of the genus Saccharomyces [86].

Recently, high-throughput sequencing methods, such as pyrosequencing, have been shown to be fast and very efficient tools for identifying members of the complex populations. In general, two approaches can be taken: diversity studies based on the sequencing of ribosomal gene (rRNA gene) amplicons, and metagenomic studies where whole-community DNA is subjected to shotgun sequencing [87]. While sequencing ribosomal amplicons is much cheaper because only one gene is being sequenced, the metagenomic approach sequences all of the DNA genes, thus revealing the functions of the microbial community [87]. A useful innovation for these two approaches is to analyze multiple samples at the same time (multiplexing), which can be accomplished using barecoded pyrosequencing or by physically separating the samples in the sequencing plates. In the barecode technique, the sequences in each sample are tagged with a unique barecode using barcoded primers during PCR amplification. The result of these high-throughput sequencing methods is several thousand sequences per sample in just few days, which must necessarily be analyzed using bioinformatics tools. Although the costs associated with these new technologies are less than for the Sanger method (considering the cost for one sequence), high-throughput sequencing methods remain an expensive approach. To date, these methods have been applied to the study of the microbial diversity of several communities, especially from human and animal guts [88-93]. Diversity analysis targeting the D1/D2 domain of the 26S rRNA gene or the internal transcribed regions (ITSs) allow yeast species to be distinguished [8], and have recently been applied to the study of some fungal communities [94, 95], and to the identification of some clinical yeast isolates [96-98]. These methods appear to be very suitable for studying the yeast biodiversity in the fish gut. This new knowledge, together with the information available in various databases, will allow both the accurate identification of new yeast isolates and the application of molecular strategies to characterize the yeast population in the fish gut.

USE OF YEAST AS PROBIOTICS IN AQUACULTURE: STIMULATION OF THE IMMUNE AND ANTIOXIDANT SYSTEMS, GUT MATURATION AND FISH GROWTH

The natural occurrence of numerous yeast species in the gastrointestinal tract of healthy fish has been well described, and yeast have been shown to constitute an important part of the microbiota of the fish gut [6]. In addition to S. cerevisiae, the halotolerant yeast D. hansenii has been considered an excellent probiotic candidate in fish aquaculture. Because, the number of experiences reporting the use of D. hansenii is increasing, this allows us to know the capacity of this yeast to enhance growth, survival, and gut maturation and to improve of the immune and antioxidant systems in fish larvae and juveniles. Yeast cells can be a hundred times larger than bacterial cells, which may explain the fact that the introduction of a low yeast population (10^4 CFU/g) through feed can induce beneficial effects in the host.

The effects of probiotics differ greatly depending on the microbial species, isolation source, experimental concentration and finally, the rearing conditions of the fish. However, the improvement of the immune response is one of the most encountered side effects in the host because immune system stimulation or immunomodulation are considered important mechanisms supporting probiosis. Yeast have immunostimulatory properties because they possess components such as β-glucan, mannoproteins, chitin (as a minor component) and nucleic acids [99].

Recent studies have shown the beneficial effect of dietary administered Saccharomyces cerevi- siae in fish. Yeast supplemented diets stimulate growth, feed efficiency, blood biochemistry, survival rate, and non- specific immune responses in Uronema marinum-infected olive flounder (Paralichthys olivaceus) [100]. A diet supplemented with S. cerevisiae treated with beta-mercap- toethanol was better than whole cell yeast and n-3 highly unsaturated fatty acids

(HUFA)- enriched yeast as an immune system and growth stimulator in juvenile rainbow trout challenged with Yersinia ruckeri [101]. Similarly, the dietary administration of the probiotic S. cerevisiae P13 at a minimum level of 105 CFU/kg enhanced the growth, innate immune responses and disease resistance of grouper (Epinephelus coioides) [102]. Cellular yeast components also stimulate the immune system: an improvement in gut mucus lysozyme activity was observed in European sea bass (Dicentrarchus labrax) fed mannan oligosaccharides (MOS) derived from the outer cell wall of a select strain of S. cerevisiae (Bio-Mos, Alltech Inc, USA) [103]. Furthermore, channel catfish (Ictalurus punctatus) juveniles fed diets supplemented with whole-cell S. cerevisiae (Levucell SB20®) or yeast subcomponents such as commercial preparations of β-glucan (MacroGard ® and Betagard A ®) had a significantly higher survival rate after Edwardsiella ictaluri challenge than did catfish fed with a controlled diet [104].

As previously described, most published studies have been performed using S. cerevisiae; however, promising results have also been obtained with Debaryomyces hansenii. A diet supplemented with D. hansenii stimulates the immune system of juvenile leopard grouper, Mycteroperca rosacea, by increasing IgM and superoxide dismutase (SOD) activity and enhances the resistance of the fish to infection by the dinoflagellate, Amyloodinium ocellatum [105]. Additionally, other studies have demonstrated the immune system improvement of M. rosacea when the fish were fed for 4 weeks with a compound diet enriched with 1.1% D. hansenii. After the 4 weeks, the fish were challenged with the pathogenic bacteria Aeromonas hydrophila strain Ah-315, resulting in an increase in IgM levels as well as catalase (CAT) and SOD activities in those fish fed yeast diets. Improvements were also observed at the molecular level, where CAT and HSP70 expression levels were enhanced in M. rosacea fed with D. hansenii, and challenged with A. hydrophila [106].

D. hansenii administration to the gilthead seabream (Sparus aurata L.) significantly enhances leukocyte peroxidase and respiratory burst activity by week 4 of feeding with yeast. Yeast feeding causes an up-regulation in the expression of the immune-associated genes Hep, IgM, TCR-ß, NCCRP-1, MHC-IIa, CSF-1R, C3, TNF-α and IL-1ß in the head-kidney: C3 expression was only stimulated in the liver, whereas the expression of TCR-ß, TNF-α and C3 was stimulated in the intestine of S. aurata [107].

When D. hansenii was administered at 1.1% in a compound diet to D. labrax larvae, the yeast stimulated the antioxidant status [108]. The group fed with yeast showed lower glutathione peroxidase (GPX) and SOD activity compared to fish fed the control diet, suggesting a possible involvement of superoxide anion retention in fish larvae, which could represent importance to the host to increase cell or tissue responsiveness to growth- and/or differentiation-enhancing factors [109]. The group fed the control diet showed oxidative stress represented by an increase in GPX activity at 48 days post hatching (dph) and gene expression levels for both GPX and SOD at 23 dph.

The ontogeny of the digestive tract of fish larvae has been the subject of many studies in the last decades with the purpose of increasing production rates by reducing the bottlenecks in larviculture. In this sense, the number of reports concerning the use of yeast to enhance gut maturation and digestive enzyme activity in fish are also increasing. The activity and expression of digestive enzyme-related genes during fish development provides an excellent marker of digestive development in fish larvae. The enzymes secreted from the pancreas (trypsin, lipase, and amylase) as well as those encountered in the intestinal brush border membranes (BBM) (leucine aminopeptidase N, alkaline phosphatase, maltase, γ-glutamyl-transpeptidase, and the cytosolic leucine-alanine peptidase), are the most common indicators of digestive system maturation in fish larvae. The degree of enterocyte maturation is described by increasing ratios of activities of BBM enzymes vs cytosolic leucine alanine peptidase; in the case of pancreatic enzymes, a decrease in amylase activity or expression level with the concomitant increase in trypsin or lipase activities characterizes the maturation of the exocrine pancreas.

Given this previous work, the effect of D. hansenii-enriched microparticulated diets were tested in European sea bass larvae. The authors observed an increase in intestinal brush border and pancreatic enzyme activities at 27 dph, indicating an achievement of gut morphology in this larvae stage compared with the control diet lacking yeast [110]. In a second feeding trial, where the introduction of yeast to the microparticulated diet was improved, larvae fed 1% D. hansenii matured earlier than fish fed a control diet after day 26, as revealed by lower amylase expression, higher lipase and trypsin expression, and high levels of the BBM enzymes, aminopeptidase N, maltase and alkaline phosphatase [111].

In another study [42], two yeast strains: Saccharomyces cerevisiae and S. boulardii, were evaluated as probiotics for rainbow trout (Oncorhynchus mykiss) fry to compare the cross effects of the two rearing conditions, with the intestinal microbiota and the brush border enzyme activities. Intestinal maturation at 10 dph was observed in trout fed S. boulardii, and kept in spring water, and these fish displayed the highest ratios of BBM leucine aminopeptidase N vsleucine-alanine peptidase, compared with those fish fed S. cerevisiae, and kept in river water.

Overall, yeast has been added directly to the water, administered as an additive in micropar- ticulated diets, and has been used alive to feed live food (rotifers or Artemia) as a possible vector to deliver yeast into the gut of fish larvae. Rotifers have been established as the most common live prey to feed larval fish in hatcheries around the world, and baker's yeast (S. cerevisiae) is the most common nutrient source for culturing rotifers in addition to algae, emulsified oils or bacteria. Currently, efforts are being made to introduce D. hansenii into Brachionus rotundifor- mis to deliver yeast into the intestine of L. guttatus to accelerate digestive maturation [112].

In this regard, the use of Levucell® (S. boulardii), Bactocell® (Pediococcus acidilactici) and live yeast (S. cerevisiae) produced no significant effect on trypsin, lipase, and leucine aminopepti- dase activities in California halibut larvae, Paralichtys californicus, at 46 dph. Contrary to this, an increase in pepsin and chymotrypsin activity was only observed in fish larvae fed Bactocell® at the final endpoint of the experiment (46 days), suggesting a potential use of these probiotics once metamorphosis is completed [113].

In fish aquaculture, the most utilized growth-promoting additives are hormones, antibiotics, ionophores, and salts [114].

The use of probiotics as growth-promoters has been recognized in the last decade with a number of studies related to this topic being published. Probiotics can be used as an alternative to avoid the use of antibiotics for growth promotion, thus eliminating the possibility of generating antibiotic-resistant bacteria in the aquaculture systems. When yeast probiotics have been used in the earliest developmental stages of fish larvae, enhanced growth and survival have been observed. Several yeast species have been documented to enhance growth following artificial colonization, particularly S. cerevisiae and D. hansenii either alone, or in synergic association with bacteria. One study [115] observed that pollock (Pollachius pollachius) larvae grew better when Artemia nauplii was first treated with S. cerevisiae var. boulardii CNCM I-1079 and then with Pediococcus acidilactici MA185 M than did larvae fed with one or no probiotic. When S. cerevisiae was used alone in feeding trials, it improved feed

efficiency in Israeli carp and Nile tilapia [116, 117]. In addition to growth enhancement, an improvement in the conformation of the larvae was also observed in D. labrax fed with D. hansenii; a reduction in spinal deformity from 13.6% to 1.1% in fish fed yeast vs the control group was observed [111].

The use of probiotics in well-established fish industry is easier of because the existence of many individuals to experiment upon (S. salar, D. labrax, S. aurata, S. maximus, O. mykiss, etc). Nevertheless, it is advisable to use probiotics in species with the potential for exploitation to optimize the results for growth, survival and health. In this context, several studies have been performed in fish species with emerging aquaculture potential to contribute to the establish- ment of a continuous production line for experimental purposes. A commercial preparation of live yeast (S. cerevisiae and Lactobacillus coagulans) was tested on Indian carp fry, Labeo rohita, with no conclusive effects on growth [118]. D. hansenii has been observed to function as a growth promoter in Mycteroperca rosacea juveniles because after 4 weeks of continuous feeding, a weight gain (33%) and condition factor were observed in those fish fed a micropar- ticulated diet enriched with yeast, compared with those fish fed without yeast [106]. In spotted sand bass larvae, Paralabrax maculatofasciatus, the highest survival (13.0%) was obtained with D. hansenii enriched microparticulated diets, but no effects on growth were observed with the use of probiotics [119].

At present, much of the existing evidence indicates that yeast promotes growth, and survival because of gut maturation, conformation of the larvae, and stimulation of the immune system by a possible involvement of endoluminal yeast-secreted polyamines in the host. As was earlier demonstrated, D. hansenii produces more polyamines (putrescine, spermidine and spermine) than S. cerevisiae and S. boulardii [111]. Polyamines participate in several physiological proc- esses such as cell proliferation and differentiation and appear to have a broad influence on digestive tract maturation. In particular, the roles of dietary spermine and spermidine have been previously described [120]. These molecules enter enterocytes, where they induce a hormonal cascade that affects organs such as the pancreas and liver. Recently, the production of polyamines in 13 strains of D. hansenii, isolated from different sources, using high-pressure liquid chromatography (HPLC) has been reported [121]. In this study, they found that the L2, and CBS004 strains isolated from citrus fruit and marine water, respectively, were the main polyamine-secreting yeasts. Later, L2 strain was shown to have a probiotic effect because it enhanced the immune status, and intestinal function of gilthead seabream, Sparus aurata [122].

Finally, evidence of polyamine contribution to larviculture performance was reported when the spotted sand bass larvae, P. maculatofasciatus fed with D. hansenii with un-inhibited ornithine decarboxylase (ODC) activity had precocious digestive maturation compared to those larvae fed ODC-inhibited (with α-difluoromethylornithine (DFMO)) yeast [119]. Ornithine decarboxylase, which catalyzes the formation of putrescine, is the rate-limiting enzyme in the biosynthesis of polyamines in cells.

YEAST β-GLUCANS: STRUCTURE, MECHANISMS OF ACTION AND ITS APPLICATION AS IMMUNOSTIMULANT IN AQUACULTURE

The glucose polymer β-glucan is a major structural component of the cell wall of some plants (such as the cereals oat and barley), seaweeds, and the outer cell wall of bacteria, fungi and yeast. Different β-glucans vary in structure, size, branching frequency, structural modifications, conformation and solubility, which may influence their physiological functions. Glucose molecules, in all β-glucan polymers, are linked together by a β-(1→3) linear β-glycosidic chain core, but differ in their length and branching structures. For example, the β-glucans from oat and barley are linear with β-(1→4) linkages, and shorter stretches of β-(1→3) linkages, while the structure of yeast β-glucans is composed of β-(1→3)-D-glucans with β-(1→6)-glycosidic linked branches, which apparently corresponds to the most active form of β-glucan. The relationship between structure and biological activity is controversial, but it appears that large molecular weight β-glucans are the most active compared with small β-glucans below 5,000-10,000 Da that are generally inactive. The solubility of β-glucans also influences their biological activity, with soluble β-glucans appearing to be more active.

The consumption of β-glucans has been associated with beneficial health effects in humans, including anticancer properties [123], metabolic syndrome prevention [124, 125], cholesterollowering effects [126], anti-atherogenic properties [127] and skin health promotion [128]. In vitro and in vivo studies in animals and humans show that the β-glucans derived from fungi and yeasts in particular, have interesting immune modulating properties [129-132]. This immune stimulation can be achieved when β-glucans are administered by a parenteral or an oral (dietary) route.

Despite their structural versatility, β-glucans are highly conserved structural components and belong to a group of physiologically active compounds called biological response modifiers [133]. Because of their large molecular weight, they cannot penetrate the cell membrane and therefore they must interact with cell-surface receptors; it has been shown that β-glucans are recognized by several receptors found on neutrophils, macrophages, and dendritic cells [129, 134]. Additionally, β-glucans belong to the group of non-self-molecules called pathogenassociated molecular patterns (PAMPS), which are recognized by pattern recognition receptors (PRRs) on the cell surface [135]. The principal β-glucans PRRs are dectin-1 and the toll-like receptors (TLRs), but other receptors are suggested to be involved, such as scavenger receptors, complement receptor 3, and lactosylceramide [136, 137]. Dectin-1 specifically recognizes β- (1→3)(1→6) glucans from fungi, plants, and bacteria [138], but it is not reactive toward β-(1→4) glucans or α-mannan [139]. The stimulation of dectin-1 activates the innate immune response, ROS and inflammatory cytokine production [140] through the activation of phospholipase C [141], the PI3K/Akt pathway, MAPK, NFAT, and NF-κB [142]. The interaction of β-glucans with TLRs results in activation of NF-κB and MAPK signaling [143]. Zymosan (β-glucans from the yeast Saccharomyces cerevisiae) binds to TLR2 and TLR4 (as well as dectin-1) found on macrophages, leading to an increase in cytokine production, such as TNF-α and IL-12, through the NF-κB pathway [143]. The β-glucan (β-(1→3) (1→6)-D-glucan) from Aureobasidium pullulans ADK-34 stimulates intestinal Peyer›s patch cells both in vitro and in vivo as reflected by an increase in IL-5, IL-6, and IgA [129]. The interaction of β-glucans with specific receptors on macrophages and dendritic cells can induce the production of several cytokines, which in turn activate other immune cells such as B and T cells, thus activating the systemic immune response.

Yeast β-glucans have been applied in aquaculture to modulate the innate immune system of fish to improve their survival until adaptive immune responses are sufficiently developed to mount effective responses against pathogens [144, 145]. If the β-glucans are administered as feed additives, they can exert their primary effects at the intestinal level through the induction of cytokines, which in turn affect the systemic immune response in fish. Different sources of β-glucans have been evaluated, although the most frequent sources are those obtained from the baker's yeast, Saccharomyces cerevisiae. Some commercial preparations of β-glucan (MacroGard ®, Betagard A ®, EcoActivaTM, Nutriferm TM, BG, Fibosel®, etc) are available on the market that can be used in aquaculture. Many studies have explored the in vitro response of the macrophage to β-glucan [146, 147], while other studies have addressed the in vivo effect of β-glucans in different fish species (Table 3). The β-glucans from several sources have been administered to fish via the oral or intra-peritoneal route with different effects. Many studies have shown the immune effects of β-glucans specifically on antibody production, expression of immune system genes, survival, resistance to infectious diseases, and improvement in stress resistance. The growth enhancement of fish has also been observed as another beneficial effect of β-glucans (Table 3).

Recently, juvenile channel catfish (Ictalurus punctatus) fed diets supplemented with wholecell Saccharomyces cerevisiae (Levucell SB20®) or yeast subcomponents such as commercial preparations of β-glucan (MacroGard ® and Betagard A ®) had a significantly higher survival rate after challenge with Edwardsiella ictaluri than did catfish fed with a control diet [104]. Atlantic cod (Gadus morhua L.) were fed for 5 weeks with a purified β-glucan product [148], after which the fish were bath-challenged with the bacterial pathogen, Vibrio anguillarum. The transcription of selected cytokines (proinflammatory: IL1-β, IL-8, IFNγ; anti-inflammatory: IL-10) in different intestinal segments was analyzed using qPCR, and the β-glucan product was found to have a differential effect on the expression of the cytokine genes. In the anterior intestine and rectum, the β-glucan significantly elevated the expression of IL-1β when challenged with V. anguillarum. Moreover, an effect on the anti-inflammatory cytokine IL-10 was also visible in the rectum after the pathogen challenge. The differential responses of cytokines in the intestine of fish upon exposure to V. anguillarum suggests that β-glucans impact the ability of Atlantic cod to respond to the pathogen [148].

In another recently study, different concentrations of the yeast β-glucan preparation Macro- Gard ® (0.1%, 1% or 2%) were orally administered to mirror carp (Cyprinus carpio L.) for 8 weeks [149]. Fish fed diets containing 1% and 2% MacroGard® showed significant improve- ments in weight gain, specific growth rate and feed conversion ratio compared to fish fed both the control and the 0.1% MacroGard® containing diet. At the end of the experiment, the haematocrit value was significantly elevated in fish fed the 2% MacroGard® diet, compared to the control fed fish, with the blood monocyte fraction significantly higher in fish fed the 1% and 2% MacroGard® diets [149].

Zebrafish (Danio rerio) have been suggested as a model aquacultured fish, especially for genetic [150], nutritional and comparative growth studies [151]. Furthermore, zebrafish have been suggested as a model for pathogen studies in finfish [152]. Yeast β-glucans have also been evaluated in the zebrafish (Danio rerio) model with promising results [153]. In the study, a 5 mg/ml β-glucan preparation derived from S. cerevisiae was injected intra-peritoneally into adult zebrafish, leading to a significant reduction in mortality after challenge with Aeromonas hydrophila. In zebrafish treated with β-glucan, the ability of kidney cells to kill A. hydrophila was enhanced. Moreover, the myelomonocytic cell population in the kidney at 6 h postchallenge with A. hydrophila was increased. The β-glucan also appears to modulate the expression of IFN-γ and chemokines in the kidney [153].

Recently, the effect of β-glucan (derived from yeast, Fibosel® (Lallemand) on the growth performance and antioxidant enzyme activity in red snapper (Lutjanus peru), before and after exposure to lipopolysaccharides (LPS), was investigated. The fish were fed commercial diets with 0.0%, 0.1% and 0.2% Fibosel® for 6 weeks, after which, LPS was injected intra-peritoneally. The results showed a significant increase in growth performance after 6 weeks of β-glucan feeding; the SOD activity was also significantly higher in diets containing 0.1%β-glucanin weeks 4 and 6 with respect to the control group. At 72 h after injection of LPS, samples showed a significant increase in CAT activity in fish fed diets supplemented with 0.2% β-glucan and SOD activity increased under diets containing 0.1% and 0.2% β-glucan compared to controls [154]. To explain the enhanced growth, the authors suggested that some bacterial populations modify the host's digestive enzyme activity through their ability to produce and liberate exogenous digestive enzymes, as was previously observed [155]. Other authors reported that polysaccharides used as prebiotics can stimulate the growth of beneficial microbiota in fish [156].

Table 3. Biological effect of different β-glucans in fish

β-glucan sources	Administration route	Fish species	Biological effects	Ref.
β-glucan (Aqua-In-Tech, Inc.)	Oral	Nile tilapia (*Oreochromis niloticus*)	no effect	[157]
Betagard A*	oral	Nile tilapia (*Oreochromis niloticus*)	Immune modulation	[158]
S. cerevisiae (Hang Zhou Bio-Technology Co)	oral	Nile tilapia (*Oreochromis niloticus*)	Immune modulation	[159]
MacroGard® Betagard A®	oral	Channel catfish (*Ictalurus punctatus*)	improvement in stress resistance	[160]
MacroGard® Betagard A®	oral	Channel catfish (*Ictalurus punctatus*)	Immune modulation	[104]
Saccharomyces cerevisiae	oral	Large yellow croaker (*Pseudosciaena crocea*)	Immune modulation growth enhancement	[161]
MacroGard® Zymosan GY (Sigma)* GB (Sigma)**	oral	Fathead minnows (*Pimephales promelas*)	Immune modulation	[162]
Saccharomyces cerevisiae	intra-peritoneal	Atlantic salmon (*Salmo salar* L)	Immune modulation	[163]
Saccharomyces cerevisiae	oral	Atlantic salmon (*Salmo salar* L)	Enhancement of salmon lice resistance	[164]
marine diatom *Chaetoceros mülleri*	oral	Atlantic cod (*Gadus morhua* L.)	Survival and growth enhancement	[165]
β-1,3/1,6 glucan: BG (Biorigin Europe, Oslo, Norway)	oral	Atlantic cod (*Gadus morhua* L.)	Immune modulation	[148]
Barley	intra-peritoneal	Rohu (*Labeo rohita)*	Immune modulation	[166]
β-glucan (Sigma)	oral	Rohu (*Labeo rohita)*	Immune modulation	[167]
β-1,3 glucan (Sigma)	oral	Rohu (*Labeo rohita*)	Immune modulation	[168]
β-1,3 glucan (Sigma)	oral	Rohu (*Labeo rohita*)	Immune modulation	[169]
Yeast cell wall preparation from *Saccharomyces cerevisiae* (Nutriferm ™)	oral	Rohu (*Labeo rohita*)	Immune modulation	[170]
β-1,3 glucan (Sigma)	oral	Asian catfish (*Clarias batrachus*)	Immune modulation	[171, 172]
Glucan (Taito Co.Ltd., Tokyo, Japan)	oral	Rainbow trout (*Oncorhynchus mykiss*)	Decrease in plasmatic cortisol	[173]
MacroGard®	oral	Rainbow trout (*Oncorhynchus mykiss*)	Immune modulation	[174]
MacroGard®	oral	Rainbow trout (*Oncorhynchus mykiss*)	Immune modulation	[175]
Barley	oral	Rainbow trout (*Oncorhynchus mykiss*)	Immune modulation	[176]
β(1,3)-D-glucan (laminaran) from *Laminaria hyperborea*	intra-peritoneal	Rainbow trout (*Oncorhynchus mykiss*)	Immune modulation	[177]
β(1,3)-D-glucan (laminaran) from *Laminaria hyperborea*	inmersion	Rainbow trout (*Oncorhynchus mykiss*)	Immune modulation	[178]

Saccharomyces cerevisiae	intra-peritoneal oral	Rainbow trout (*Oncorhynchus mykiss*)	Immune modulation	[179]
Saccharomyces cerevisiae	intra-peritoneal	Carp (*Cyprinus carpio*)	Immune modulation	[180, 181]
Saccharomyces cerevisiae	intra-peritoneal oral	Carp (*Cyprinus carpio*)	Immune modulation	[182]
Saccharomyces cerevisiae	oral	Carp (*Cyprinus carpio*)	Immune modulation	[183]
MacroGard®	oral	Carp (*Cyprinus carpio*)	Immune modulation	[184]
MacroGard®	oral	Carp (*Cyprinus carpio*)	apoptosis modulation	[185]
MacroGard®	oral	Carp (*Cyprinus carpio*)	Growth enhancement	[186]
Saccharomyces cerevisiae	intra-peritoneal	Zebrafish (*Danio rerio*)	Immune modulation	[153]
MacroGard®	oral	Sea bass (*Dicentrarchus labrax*)	Immune modulation	[187]
MacroGard®	oral	Sea bass (*Dicentrarchus labrax*)	Immune modulation	[188]
MacroGard®	oral	European sea bass (*Dicentrarchus labrax*)	Immune modulation	[189]
EcoActiva™	oral	Pink snapper (*Pagrus auratus*)	growth enhancement Immune modulation	[190]
oyster mushroom (*Pleurotus florida*)	intra-peritoneal	Catla (*Catla catla*)	Immune modulation	[191]
Poria cocos	oral	*Ctenopharyngodon idella*	Immune modulation	[192]
Fibosel® (Lallemand) *Saccharomyces cerevisiae*	oral	*Lutjanus peru*	growth enhancement Immune modulation	

*GY: β-1,3-glucan from baker's yeast
**GB: β -1,3-glucan from barley

CONCLUSIONS

It is interesting to note that even after several decades of investigation, the potential of yeast, especially those of marine origin, has not yet been fully exploited. Yeasts can be part of the gut microbiota of wild and cultivated fish; however, more information derived using molecular approaches, is needed regarding the yeast composition in fish. Although much has been reported on the molecular aspects of yeasts, the exploration of the complete yeast community through the analysis of yeast DNA or RNA is lacking. The application of such methodologies will provide us with an overview of the non-cultivated yeasts, which could play a major role in the fish host. Different enzymes can be synthesized by yeast that have biotechnological potential, but the direct contribution of this potential to fish nutrition must be explored. In contrast, several publications confirm the beneficial probiotic effects of yeast in aquaculture, but the majority of these studies are focused on two species: S. cerevisiae and D. hansenii. The identification of new yeast species/strains from other cultured fish species is required to explore new beneficial properties to improve fish health and nutrition for a more sustainable aquaculture.

ACKNOWLEDGEMENTS

This work has been supported by FONDECYT grant 11110414 from CONICYT-Chile. D. Tovar, thanks to the Mexican Council of Science and Technology (CONACYT) for the grant CB-2010 157763.

REFERENCES

1. FAO. http://www.fao.org/fishery/aquaculture/en (accessed 5 April 2013).
2. Rumsey GL, Siwicki AK, Anderson DP, Bowser PR. Effect of soybean protein on serological response, non-specific defense mechanisms, growth, and protein utilization in rainbow trout. Vet Immunol Immunopath 1994; 41(3–4) 323-39.
3. Van den Ing T, Olli JJ, Krogdahl A. Effects of soybean-containing diets on the proximal and distal intestine in Atlantic salmon (Salmo salar): a morphological study. Aquaculture 1991;94 297-305.
4. Sealey WM, Barrows FT, Smith CE, Overturf K, LaPatra SE. Soybean meal level and probiotics in first feeding fry diets alter the ability of rainbow trout Oncorhynchus mykiss to utilize high levels of soybean meal during grow-out. Aquaculture 2009;293(3-4) 195-203.
5. Rawls JF, Samuel BS, Gordon JI. Gnotobiotic zebrafish reveal evolutionarily conserved responses to the gut microbiota. P Natl Acad Sci USA 2004;101(13) 4596-601.
6. Gatesoupe FJ. Live yeasts in the gut: Natural occurrence, dietary introduction, and their effects on fish health and

development. Aquaculture 2007;267(1-4) 20-30.

7. Kurtzman CP. Molecular taxonomy of the yeasts. Yeast 1994;10(13) 1727-40.

8. Suh SO, Blackwell M, Kurtzman CP, Lachance MA. Phylogenetics of Saccharomycetales, the ascomycete yeasts. Mycologia 2006;98(6) 1006-17.

9. Scorzetti G, Fell JW, Fonseca A, Statzell-Tallman A. Systematics of basidiomycetous yeasts: a comparison of large subunit D1/D2 and internal transcribed spacer rDNA regions. FEMS yeast research 2002;2(4) 495-517.

10. Lachance MA, Starmer WT. Ecology and yeasts. In: Kurtzman CP, Fell JW, editors. The Yeasts, a Taxonomic Study. Amsterdam: Elsevier; 1998. p. 21-30.

11. Raghukumar S. The Role of Fungi in Marine Detrital Processes. In: Ramaiah N, editor. Marine Microbiology: Facets & Opportunities: National Institute of Oceanography, Goa; 2004. p. 91-101.

12. Rhishipal R, Philip R. Selection of marine yeasts for the generation of single cell protein from prawn-shell waste. Bioresource Technol 1998;65(3) 255-6.

13. Rocha LL, Cordeiro RDA, Cavalcante RM, do Nascimento RF, Martins SCS, Santaella ST, et al. Isolation and characterization of phenol-degrading yeasts from an oil refinery wastewater in Brazil. Mycopathologia 2007;164(4) 183-8.

14. Burgaud G, Arzur D, Durand L, Cambon-Bonavita MA, Barbier G. Marine culturable yeasts in deep-sea hydrothermal vents: species richness and association with fauna. Fems Microbiol Ecol 2010;73(1) 121-33.

15. Chi ZM, Chi Z, Zhang T, Liu GL, Li J, Wang XH. Production, characterization and gene cloning of the extracellular enzymes from the marine-derived yeasts and their potential applications. Biotechnol Adv 2009;27(3) 236-55.

16. Lee JK, Park KS, Park S, Park H, Song YH, Kang SH, et al. An extracellular ice-binding glycoprotein from an Arctic psychrophilic yeast. Cryobiology 2010;60(2) 222-8.

17. Li J, Chi ZM, Wang XH, Peng Y, Chi Z. The selection of alkaline protease-producing yeasts from marine environments and evaluation of their bioactive peptide production. Chin J Oceanol Limn 2009;27(4) 753-61.

18. Peng Y, Chi ZM, Wang XH, Li J. Purification and molecular characterization of exobeta-1,3-glucanases from the marine yeast Williopsis saturnus WC91-2. Appl Microbiol Biot 2009;85(1) 85-94.

19. Song CL, Chi ZM, Li J, Wang XH. beta-Galactosidase production by the psychrotolerant yeast Guehomyces pullulans 17-1 isolated from sea sediment in Antarctica and lactose hydrolysis. Bioproc Biosyst Eng 2010;33(9) 1025-31.

20. Song CL, Liu GL, Xu JL, Chi ZM. Purification and characterization of extracellular beta-galactosidase from the psychrotolerant yeast Guehomyces pullulans 17-1 isolated from sea sediment in Antarctica. Process Biochem 2010;45(6) 954-60.

21. Kutty SN, Philp R. Marine yeasts - a review. Yeast 2008;25(7) 465-83.

22. Kandasamy K, Alikunhi N, Subramanian M. Yeasts in marine and estuarine environments. Journal of Yeast and Fungal Research 2012;3(6) 74-82.

23. Lodder J, Kreger-van rij JW. The Yeasts. A taxonomic study. Amsterdam, Holland.: North Holland Publ. Co.,; 1952.

24. Wickerham L. Taxonomy of yeasts. U S Dept Agr Tech Bull 1951;1029 1-55.

25. Roth F, Ahean D, Fell JW, Meyers S, Meyer S. Ecology and taxonomy of yeasts isolated from various marine substrates. Limnology and Oceanography 1962;7(2) 178-85.

26. Lodder J. The yeasts. A taxonomic study. Amsterdam: North- Holland Publishing Co.; 1970.

27. Lazarus C, Koburger J. Identification of Yeasts From the Suwannee River Florida Estuary. Applied Microbiology 1974;27(6) 1108-11.

28. Hernandez-Saavedra NY, Hernandez-Saavedra D, Ochoa JL. Distribution of Sporobolomyces (Kluyver-Et-Vanniel) Genus in the Western Coast of Baja-California-Sur, Mexico. Syst Appl Microbiol 1992;15(2) 319-22.

29. Nagahama T, Hamamoto M, Nakase T, Horikoshi K. Kluyveromyces nonfermentans sp nov., a new yeast species isolated from the deep sea. International journal of systematic bacteriology 1999;49 1899-905.

30. Nagahama T, Hamamoto M, Nakase T, Takami H, Horikoshi K. Distribution and identification of red yeasts in deep-sea environments around the northwest Pacific Ocean. Anton Leeuw Int J G 2001;80(2) 101-10.

31. Nagahama T, Hamamoto M, Nakase T, Horikoshi K. Rhodotorula lamellibrachii sp nov., a new yeast species from a tubeworm collected at the deep-sea floor in Sagami Bay and its phylogenetic analysis. Anton Leeuw Int J G 2001;80(3-4) 317-23.

32. Nagahama T, Hamamoto M, Nakase T, Horikoshi K. Rhodotorula benthica sp nov and Rhodotorula calyptogenae sp nov., novel yeast species from animals collected from the deep-sea floor, and Rhodotorula lysiniphila sp nov., which is related phylogenetically. International journal of systematic and evolutionary microbiology 2003;53 897-903.

33. Gadanho M, Sampaio JP. Occurrence and diversity of yeasts in the Mid-Atlantic ridge hydrothermal fields near the Azores Archipelago. Microb Ecol 2005;50(3) 408-17.

34. Nagahama T, Hamamoto M, Horikoshi K. Rhodotorula pacifica sp nov., a novel yeast species from sediment collected on the deep-sea floor of the north-west Pacific Ocean. International journal of systematic and evolutionary microbiology 2006;56 295-9.

35. Bass D, Howe A, Brown N, Barton H, Demidova M, Michelle H, et al. Yeast forms dominate fungal diversity in the deep oceans. P R Soc B 2007;274(1629) 3069-77.

36. Nagahama T, Abdel-Wahab MA, Nogi Y, Miyazaki M, Uematsu K, Hamamoto M, et al. Dipodascus tetrasporeus sp nov., an ascosporogenous yeast isolated from deep-sea sediments in the Japan Trench. International journal of systematic and evolutionary microbiology 2008;58

1040-6.

37. Chen YS, Yanagida F, Chen LY. Isolation of marine yeasts from coastal waters of northeastern Taiwan. Aquat Biol 2009;8(1) 55-60.

38. Kutty SN, Philip R, Damodaran R. Marine yeasts in the slope sediments of Arabian sea and bay of Bengal. European Journal of Experimental Biology 2013;3(3) 311- 27.

39. van Uden N, Branco C. Distribution and population densities of yeast species in Pacific water, air, animals and kelp off southern California. Limnology and Oceanography 1963;8 323-9.

40. Slavikova E, Vadkertiova R. Yeasts and yeast-like organisms isolated from fish-pond waters. Acta microbiologica Polonica 1995;44(2) 181-9.

41. Andlid T, Vazquez-Juarez R, Gustafsson L. Yeasts isolated from the intestine of rainbow trout adhere to and grow in intestinal mucus. Mol Mar Biol Biotech 1998;7(2) 115-26.

42. Wache Y, Auffray F, Gatesoupe FJ, Zambonino J, Gayet V, Labbe L, et al. Cross effects of the strain of dietary Saccharomyces cerevisiae and rearing conditions on the onset of intestinal microbiota and digestive enzymes in rainbow trout, Onchorhynchus mykiss, fry. Aquaculture 2006;258(1-4) 470-8.

43. Vázquez-Juárez R, Andlid T, Gustafsson L. Cell surface hydrophobicity and its relation to adhesion of yeasts isolated from fish gut. Colloids and Surfaces B: Biointerfaces 1994;2(1-3) 199-209.

44. Vazquez-Juarez R, Andlid T, Gustafsson L. Adhesion of yeast isolated from fish gut to crude intestinal mucus of rainbow trout, Salmo gairdneri. Mol Mar Biol Biotechnol 1997;6(1) 64-71.

45. Andlid T, Juarez RV, Gustafsson L. Yeast Colonizing the Intestine of Rainbow-Trout (Salmo-Gairdneri) and Turbot (Scophtalmus-Maximus). Microb Ecol 1995;30(3) 321-34.

46. van Uden N, Farinha M. On the significance of temperature relations and vitamin deficiency patterns in the delimitation of yeast species. Port Acta Biol, Sér B 1958;6 161-78.

47. Kreger-van rij JW. A Taxonomic study of the Yeast genera Endomycopsis, Pichia and Debaryomyces. . Kerkwerve, Netherlands: Backhuys Biological Books 1964.

48. Bruce J, Morris EO. Psychrophilic yeasts isolated from marine fish. Antonie van Leeuwenhoek 1973;39(1) 331-9.

49. Aubin J, Gatesoupe FJ, Labbe L, Lebrun L. Trial of probiotics to prevent the vertebral column compression syndrome in rainbow trout (Oncorhynchus mykiss Walbaum). Aquaculture Research 2005;36(8) 758-67.

50. Barnett JA, Payne RW, Yarrow D. Yeast: Characteristics and Identification. 2nd ed. Cambridge: Cambridge Univ. Press.; 1990.

51. Laconi S, Pompei R. Study and characterization of intestinal yeasts of mullet (Mugil spp.) for potential probiotic use. J Food Agric Environ 2007;5(3-4) 475-80.

52. Breuer U, Harms H. Debaryomyces hansenii - an extremophilic yeast with biotechnological potential. Yeast 2006;23(6) 415-37.

53. Gao LM, Chi ZM, Sheng J, Wang L, Li J, Gong F. Inulinase-producing marine yeasts: Evaluation of their diversity and inulin hydrolysis by their crude enzymes. Microb Ecol 2007;54(4) 722-9.

54. Bolumar T, Sanz Y, Aristoy MC, Toldra F. Purification and characterisation of proteases a and D from Debaryomyces hansenii. Int J Food Microbiol 2008;124(2) 135-41.

55. Garcia-Gonzalez A, Lotz M, Ochoa JL. Anti-inflammatory activity of superoxide dismutase obtained from Debaryomyces hansenii on type II collagen induced arthritis in rats. Rev Invest Clin 2009;61(3) 212-20.

56. Takac S, Sengel BS. Extracellular lipolytic enzyme activity of a newly isolated Debaryomyces hansenii. Preparative biochemistry & biotechnology 2010;40(1) 28-37.

57. Segal-Kischinevzky C, Rodarte-Murguia B, Valdes-Lopez V, Mendoza-Hernandez G, Gonzalez A, Alba-Lois L. The Euryhaline Yeast Debaryomyces hansenii has Two Catalase Genes Encoding Enzymes with Differential Activity Profile. Curr Microbiol 2011;62(3) 933-43.

58. Viana PA, de Rezende ST, Alves AD, Manfrini RM, Alves RJ, Bemquerer MP, et al. Activity of Debaryomyces hansenii UFV-1 alpha-galactosidases against alpha-D-galactopyranoside derivatives. Carbohyd Res 2011;346(5) 602-5.

59. Viana PA, de Rezende ST, Falkoski DL, Leite TD, Jose IC, Moreira MA, et al. Hydrolysis of oligosaccharides in soybean products by Debaryomyces hansenii UFV-1 alphagalactosidases. Food Chem 2007;103(2) 331-7.

60. Viana PA, de Rezende ST, Marques VM, Trevizano LM, Passos FML, Oliveira MGA, et al. Extracellular alpha-galactosidase from Debaryomyces hansenii UFV-1 and its use in the hydrolysis of raffinose oligosaccharides. Journal of Agricultural and Food Chemistry 2006;54(6) 2385-91.

61. Viana PA, de Rezende ST, Passos FML, Machado SG, Maitan GP, Coelho VTD, et al. alpha-Galactosidases production by Debaryomyces hansenii UFV-1. Food Sci Biotechnol 2011;20(3) 601-6.

62. Viana PA, de Rezende ST, Passos FML, Oliveira JS, Teixeira KN, Santos AMC, et al. Debaryomyces hansenii UFV-1 Intracellular alpha-Galactosidase Characterization and Comparative Studies with the Extracellular Enzyme. Journal of Agricultural and Food Chemistry 2009;57(6) 2515-22.

63. Barbosa AM, Giese EC, Dekker RFH, Borsato D, Perez AIB, Iranzo JFU. Extracellular beta-glucosidase production by the yeast Debaryomyces pseudopolymorphus UCLMNS7A: optimization using response surface methodology. New Biotechnol 2010;27(4) 374-81.

64. Ragon M, Aumelas A, Chemardin P, Galvez S, Moulin G, Boze H. Complete hydrolysis of myo-inositol hexakisphosphate by a novel phytase from Debaryomyces castellii CBS 2923. Appl Microbiol Biot 2008;78(1) 47-53.

65. Dujon B, Sherman D, Fischer G, Durrens P, Casaregola S, Lafontaine I, et al. Genome evolution in yeasts. Nature 2004;430(6995) 35-44.

66. White TJ, Bruns T, Lee S, Taylor S. Amplification and direct sequencing of fungal ribosomal RNA genes for phylogenetics. In: Innis MAG, P.H.; Sninsky, J.J.; White, T.J. , editor. PCR Protocols. San Diego, USA: Academic Press Inc; 1990. p. 315-22.

67. The Centraal bureau voor Schimmelcultures (CBS) database. http:// www.cbs.knaw.nl/collections/BioloMICSSequences.aspx (accessed 5 April 2013)

68. The Centraal bureau voor Schimmelcultures (CBS) database. http:// www.cbs.knaw.nl/collections/BiolomicsID.aspx?IdentScenario=Yeasts%20species%20online%20ID (accessed 5 April 2013).

69. Amann RI, Ludwig W, Schleifer KH. Phylogenetic Identification and in-Situ Detection of Individual Microbial-Cells without Cultivation. Microbiol Rev 1995;59(1) 143-69.

70. Urubschurov V, Janczyk P. Biodiversity of Yeasts in the Gastrointestinal Ecosystem with Emphasis on Its Importance for the Host. In: Grillo O, Venora G, editors. The Dynamical Processes of Biodiversity - Case Studies of Evolution and Spatial Distribution: INTECH; 2011.

71. Hovda MB, Lunestad BT, Fontanillas R, Rosnes JT. Molecular characterisation of the intestinal microbiota of farmed Atlantic salmon (Salmo salar L.). Aquaculture 2007;272 581-8.

72. Kim DH, Brunt J, Austin B. Microbial diversity of intestinal contents and mucus in rainbow trout (Oncorhynchus mykiss). J Appl Microbiol 2007;102(6) 1654-64.

73. Navarrete P, Espejo RT, Romero J. Molecular analysis of microbiota along the digestive tract of juvenile Atlantic salmon (Salmo salar L.). Microb Ecol 2009;57(3) 550-61.

74. Navarrete P, Magne F, Mardones P, Riveros M, Opazo R, Suau A, et al. Molecular analysis of intestinal microbiota of rainbow trout (Oncorhynchus mykiss). Fems Microbiol Ecol 2010;71(1) 148-56.

75. Pond MJ, Stone DM, Alderman DJ. Comparison of conventional and molecular techniques to investigate the intestinal microflora of rainbow trout (Oncorhynchus mykiss). Aquaculture 2006;261(1) 194-203.

76. Romero J, Navarrete P. 16S rDNA-based analysis of dominant bacterial populations associated with early life stages of coho salmon (Oncorhynchus kisutch). Microb Ecol 2006;51(4) 422-30.

77. Navarrete P, Magne F, Araneda C, Fuentes P, Barros L, Opazo R, et al. PCR-TTGE analysis of 16S rRNA from rainbow trout (Oncorhynchus mykiss) gut microbiota reveals host-specific communities of active bacteria. Plos One 2012;7(2).

78. Muyzer G, Dewaal EC, Uitterlinden AG. Profiling of Complex Microbial-Populations by Denaturing Gradient Gel-Electrophoresis Analysis of Polymerase Chain ReactionAmplified Genes-Coding for 16s Ribosomal-Rna. Appl Environ Microb 1993;59(3) 695-700.

79. Lachance MA, Daniel HM, Meyer W, Prasad GS, Gautam SP, Boundy-Mills K. The D1/D2 domain of the large-subunit rDNA of the yeast species Clavispora lusitaniae is unusually polymorphic. FEMS yeast research 2003;4(3) 253-8.

80. Cocolin L, Aggio D, Manzano M, Cantoni C, Comi G. An application of PCR-DGGE analysis to profile the yeast populations in raw milk. Int Dairy J 2002;12(5) 407-11.

81. Cocolin L, Manzano M, Rebecca S, Comi G. Monitoring of yeast population changes during a continuous wine fermentation by molecular methods. Am J Enol Viticult 2002;53(1) 24-7.

82. Gatto V, Torriani S. Microbial population changes during sourdough fermentation monitored by DGGE analysis of 16S and 26S rRNA gene fragments. Ann Microbiol 2004;54(1) 31-42.

83. Molnar O, Wuczkowski M, Prillinger H. Yeast biodiversity in the guts of several pests on maize; comparison of three methods: classical isolation, cloning and DGGE. Mycol Prog 2008;7(2) 111-23.

84. Prakitchaiwattana CJ, Fleet GH, Heard GM. Application and evaluation of denaturing gradient gel electrophoresis to analyse the yeast ecology of wine grapes. FEMS yeast research 2004;4(8) 865-77.

85. Valmorri S, Tofalo R, Settanni L, Corsetti A, Suzzi G. Yeast microbiota associated with spontaneous sourdough fermentations in the production of traditional wheat sourdough breads of the Abruzzo region (Italy). Anton Leeuw Int J G 2010;97(2) 119-29.

86. Ubeda JF, Fernandez-Gonzales M, Briones AI. Application of PCR-TTGE and PCRRFLP for Intraspecific and Interspecific Characterization of the Genus Saccharomyces Using Actin Gene (ACT1) Primers. Curr Microbiol 2009;58(1) 58-63.

87. Hamady M, Knight R. Microbial community profiling for human microbiome projects: Tools, techniques, and challenges. Genome research 2009;19(7) 1141-52.

88. Beloshapka AN, Dowd SE, Suchodolski JS, Steiner JM, Duclos L, Swanson KS. Fecal microbial communities of healthy adult dogs fed raw meat-based diets with or without inulin or yeast cell wall extracts as assessed by 454 pyrosequencing. Fems Microbiol Ecol 2013;84(3) 532-41.

89. Delgado S, Cabrera-Rubio R, Mira A, Suarez A, Mayo B. Microbiological Survey of the Human Gastric Ecosystem Using Culturing and Pyrosequencing Methods. Microb Ecol 2013;65(3) 763-72.

90. Marzorati M, Maignien L, Verhelst A, Luta G, Sinnott R, Kerckhof FM, et al. Barcoded pyrosequencing analysis of the microbial community in a simulator of the human gastrointestinal tract showed a colon region-specific microbiota modulation for two plant-derived polysaccharide blends. Anton Leeuw Int J G 2013;103(2) 409-20.

91. Oikonomou G, Teixeira AGV, Foditsch C, Bicalho ML, Machado VS, Bicalho RC. Fecal Microbial Diversity in Pre-Weaned Dairy Calves as Described by Pyrosequencing of Metagenomic 16S rDNA.

Associations of Faecalibacterium Species with Health and Growth. Plos One 2013;8(4).

92. Roos S, Dicksved J, Tarasco V, Locatelli E, Ricceri F, Grandin U, et al. 454 Pyrosequencing Analysis on Faecal Samples from a Randomized DBPC Trial of Colicky Infants Treated with Lactobacillus reuteri DSM 17938. Plos One 2013;8(2).

93. Zened A, Combes S, Cauquil L, Mariette J, Klopp C, Bouchez O, et al. Microbial ecology of the rumen evaluated by 454 GS FLX pyrosequencing is affected by starch and oil supplementation of diets. Fems Microbiol Ecol 2013;83(2) 504-14.

94. Heinrichs G, Hubner I, Schmidt CK, de Hoog GS, Haase G. Analysis of Black Fungal Biofilms Occurring at Domestic Water Taps (I): Compositional Analysis Using TagEncoded FLX Amplicon Pyrosequencing. Mycopathologia 2013;175(5-6) 387-97.

95. Monard C, Gantner S, Stenlid J. Utilizing ITS1 and ITS2 to study environmental fungal diversity using pyrosequencing. Fems Microbiol Ecol 2013;84(1) 165-75.

96. Pannanusorn S, Elings MA, Romling U, Fernandez V. Pyrosequencing of a hypervariable region in the internal transcribed spacer 2 to identify clinical yeast isolates. Mycoses 2012;55(2) 172-80.

97. Heinrichs G, de Hoog GS, Haase G. Barcode Identifiers as a Practical Tool for Reliable Species Assignment of Medically Important Black Yeast Species. J Clin Microbiol 2012;50(9) 3023-30.

98. Kim JY, Hahn HJ, Choe YB, Lee YW, Ahn KJ, Moon KC. Molecular Biological Identification of Malassezia Yeasts Using Pyrosequencing. Ann Dermatol 2013;25(1) 73-9.

99. Ortuno J, Cuesta A, Rodriguez A, Esteban MA, Meseguer J. Oral administration of yeast, Saccharomyces cerevisiae, enhances the cellular innate immune response of gilthead seabream (Sparus aurata L.). Vet Immunol Immunopathol 2002;85(1-2) 41-50.

100. Harikrishnan R, Kim MC, Kim JS, Balasundaram C, Heo MS. Immunomodulatory effect of probiotics enriched diets on Uronema marinum infected olive flounder. Fish Shellfish Immunol 2011;30(3) 964-71.

101. Tukmechi A, Rahmati Andani HR, Manaffar R, Sheikhzadeh N. Dietary administration of beta-mercapto-ethanol treated Saccharomyces cerevisiae enhanced the growth, innate immune response and disease resistance of the rainbow trout, Oncorhynchus mykiss. Fish Shellfish Immunol 2011;30(3) 923-8.

102. Chiu CH, Cheng CH, Gua WR, Guu YK, Cheng W. Dietary administration of the probiotic, Saccharomyces cerevisiae P13, enhanced the growth, innate immune responses, and disease resistance of the grouper, Epinephelus coioides. Fish Shellfish Immunol 2010;29(6) 1053-9.

103. Torrecillas S, Makol A, Benitez-Santana T, Caballero MJ, Montero D, Sweetman J, et al. Reduced gut bacterial translocation in European sea bass (Dicentrarchus labrax) fed mannan oligosaccharides (MOS). Fish Shellfish Immunol 2011;30(2) 674-81.

104. Welker TL, Lim C, Yildirim-Aksoy M, Klesius PH. Effect of short-term feeding duration of diets containing commercial whole-cell yeast or yeast subcomponents on im-mune function and disease resistance in channel catfish, Ictalurus punctatus. Journal of animal physiology and animal nutrition 2012;96(2) 159-71.

105. Reyes-Becerril M, Tovar-Ramirez D, Ascencio-Valle F, Civera-Cerecedo R, Gracia-Lopez V, Barbosa-Solomieu V. Effects of dietary live yeast Debaryomyces hansenii on the immune and antioxidant system in juvenile leopard grouper Mycteroperca rosacea exposed to stress. Aquaculture 2008;280(1-4) 39-44.

106. Reyes-Becerril M, Tovar-Ramirez D, Ascencio-Valle F, Civera-Cerecedo R, Gracia-Lopez V, Barbosa-Solomieu V, et al. Effects of dietary supplementation with probiotic live yeast Debaryomyces hansenii on the immune and antioxidant systems of leopard grouper Mycteroperca rosacea infected with Aeromonas hydrophila. Aquaculture Research 2011;42(11) 1676-86.

107. Reyes-Becerril M, Salinas I, Cuesta A, Meseguer J, Tovar-Ramirez D, Ascencio-Valle F, et al. Oral delivery of live yeast Debaryomyces hansenii modulates the main innate immune parameters and the expression of immune-relevant genes in the gilthead seabream (Sparus aurata L.). Fish Shellfish Immunol 2008;25(6) 731-9.

108. Tovar-Ramirez D, Mazurais D, Gatesoupe JF, Quazuguel P, Cahu CL, Zambonino-Infante JL. Dietary probiotic live yeast modulates antioxidant enzyme activities and gene expression of sea bass (Dicentrarchus labrax) larvae. Aquaculture 2010;300(1-4) 142-7.

109. Buetler TM, Krauskopf A, Ruegg UT. Role of superoxide as a signaling molecule. News Physiol Sci 2004;19 120-3.

110. Tovar D, Zambonino J, Cahu C, Gatesoupe FJ, Vazquez-Juarez R, Lesel R. Effect of live yeast incorporation in compound diet on digestive enzyme activity in sea bass (Dicentrarchus labrax) larvae. Aquaculture 2002;204(1-2) 113-23.

111. Tovar-Ramirez D, Infante JZ, Cahu C, Gatesoupe FJ, Vazquez-Juarez R. Influence of dietary live yeast on European sea bass (Dicentrarchus labrax) larval development. Aquaculture 2004;234(1-4) 415-27.

112. Tovar-Ramirez D. El efecto probiótico de la levadura Debaryomyces hansenii CBS 8339 sobre el desarrollo larvario del pargo Lunarejo Lutjanus guttatus. La Paz, BCS, México: Personal Communication; 2013.

113. Zacarias-Soto M, Lazo JP, Viana MT. Effect of three probiotics administered through live feed on digestive enzyme activity in California halibut, Paralichthys californicus, larvae. J World Aquacult Soc 2011;42(3) 321-31.

114. Klaenhammer TR, Kullen MJ. Selection and design of probiotics. Int J Food Microbiol 1999;50(1-2) 45-57.

115. Gatesoupe FJ. Probiotic and formaldehyde treatments of Artemia nauplii as food for larval pollack, Pollachius pollachius. Aquaculture 2002;212(1-4) 347-60.

116. Lara-Flores M, Olvera-Novoa MA, Guzman-Mendez BE, Lopez-Madrid W. Use of the bacteria Streptococcus faecium and Lactobacillus acidophilus, and the yeast

Saccharomyces cerevisiae as growth promoters in Nile tilapia (Oreochromis niloticus). Aquaculture 2003;216(1-4) 193-201.

117. Noh SH, Han K, Won TH, Choi YJ. Effect of antibiotics, enzyme, yeast culture and probiotics on the growth performance of Israeli carp. Korean Journal of Animal Science 1994;36(480-486).

118. Mohanty SN, Swain SK, Tripathi SD. Growth and survival of rohu spawn fed on liver based diet. Journal of Inland Fisheries Society of India 1993;25(2) 41-5.

119. Guzmán-Villanueva L, Tovar-Ramírez D, Civera-Cerecedo R, editors. Effect of wild and ornithine decarboxylase deficient Debaryomyces hansenii, on Paralabrax maculatofasciatus larvae development. Caribbean and Latin American Aquaculture; 2007 6–9 November; San Juan Puerto Rico.

120. Dufour C, Dandrifosse G, Forget P, Vermesse F, Romain N, Lepoint P. Spermine and spermidine induce intestinal maturation in the rat. Gastroenterology 1988;95(1) 112-6.

121. Reyes-Becerril M, Ángeles Esteban MA, Tovar-Ramírez D, Ascencio-Valle F. Polyamine determination in different strains of the yeast Debaryomyces hansenii by high pressure liquid chromatography. . Food Chem 2011;127 1862-5.

122. Reyes-Becerril M, Ascencio-Valle F, Meseguer J, Tapia-Paniagua ST, Moriñigo MA, Esteban MÁ. Debaryomyces hansenii L2-enriched diet enhances the immunity status, gene expression and intestine functionality in gilthead seabream (Sparus aurata L.). Aquaculture Research 2012;43 1107-18.

123. Aleem E. beta -Glucans and their Applications in Cancer Therapy: Focus on human studies. Anti-cancer agents in medicinal chemistry 2013;13(5) 709-19.

124. Cloetens L, Ulmius M, Johansson-Persson A, Akesson B, Onning G. Role of dietary beta-glucans in the prevention of the metabolic syndrome. Nutrition reviews 2012;70(8) 444-58.

125. Tosh SM. Review of human studies investigating the post-prandial blood-glucose lowering ability of oat and barley food products. Eur J Clin Nutr 2013;67(4) 310-7.

126. Othman RA, Moghadasian MH, Jones PJH. Cholesterol-lowering effects of oat betaglucan. Nutrition reviews 2011;69(6) 299-309.

127. Andersson KE, Hellstrand P. Dietary oats and modulation of atherogenic pathways. Molecular nutrition & food research 2012;56(7) 1003-13.

128. Du B, Bian Z, Xu B. Skin Health Promotion Effects of Natural Beta-Glucan Derived from Cereals and Microorganisms: A Review. Phytotherapy research : PTR 2013.

129. Tanioka A, Tanabe K, Hosono A, Kawakami H, Kaminogawa S, Tsubaki K, et al. Enhancement of intestinal immune function in mice by β-D-glucan from Aureobasidium pullulans ADK-34. Scand J Immunol 2013;78(1) 61-8.

130. Ganner A, Schatzmayr G. Capability of yeast derivatives to adhere enteropathogenic bacteria and to modulate cells of the innate immune system. Appl Microbiol Biot 2012;95(2) 289-97.

131. Muramatsu D, Iwai A, Aoki S, Uchiyama H, Kawata K, Nakayama Y, et al. beta-glucan derived from Aureobasidium pullulans is effective for the prevention of influenza in mice. Plos One 2012;7(7).

132. Volman JJ, Ramakers JD, Plat J. Dietary modulation of immune function by beta-glucans. Physiol Behav 2008;94(2) 276-84.

133. Novak M, Vetvicka V. beta-glucans, history, and the present: Immunomodulatory aspects and mechanisms of action. J Immunotoxicol 2008;5(1) 47-57.

134. Goodridge HS, Wolf AJ, Underhill DM. beta-glucan recognition by the innate immune system. Immunol Rev 2009;230 38-50.

135. Brown GD, Gordon S. Immune recognition of fungal beta-glucans. Cell Microbiol 2005;7(4) 471-9.

136. Dalmo RA, Bogwald J. beta-glucans as conductors of immune symphonies. Fish Shellfish Immun 2008;25(4) 384-96.

137. Zimmerman JW, Lindermuth J, Fish PA, Palace GP, Stevenson TT, DeMong DE. A novel carbohydrate-glycosphingolipid interaction between a beta-(1-3)-glucan immunomodulator, PGG-glucan, and lactosylceramide of human leukocytes. J Biol Chem 1998;273(34) 22014-20.

138. Palma AS, Feizi T, Zhang YB, Stoll MS. Ligands for the beta-glucan receptor, Dectin-1, assigned using "designer" microarrays of oligosaccharide probes (neoglycolipids) generated from glucan polysaccharides. (vol 281, pg 5771, 2006). J Biol Chem 2006;281(34) 24999-.

139. Brown GD, Taylor PR, Reid DM, Willment JA, Williams DL, Martinez-Pomares L, et al. Dectin-1 is a major beta-glucan receptor on macrophages. J Exp Med 2002;196(3) 407-12.

140. Grunebach F, Weck MM, Reichert J, Brossart P. Molecular and functional characterization of human Dectin-1. Exp Hematol 2002;30(11) 1309-15.

141. Xu S, Huo J, Lee KG, Kurosaki T, Lam KP. Phospholipase C gamma 2 Is Critical for Dectin-1-mediated Ca(2+) Flux and Cytokine Production in Dendritic Cells. J Biol Chem 2009;284(11) 7038-46.

142. Shah VB, Ozment-Skelton TR, Williams DL, Keshvaraa L. Vav1 and PI3K are required for phagocytosis of beta-glucan and subsequent superoxide generation by microglia. Mol Immunol 2009;46(8-9) 1845-53.

143. Gantner BN, Simmons RM, Canavera SJ, Akira S, Underhill DM. Collaborative induction of inflammatory responses by dectin-1 and toll-like receptor 2. J Exp Med 2003;197(9) 1107-17.

144. Bricknell I, Dalmo RA. The use of immunostimulants in fish larval aquaculture. Fish Shellfish Immun 2005;19(5) 457-72.

145. Meena DK, Das P, Kumar S, Mandal SC, Prusty AK, Singh SK, et al. Beta-glucan: an ideal immunostimulant in aquaculture (a review). Fish Physiol Biochem 2013;39(3) 431-57.

146. Jorgensen JB, Sharp GJE, Secombes CJ, Robertsen B.

Effect of a Yeast-Cell-Wall Glucan on the Bactericidal Activity of Rainbow-Trout Macrophages. Fish Shellfish Immun 1993;3(4) 267-77.

147. Paulsen SM, Engstad RE, Robertsen B. Enhanced lysozyme production in Atlantic salmon (Salmo salar L.) macrophages treated with yeast beta-glucan and bacterial lipopolysaccharide. Fish Shellfish Immun 2001;11(1) 23-37.

148. Lokesh J, Fernandes JMO, Korsnes K, Bergh O, Brinchmann MF, Kiron V. Transcriptional regulation of cytokines in the intestine of Atlantic cod fed yeast derived mannan oligosaccharide or beta-Glucan and challenged with Vibrio anguillarum. Fish Shellfish Immun 2012;33(3) 626-31.

149. Kühlwein H, Merrifield D, Rawling M, Foey A, Davies S. Effects of dietary β-(1,3) (1,6)-D-glucan supplementation on growth performance, intestinal morphology and haemato-immunological profile of mirror carp (Cyprinus carpio L.). Journal of Animal Physiology and Animal Nutrition (Berl) 2013;doi: 10.1111/jpn.12078.

150. Dahm R, Geisler R. Learning from small fry: The zebrafish as a genetic model organism for aquaculture fish species. Mar Biotechnol 2006;8(4) 329-45.

151. Ulloa PE, Iturra P, Neira R, Araneda C. Zebrafish as a model organism for nutrition and growth: towards comparative studies of nutritional genomics applied to aquacultured fishes. Rev Fish Biol Fisher 2011;21(4) 649-66.

152. Sullivan C, Kim CH. Zebrafish as a model for infectious disease and immune function. Fish Shellfish Immun 2008;25(4) 341-50.

153. Rodriguez I, Chamorro R, Novoa B, Figueras A. beta-Glucan administration enhances disease resistance and some innate immune responses in zebrafish (Danio rerio). Fish Shellfish Immun 2009;27(2) 369-73.

154. Guzmán-Villanueva LT, Tovar-Ramírez D, Ascencio-Valle F, Macías-Rodríguez ME, Esteban MA, Gisbert E, editors. Effects of β-Glucan 1,3/1,6 on antioxidant enzyme activity in Lutjanus peru exposed to lipopolysaccharides. . Aquaculture 2012; 2012 1-5 September; Prague, Czech Republic: World Aquaculture Society.

155. Dimitroglou A, Merrifield DL, Carnevali O, Picchietti S, Avella M, Daniels C, et al. Microbial manipulations to improve fish health and production - A Mediterranean perspective. Fish Shellfish Immun 2011;30(1) 1-16.

156. Ringo E, Olsen RE. The effect of diet on aerobic bacterial flora associated with intestine of Arctic charr (Salvelinus alpinus L.). J Appl Microbiol 1999;86(1) 22-8.

157. Whittington R, Lim C, Klesius PH. Effect of dietary beta-glucan levels on the growth response and efficacy of Streptococcus iniae vaccine in Nile tilapia, Oreochromis niloticus. Aquaculture 2005;248(1-4) 217-25.

158. R. S, C. L, M. Y-A, Welker TL, Klesius PH. Effect of yeast oligosaccharide diet supplements on growth and disease resistance in juvenile Nile tilpia (Oreochromis niloticus). Journal of applied aquaculture 2009;21 61-71.

159. El-Boshy ME, El-Ashram AM, AbdelHamid FM, Gadalla HA. Immunomodulatory effect of dietary Saccharomyces cerevisiae, beta-glucan and laminaran in mercuric chloride treated Nile tilapia (Oreochromis niloticus) and experimentally infected with Aeromonas hydrophila. Fish Shellfish Immun 2010;28(5-6) 802-8.

160. Welker TL, Lim C, Yildirim-Aksoy M, Shelby R, Klesius PH. Immune response and resistance to stress and Edwardsiella ictaluri challenge in channel catfish, Ictalurus punctatus, fed diets containing commercial whole-cell yeast or yeast subcomponents. J World Aquacult Soc 2007;38(1) 24-35.

161. Ai QH, Mai KS, Zhang L, Tan BP, Zhang WB, Xu W, et al. Effects of dietary beta-1,3 glucan on innate immune response of large yellow croaker, Pseudosciaena crocea. Fish Shellfish Immun 2007;22(4) 394-402.

162. Palic D, Andreasen C, Herolt D, Menzel B, Roth J. Immunomodulatory effects of bglucan on neutrophil function in fathead minnows (Pimephales promelas Rafinesque, 1820). Dev Comp Immunol 2006;30 817-30.

163. Rorstad G, Aasjord PM, Robertsen B. Adjuvant Effect of a Yeast Glucan in Vaccines against Furunculosis in Atlantic Salmon (Salmo salar L). Fish Shellfish Immun 1993;3(3) 179-90.

164. Refstie S, Baeverfjord G, Seim RR, Elvebo O. Effects of dietary yeast cell wall betaglucans and MOS on performance, gut health, and salmon lice resistance in Atlantic salmon (Salmo salar) fed sunflower and soybean meal. Aquaculture 2010;305(1-4) 109-16.

165. Skjermo J, Storseth TR, Hansen K, Handa A, Oie G. Evaluation of beta-(1 -> 3, 1 -> 6)- glucans and high-M alginate used as immunostimulatory dietary supplement during first feeding and weaning of Atlantic cod (Gadus morhua L.). Aquaculture 2006;261(3) 1088-101.

166. Misra CK, Das BK, Mukherjee SC, Pattnaik P. Effect of multiple injections of betaglucan on non-specific immune response and disease resistance in Labeo rohita fingerlings. Fish Shellfish Immunol 2006;20(3) 305-19.

167. Misra CK, Das BK, Mukherjee SC, Pattnaik P. Effect of long term administration of dietary beta-glucan on immunity, growth and survival of Labeo rohita fingerlings. Aquaculture 2006;255(1-4) 82-94.

168. Sahoo PK, Mukherjee SC. Effect of dietary beta-1,3 glucan on immune responses and disease resistance of healthy and aflatoxin B-1-induced immunocompromised rohu (Labeo rohita Hamilton). Fish Shellfish Immun 2001;11(8) 683-95.

169. Sahoo PK, Mukherjee SC. The effect of dietary immunomodulation upon Edwardsiella tarda vaccination in healthy and immunocompromised Indian major carp (Labeo rohita). Fish Shellfish Immun 2002;12(1) 1-16.

170. Pal D, Joardar SN, Roy B. Immunostimulatory effects of a yeast (Saccharomyces cerevisiae) cell wall feed supplement on rohu (Labeo rohita), an Indian major carp. Isr J Aquacult-Bamid 2007;59(3) 175-81.

171. Kumari J, Sahoo PK. Dietary immunostimulants influence specific immune response and resistance of healthy and immunocompromised Asian catfish Clarias batrachus to Aeromonas hydrophila infection. Dis Aquat Organ 2006;70(1-2) 63-70.

172. Kumari J, Sahoo PK. Dietary beta-1,3 glucan potentiates innate immunity and disease resistance of Asian catfish, Clarias batrachus (L.). J Fish Dis 2006;29(2) 95-101.

173. Jeney G, Galeotti M, Volpatti D, Jeney Z, Anderson DP. Prevention of stress in rainbow trout (Oncorhynchus mykiss) fed diets containing different doses of glucan. Aquaculture 1997;154(1) 1-15.

174. Verlhac V, Obach A, Gabaudan J, Schuep W, Hole R. Immunomodulation by dietary vitamin C and glucan in rainbow trout (Oncorhynchus mykiss). Fish Shellfish Immun 1998;8(6) 409-24.

175. Siwicki A, Kazuñ K, Glabski E, Terech-Majewska E, Baranowski P, Trapkowska S. The effect of beta-1.3/1.6 – glucan in diets on the effectiveness of anti-Yersinia ruckeri vaccine – an experimental study in rainbow trout (Oncorhynchus mykiss). Polish journal of food and nutrition sciences 2004;13(54) 59-61.

176. Sealey WM, Barrows FT, Hang A, Johansen KA, Overturf K, LaPatra SE, et al. Evaluation of the ability of barley genotypes containing different amounts of beta-glucan to alter growth and disease resistance of rainbow trout Oncorhynchus mykiss. Anim Feed Sci Tech 2008;141(1-2) 115-28.

177. Lovoll M, Fischer U, Mathisen GS, Bogwald J, Ototake M, Dalmo RA. The C3 subtypes are differentially regulated after immunostimulation in rainbow trout, but head kidney macrophages do not contribute to C3 transcription. Vet Immunol Immunopathol 2007;117(3-4) 284-95.

178. Zhang Z, Swain T, Bogwald J, Dalmo RA, Kumari J. Bath immunostimulation of rainbow trout (Oncorhynchus mykiss) fry induces enhancement of inflammatory cytokine transcripts, while repeated bath induce no changes. Fish Shellfish Immunol 2009;26(5) 677-84.

179. Kunttu HMT, Valtonen ET, Suomalainen LR, Vielma J, Jokinen IE. The efficacy of two immunostimulants against Flavobacterium columnare infection in juvenile rainbow trout (Oncorhynchus mykiss). Fish Shellfish Immun 2009;26(6) 850-7.

180. Selvaraj V, Sampath K, Sekar V. Administration of yeast glucan enhances survival and some non-specific and specific immune parameters in carp (Cyprinus carpio) infected with Aeromonas hydrophila. Fish Shellfish Immun 2005;19(4) 293-306.

181. Selvaraj V, Sampath K, Sekar V. Use of glucan from Saccharomyces cerevisiae as an immunostimulant in carp: Impact on hematology, phagocyte function, and infection with Aeromonas hydrophila. Isr J Aquacult-Bamid 2005;57(1) 39-48.

182. Selvaraj V, Sampath K, Vaithilingam S. Adjuvant and immunostimulatory effects of b-glucan administration in combination with lipopolysaccharide enhances survival and some immune parameters in carp challenged with Aeromonas hydrophila. Veterinary Immunology and Immunopathology 2006;114 15-24.

183. Lin S, Pan Y, Luo L, Luo L. Effects of dietary beta-1,3-glucan, chitosan or raffinose on the growth, innate immunity and resistance of koi (Cyprinus carpio koi). Fish Shellfish Immunol 2011;31(6) 788-94.

184. Falco A, Frost P, Miest J, Pionnier N, Irnazarow I, Hoole D. Reduced inflammatory response to Aeromonas salmonicida infection in common carp (Cyprinus carpio L.) fed with beta-glucan supplements. Fish Shellfish Immunol 2012;32(6) 1051-7.

185. Miest JJ, Falco A, Pionnier NP, Frost P, Irnazarow I, Williams GT, et al. The influence of dietary beta-glucan, PAMP exposure and Aeromonas salmonicida on apoptosis modulation in common carp (Cyprinus carpio). Fish Shellfish Immunol 2012;33(4) 846-56.

186. Kuhlwein H, Merrifield DL, Rawling MD, Foey AD, Davies SJ. Effects of dietary beta-(1,3)(1,6)-D-glucan supplementation on growth performance, intestinal morphology and haemato-immunological profile of mirror carp (Cyprinus carpio L.). Journal of animal physiology and animal nutrition 2013.

187. Bagni M, Archetti L, Amadori M, Marino G. Effect of long-term oral administration of an immunostimulant diet on innate immunity in sea bass (Dicentrarchus labrax). J Vet Med B 2000;47(10) 745-51.

188. Bagni M, Romano N, Finoia MG, Abelli L, Scapigliati G, Tiscar PG, et al. Short- and long-term effects of a dietary yeast beta-glucan (Macrogard) and alginic acid (Ergosan) preparation on immune response in sea bass (Dicentrarchus labrax). Fish Shellfish Immunol 2005;18(4) 311-25.

189. Bonaldo A, Thompson KD, Manfrin A, Adams A, Murano E, Mordenti AL, et al. The influence of dietary beta-glucans on the adaptive and innate immune responses of European sea bass (Dicentrarchus labrax) vaccinated against vibriosis. Ital J Anim Sci 2007;6(2) 151-64.

190. CooK MT, Hayball PJ, Hutchinson W, Nowak BF, Hayball JD. Administration of a commercial immunostimulant preparation, EcoActiva (TM) as a feed supplement enhances macrophage respiratory burst and the growth rate of snapper (Pagrus auratus, Sparidae (Bloch and Schneider)) in winter. Fish Shellfish Immun 2003;14(4) 333-45.

191. Kamilya D, Maiti TK, Joardar SN, Mal BC. Adjuvant effect of mushroom glucan and bovine lactoferrin upon Aeromonas hydrophila vaccination in catla, Catla catla (Hamilton). J Fish Dis 2006;29(6) 331-7.

192. Y. K, Q. Z. Effect of b-glucan on activity of antioxidant enzymes and Mx gene expression in virus infected grass carp. Fish & Shellfish Immunology 2009;27 336-40.

CHAPTER 8

ORAL ADMINISTRATION OF BAKER'S YEAST (SACCHAROMYCES CEREVISIAE) ACTS AS A GROWTH PROMOTER AND IMMUNOMODULATOR IN LABEO ROHITA (HAM.)

ArupTewary and Bidhan C Patra

Keywords: Immunostimulant, *Labeo rohita*, *Saccharomyces cerevisiae*

INTRODUCTION

Several whole microorganisms, live or not, such as bacteria, fungi or algae, increase disease resistance in mammals and fish [1,2]. In fish, as in other aquatic organisms, the whole microorganisms administered have mainly been bacterial species, which in the form of feed additives, have been shown to improve the intestinal microbial balance and increase the health status of fish, seemingly by colonising the gut and acting as antagonists to pathogens and so increasing resistance to pathogens [1,3,4].More recently, other whole microorganisms have been tested for their possible immunostimulant properties in fish. Thus, the oral administration or injection of the yeasts Saccharomyces cerevisiae or Candida utilis has been shown to increase both humoral (myeloperoxidase and antibody titer) and cellular (phagocytosis, respiratory burst and cytotoxicity) immune responses, and to increase or confer resistance against pathogenic bacteria in channel catfish, rainbow trout or gilthead seabream [5-8]. Possible use of baker's yeast in fish diets has many advantages. Firstly, they can be produced rapidly, easily and inexpensively and, at the same time, they are very stable and can be recycled from other industries. They are also natural substances so no negative effects may be expected either to the animals or to the environment. Moreover, there is no need to isolate their components, which consists mainly of cell wall sugars (β-glucans, mannoproteins and chitin), all are well-proved immunostimulant compounds.

The yeasts (Saccharomyces cerevisiae) has been used in gilthead seabream was Lyophilized form, which is not easily available to the farmer and also very costly but the baker's yeast, which is directly used in the bakery industry is low cost material. With this aim, the present paper discusses the effects of the dietary intake of the baker's yeast as supplementary feed and possible difference with the use of control diets are also established.

MATERIAL AND METHODS

Fingerlings of the species were obtained from a carp culture farm at the vicinity of Midnapore town has an initial measurement of 12.0±0.2g. Fish were (12.2 ± 0.22g) released into continuous flow glass aquaria (76 X 41 X 41 cm^3 area; 200.00 l capacity) after acclimatization for 15 days to prevailing laboratory condition of water temperature (31-33 °C) and pH (7.42 –7.53). Studies were conducted at room temperature for 60 days. The water quality (pH, DO, Alkalinity, Ammonia) of the experimental aquaria was monitored periodically once in a week following the methods of APHA (1998) and maintained at normal level.

Citation: Tewary A, Patra BC (2011) Oral administration of baker's yeast (Saccharomyces cerevisiae) acts as a growth promoter and immunomodulator in Labeo rohita (Ham.). J Aquac Res Development 2:109. doi:10.4172/2155- 9546.1000109.

Preparation of Experimental Feeds

The four prepared feeds (Cont., Exp-1, Exp-2, Exp-3) were formulated using locally available ingredients (mustard oil cake, rice polish, fish meal and tapioca powder). Feed formulation was done basically by "Square method" using determined values of protein content of the different ingredients. Proportion of each ingredient required was calculated precisely providing allowance for the premix. Locally available Baker's Yeast (Saccharomyces cerevisiae) was used as an immune stimulant. Baker's Yeast were diluted in water and supplemented @ 5%, 7.5% and 10% in the feed Exp-1, Exp-2 and Exp-3 respectively; where as control feed was not supplemented with Saccharomyces cerevisiae. Feeds were pelleted separately with local made (Kolkata, India) hand pelletiser. The pellets were dried in a thermostatic oven (M/S Modern Industrial Corporation, Mumbai, India) at 37°C and less then 10% moist [9, 10] and stored in an air tide jars at room temperature. Proximate composition of the four prepared feeds (Cont., Exp-1, Exp-2, Exp-3) were detailed in (Table 1).

Growth Performance and Conversion Ratio

Fishes were fed twice daily at 8.00 and 16.00 h with ration size maintained with 6% of their body weight in two equal portions. The net weight was recorded every 15 days with an electronic and feed quality was readjusted after every weighting period of 15 days. For evaluating the dietary performances, the nutritional indices like live weight gain (LWG), Net weight gain (NWG), Average daily growth (ADG), Protein efficiency ration (PER), Specific growth rate (SGR) and Protein efficiency ratio (PER) were used. Two fishes of each group were terminated through overdose anaesthetiasation by MS222 (Sigma chemicals, India) [11] at the end of the experiment, and stored at –20 °C until analysis.

Proximate Analysis

Proximate analyses of ingredients, feeds and body carcass were determined following the method provided in AOAC (1990). Moisture content was determined gravimetrically in a hot air oven at 100+ 10 °C for 24 h. Crude protein contents were determined by the microkjeldahl method. Crude lipid contents were determined by the soxhlet extraction method using petroleum ether (Boiling point: 40-60°C) in the electro-thermal soxhlet apparatus. After extraction of lipid the defattend samples were used for the estimation of crude fiber following Patra [12]. Ash content was estimated by inserting samples in a muffle furnace at 500+ 50°C for 10 h.

Biochemical Analysis

DNA (Deoxy-ribo nuclic acid) and RNA (Ribo nuclic acid) contents from liver (hepatopancreas) tissues were estimated as per the scheme given by Munro and Fleck (1969). Tissue was taken from fishes and homogenized with 0.25 m sucrose solutions. 250 µl homogenate and 500 µl 5% TCA mixed thoroughly centrifuged and wait for 15 minutes and then precipitate was dissolved in PCA (several time in different concentration) and centrifuged with 5000 rpm (two times). The ultimate supernatant was used for RNA and DNA estimation by UV spectrophotometer.

Enzyme Assay

Two specimen from each of the replicate were sampled both the initial stage and at the end of a 60 days experimental period. The fishes were dissected and the liver was cleaned and removed and pieces of liver was thoroughly washed in ice chilled double distilled water and subsequently collected in ice-cooled Petri dishes, weighed and cut into small pieces. A 5% homogenate was prepared in potter- Elvehjen homogenizer with neutral glass powder and ice cold 0.1 M phosphate buffer (pH-7.4) and centrifuged at 2,500 rpm for 15 minutes in a refrigerated centrifuged. The supernatant was used for the GOT and GPT activity following the method of Bernfeld [13].

Initially and after completion of the study, experimental fishes from different feeding trials were separately pooled and liver was dissected out for ACP and ALP activity estimation. After weighting the whole quantity of proportionate quantity, 10 % sucrose and 1 % tritonex mixture were added and the contents homogenized in a tissue homogenizer. The samples were then put in appendrop tube and cold centrifuged (at -4 °C) at 11000 rpm for 15 minutes. The supernatant was collected as sample for determining the ACP [14] and ALP [15] activity.

Study of Blood Parameters

Blood samples were collected by heparinized syringe from caudal vein for haematology. EDTA (Ethyline diamine tetraacetic acid) was used as anticoagulant. 1.0 mg EDTA ml^{-1} of blood or 1 drop of 1.0% solution 5 ml^{-1} of blood was used for haematology. Haematological parameters were estimated according to the method of Wintrobe [16]. MCV, MCH, MCHC were calculated by using standard formulae (Decie and Lewis 1991). Blood samples were collected in the laboratory for serological diagnosis by syringe from caudal vein and heart. Determination of the Total Serum Protein (TSP) and albumin by Gornall's biuret method [17].

TSG = TSP- TSA

(Where TSG = Total Serum globulin, TSP= Total Serum Protein, TSA= Total Serum albumin.)

Determination of Immunity Level

On day 60, blood was collected form fishes of each group. Part of the blood was heparinised and the rest was allowed to clot for serum samples, which were preserved at -20°C for further analysis. Immediately after collection, the heparinised blood samples of each group were pooled to three aliquots. Part of the blood was analysed for leucocrit value in duplicate per sample [18]. The rest of the heparinised blood was immediately used for the phagocytic assay [8, 19]. Blood was collected from the fish by using a 0.2 ml glass syringe rinsed with an anticoagulant. Then the blood was transferred into the heparinised vial and mix properly. 0.1 ml of freshly prepared NBT solution was added to 0.1 ml of the heparin mixed blood and 15 µl of stimulant solution in the incubating bottle. The bottles were incubated at 37o C for 10 minutes and at 26o C for another 10 minutes. 50-70 µl of this blood was transferred onto a clean slide and makes a thick smear with a spreader slide. The slides were air dried and stain with Wright's stain. For staining with the Wright's stain, first flood the slide with 1 ml of the staining solution for 30 seconds then 1 ml of distilled water was added and keep for another 30 seconds. The slide was then pour off the stain and dried. Then the slide was then under oil immersion lens at 100 X. The positive cells had the violet coloured formazan granules in the cytoplasm. The percentage of the positive cells gave the idea about the non-specific immune status of the organism.

Challenge Trial

Rohu fishes (Labeo rohita) of medium size (435±28g) were collected from a semi-intensive culture pond at the University campus. The intestine were gently excised & cut open with a pair of sterile scissors. The non-adherent micro floras of the intestine were isolated by three times washed with sterile solution & homogenized with 10 ml distilled water in stomacher bags. The presumptive numbers of micro flora were determined by the spread plate technique using nutrient agar. The pathogenic strain Aeromonas hydrophila (AH1) were isolated by the method of Kaneko [20], which had been cultured and maintained in the Aeromonas selective medium (M884, Hi – Media).

After 60 days feeding trial, fishes of each experimental group were released in four aseptic tanks. Different water quality parameters (Temperature pH, Dissolve Oxygen, Alkalinity, Ammonia) were maintained in normal level. Different experimental feeds and control feeds were provided twice daily according to 6 % of their body weight. The fishes in each treatment were challenged with Aeromonas hydrophila (AH1). Fishes in all replicates were immersed in a suspension of Aeromonas hydrophila (AH1), ~ 105 CFU ml – 1 according to Austin et al. [21]. This was followed by a second immersion ~ 107 CFU ml – 1 after 7 days Austin et al. [21]. The survivability of the fishes was recorded against pathogenic strain for last 10 days.

Statistical Analyses

As all the above analyses were carried out on pooled samples of a given lot, standard deviations of means were calculated. However, for evaluating the dietary performances, nutritional indices, enzymatic activity and RNA: DNA ratio, different haematological, serological, immunological parameters and challenge trials; correlation and regression test were performed through SPSS software package. Significant differences between the means of the treatments were tested by Duncan Multiple Range Test [22] through SAS software package [23].

RESULTS

Locally available ingredients (Mustard oil cake, Rich polish and Fish meal) were used for preparing experimental feeds for Labeo rohita. Except control feed (Control), other three experimental feeds (Exp-1, Exp-2 and Exp-3) were supplemented with baker's yeast (Saccharomyces cerevisiae) and it was replaced with equal amount of rich polish. All the four experimental feeds are isocaloric and isonitrogenous. The average crude protein percentage of the feeds were 32.18± 0.22 and the crude protein percentage were around 7.67±0.09 (Table 1).

After 60 days feeding trial initial and final carcass composition of Labeo rohita in relation to various feeds was presented in (Table 2). The carcass composition of the test animals revealed an apparent increase in the final carcass protein and lipid over the initial carcass protein and lipid. Significantly highest carcass protein (12.01±1.29) and lipid (3.16± 0.31) was recorded in feed Exp-1 fed fishes as compare to other experimental and control feeds. This was clearly indicating that, the enhancement of carcass composition with the increasing supplementation of yeast at a specific level (5% supplementation). Perhaps it was due to the enzymatic activity in the gut and there by nutrients are spread for the growth and it make overall well being for the fishes.

Maximum weight gain of 135.18±1.21(of 10 fishes) was obtained in feed Exp-1 fed fishes followed by Exp-2 (109.88±1.36) and Exp-3 (102.86±1.96) and control (92.82±1.29) respectively (Table 3, Figure 1). There were significant differences ($P \leq 0.05$) in the different growth parameters among the different experimental feed fed fishes (Table 3). Significantly lowest ($P \leq 0.05$) FCR (3.54±0.033) was observed in Exp- 1 fed fishes; it was also observed that significantly higher SGR was obtained from feed Exp-1 fed fishes (1.27±0.04). It indicates that Exp-1 feed shows better utilization of nutrients than the other feeds. Similarly significantly ($P \leq 0.05$) highest PER values (6.46±0.11) was obtained from Exp-1 fed fishes, which indicate better utilization of protein for growth and metabolism.

Table 1: Proximate composition (%) of different experimental diets* for Labeo rohita

Proximate composition of feeds	Diets			
	Control	Exp –1	Exp –2	Exp –3
Moisture(%)	7.18^a	7.23^a	7.19^a	7.28^a
Crude protein(%)	32.83 ± 0.13^a	32.18 ± 0.22^a	31.85 ± 0.32^a	31.53 ± 0.32^a
Crude lipid (%)	7.92 ± 0.07^a	7.67 ± 0.09^a	7.54 ± 0.11^a	7.41 ± 0.23^a
Ash(%)	11.69 ± 0.16^a	10.62 ± 0.08^a	10.08 ± 0.06^a	9.55 ± 0.19^a
Energy (KJ/g)	9.35 ± 0.004^a	9.43 ±0.002^a	9.46 ±0.003^a	9.50 ±0.008^a

* On dry matter basis.
Results are means of five separate determinations (Mean ± SEM)
Figures having same alphabets in the different rows are significantly different ($p \leq 0.05$)

Table 2: Initial and final carcass composition of fingerlings of Labeo rohita after 60 days experimental trial on four different dietary treatment

Carcass Composition (%)	Experimental tanks				
	Initial	Final			
		Control	Exp –1	Exp –2	Exp –3
Crude protein	7.75 ± 0.86^a	9.96 ± 1.05^b	12.01 ± 1.29^c	10.40 ± 0.30^b	10.25 ± 0.65^b
Crude lipid	2.83 ± 0.71^a	2.90 ± 0.29^a	3.16 ± 0.31^a	2.93 ± 0.93^a	2.99 ± 0.12^a
Ash	12.63 ±0.062^a	12.55 ± 0.089^a	12.89 ± 0.073^a	12.55 ±0.062^a	12.68 ±0.071^a

Results are means of five separate determinations (Mean ± SEM).
Figures having different alphabets (superscribed) in the same row are significantly different ($p \leq 0.05$)

Table 3: Different growth parameters, haematological parameters and serological parameters after 60 days feeding trial on L. rohita with yeast supplemented diets

Diets	NWG (g)	ADG (g)	SGR (%)	FCR	PER	Hb (g %)	TLC ($\times 10^3 mm^{-3}$)	TEC ($\times 10^3 mm^{-3}$)	Hct (%)	Leucocrit value
Control	92.82 ± 1.29^a	1.54 ± 0.002^a	0.95 ± 0.30^a	4.64 ± 0.18^a	4.35 ± 0.14^a	7.1 ± 0.91^a	13.8 ± 0.76^a	1.30 ± 0.15^a	27.21 ±1.02^a	37.10 ± 2.05^a
Exp –1	135.18 ± 1.21^b	2.25 ± 0.001^b	1.27 ± 0.41^a	3.54 ±0.33^b	6.46 ± 0.11^b	8.2 ± 0.17^b	22.2 ± 0.57^b	1.76 ± 0.11^b	35.34 ± 1.25^b	59.20 ± 0.33^b
Exp –2	109.88 ± 1.36^c	1.83 ± 0.007^c	1.13 ± 0.39^a	4.39 ± 0.34^a	5.30 ± 0.32^c	7.9 ± 0.21^c	18.3 ± 0.76^c	1.52 ± 0.06^c	28.32 ± 1.41c	47.50 ± 1.34^c
Exp –3	102.86 ± 1.96^d	1.71 ± 0.011^d	1.10 ± 0.30^a	4.36 ± 0.44^a	5.02 ± 0.19^d	7.5 ± 0.20^d	16.7 ± 0.87^d	1.43 ± 0.14^d	32.18 ±1.30^c	42.60 ±2.13^d

Figures having different alphabets (superscribed) in the same column are significantly different ($p \leq 0.05$)

Significantly ($P \leq 0.05$) highest Alkaline phosphatase (ALP) (8.62 ± 0.04) and Acid Phosphatase (1.48 ±0.02) activity was observed in Exp- 1 feed fed fishes (Table 4) following Exp-2 and Exp-3 feed fed fishes. Similarly highest GPT (0.068±0.003) and GOT (0.042 ± 0.002) values were registered in feed Exp-1 fed fishes, where as lowest in feed control fed fishes (0.055±0.001 and 0.032 ±0.003). Significantly ($P \leq 0.05$) higher RNA: DNA ratio (2.10 ± 0.003) was observed in fishes fed with Exp-1 feed and least was recorded (1.54 ±0.001) in control feed treated fishes (Table 5) after 60 days feeding trial. It was also observed that RNA: DNA ratio of fishes increased in all the treatments over the initial RNA: DNA ratio (1.02 ±0.010).

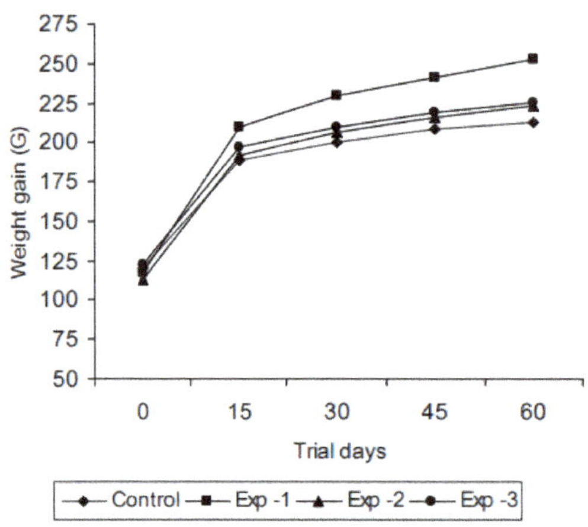

Figure 1: Growth (g) of Labeo rohita in relation to various diets.

Table 4: Activities of ALP. ACP , GOT and GPT in the intestine of Labeo rohita after 60 days feeding trial

Treatment	ALP	ACP	GOT	GPT
Control	1.08 ±0.01^a	0.40±0.001 ^{a}	0.032±0.02 ^{a}	0.055±0.001 ^{a}
Exp –1	8.62±0.04^b	1.48±0.02 ^{b}	0.042±0.06 ^{b}	0.068±0.005 ^{b}
Exp –2	5.29±0.02^c	1.01±0.001 ^{c}	0.039±0.01 ^{c}	0.062±0.002 ^{c}
Exp –3	5.10±0.03^c	0.73±0.003^d	0.035±0.04^d	0.059±0.006^d

ALP (Alkaline phosphatase); ACP (Acid Phosphatase); GOT (glutamate-oxaloacetate transaminase) (EC 2.6.1.1); GPT (glutamate-pyruvate transaminase) (EC 2.6.1.2)

Table 5: Initial and final RNA/DNA ratio in the muscle of Labeo rohita after 60 days feeding trial

Treatment	Initial	Final
Control	1.02 ± 0.010^m	1.54 ± 0.001$^{a\ n}$
Exp –1		2.10 ±0.003$^{b\ o}$
Exp –2		1.98 ±0.005$^{c\ p}$
Exp –3		1.82 ±0.000$^{c\ q}$

Results are means of three separate determinations (Mean ± SEM),
Values with the same superscript in the different rows are not significantly different ($p < 0.05$)
from each other.
RNA/DNA : Initial value = 1.02± 0.010

Table 6: Effect of Yeast (Saccharomyces cerevisiae) on A:G ratio & non specific Immunity levels of Labeo rohita after 60 days feeding trial

Treatment	Albumin: Globulin	Phagocytic Ratio	Phagocytic Index	NBT cells
Control	3.00 ± 0.04^{a}	15 ± 0.49^{a}	1.52 ± 0.12^{a}	48.5± 0.32^{a}
Exp –1	2.34 ± 0.03^{b}	63 ± 1.22^{b}	2.34 ± 0.09^{b}	57.0±0.61^{b}
Exp – 2	2.52 ± 0.06^{c}	42 ± 1.27^{c}	2.15 ± 0.09^{c}	52.5±0.24^{c}
Exp – 3	2.49 ± 0.09^{c}	18 ± 0.86^{d}	1.49 ± 0.20^{a}	49.5±0.23^{d}

Results are means of five separate determinations (Mean ± SEM)
Figures having different alphabets (superscribed) in the same column are significantly different (p ≤ 0.05)

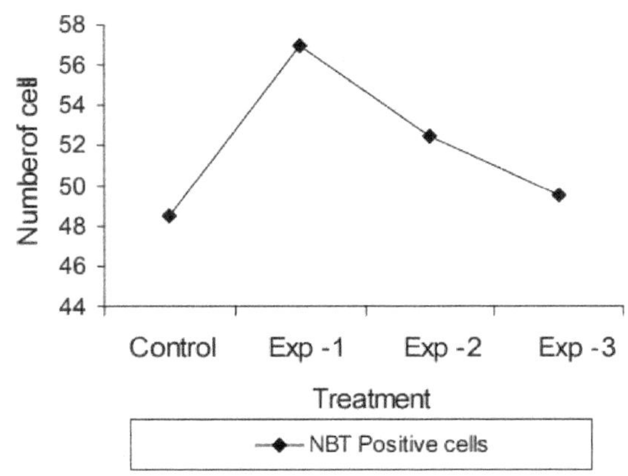

Figure 2: NBT positive cells Number of cells.

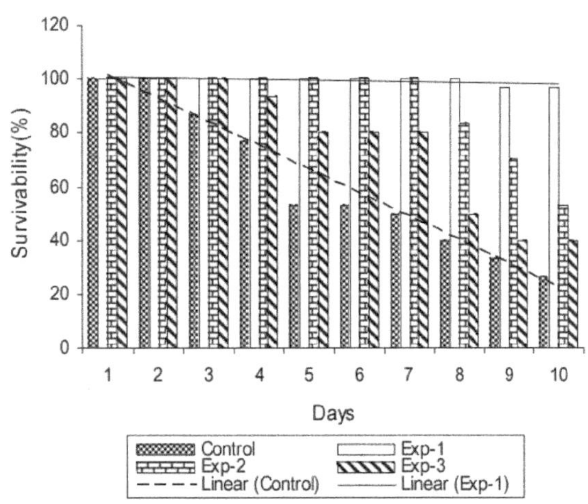

Figure 3: Survivability of Labeo rohita in relation to various diets when challenged with Aeromonas hydrophila (AH1).

Different Haematological and serological parameters of Labeo rohita after 60 days feeding trial were presented in (Table 3). Significantly (P≤0.05) highest TEC, Hb and HCT and lowest Albumin: Globulin ratio were observed in feed Exp-1 fed fishes, where as significantly (P≤0.05) lowest TEC, Hb and HCT and highest Albumin: Globulin ratio were recorded in control feed fed fishes after 60 days feeding trial.

The effect of yeast (Saccharomyces cerevisiae) on non-specific immunity was observed. Significantly (P≤0.05) highest phagocytic ratio (63 ±1.22), phagocytic index (2.34±0.09) and leucocrit value (59.20±0.57) were recorded in Exp-1 feed fed fishes, where as lowest of all these were recorded (9±0.49, 1.52 ±0.12and 37.10 ±2.05 respectively) in control feed fed fishes (Table 6). Similar trend was observed in case of number of NBT positive cells. Significantly (P≤0.05) highest number of NBT positive cells (57.0 ± 0.61) were observed in Exp-1 feed treated fishes and least one (48.5±0.32) recorded in case of control feeds of Labeo rohita (Table 6, Figure 2).

After 60 days feeding trial all treated and controlled fish faced challenge trial with Aeromonas hydrophila (AH1) for 10 days. After 10 days challenge trial significantly (P≤0.05) highest (96.66%) survivability was observed in Exp-1 feed fed fishes followed by Exp-2 (53.33%), Exp- 3 (40%) and control (26%) feed treated fishes (Figure 3).

DISCUSSION

In the last two decades, many substances have proved their usefulness in fish culture because of their properties to stimulate the immune system & increase disease resistance. Among these immunostimulants the role of isolated β-glucans

, chitin or vitamins is well documented [24-34]. However, the use of whole organisms instead of their isolated components has hardly been evaluated. In this way whole yeast cells (mainly S. cerevisiae), Which represent a major commercial source of β-glucans , have recently been described as good immiunostimulants in fish [6,8].

Charlon and Bergot [35] achieved more than 89 % survival and good growth of carp larvae; feed exclusively on dry diets having yeast powder and pork / beef freeze dried liver. Alami, Durante et al [36] reported 87 % survival and FCR of 0.62 in common carp larvae reared on liver , yeast and commercial trout starter feed . Singh et al. [37] also found that yeast increased the rate of feed intake and conversion efficiency. Mohanty et al ., reported 100 % survival using a liver based diet containing goat liver , Cod liver oil , Vitamin and mineral mixture with average growth of 133.3 mg / fry in Rohu (Labeo rohit) raised from spawn to fry .

In the present study, S. cerevisiae are also found to stimulate the digestion through the supply of digestive enzymes and certain essential nutrients to the animals. Immunostimulants, particularly Saccharomyces cerevisiae producing several enzymes, which is not, produces by the host. Similar observation was also reported by Swain et al. [38]. A complex polysaccharide including cellulose are better utilized by the host in the presence of direct feed microbes like Aspergillus oryza , Saccharomyces cerevisiae.

Furthermore, it was reported that yeast in the diet improves feed efficiency, organic phosphorus (phytic acid) utilization and fibre digestion [38]. It also reported that supplementation of immunostimulants in feed improves the nutrition by hydrolytic enzymes including amylase and proteases, the production of vitamins such as biotin and vitamin B12 [39-44] and the host immunity [45]. It indicate that a given amount of immunostimulant elicit more than one protective response by the host.

Dietary intake of immunostimulants by fish has definite advantages and it is a useful method of exposure in large scale fish culture [8], particularly lypopolysaccharides of yeast cell wall was evaluated by many investigators like Ortuno et al. [6]; Sakai [28]; Sahoo and Mukherjee [46]. It is recognized that immunostimulants enhance the host defence system against pathogens by increasing phagocytosis , antibody production , leucocrit level and reduced A:G (Sakai, 1999) . A comparative study done by Ortuno et al. [6] revealed that oral administration of whole yeast Saccharomyces cerevisiae enhances the cellular innate immune response. Notably that all kinds of cellular parameter such as phagocytic index, respiratory burst activity, number of erythrocytes, lymphocytes were also enhanced but not the humoral one.

Blood is a pathophysiological reflector of the whole body and therefore, blood parameters are important in diagnosing the status of fish health (Pecie and Lewis, 1991), particularly when some additives used in the feed. In this present study it was observed that all the blood parameters in all the treatments were similar to standard (Banerjee et al., 2002) and Exp-1 fed fishes showed superior as compare to others, indication of the blood parameters revealed the positive impact, but also demonstrated a stable physiological reflection of the whole body [47].

In this study, a superior growth performance in terms of weight gain percentage, specific growth rate was recorded in L. rolita fishes fed with Exp-1 feed as compare to other feed fed fishes. These observations determine the optimum doses of yeast supplementation in feed. This might be helpful for optimum dietary utilization. In this study, although all the feeds were isonitrogenous but the concentration of immunostimulant i.e. Baker's yeast (S. cerevisoae) in Exp-1 feed might be helpful for proper nutrient utilization. Whole body carcass composition was higher in Exp-1 feed fed fishes as compare to control one, which could have also noted to the overall low feed utilization level. RNA: DNA is known to provide dependable indication of growth trend [48-50]. The ratio was greatest in the fish fed Exp –1 feed with higher dietary utilization and best growth. Bazaz and keshavnath (1993) found higher RNA: DNA in better growing fish fed with oil supplemented diets using equal level of crude protein. The present study also reports such a finding, where all the feeds are isocaloric and isonitrogenous but 5% supplementation of Baker's Yeast (S. cerevisiae) incorporated feed (feed Exp-1) exhibit better growth as well as better RNA:DNA ratio. The highest level of GOT and GPT as well as ALP and ACP were found in Exp-1 feed fed fishes and lowest in control. Most of the amino acids normally found in protein undergo transamination reaction and transaminases are localized in both citosol and mitochondria (Wada and Marino, 1964), which is induced by high protein diet [51], thus a positive correlation between the immunostimulant concentration at a specific level and the GOT, GPT level in the liver could be observed.

The result obtain in this study not only support the use of Baker's yeast for better growth, and proper nutrient utilization but also it act as an immunostinunlant by stimulating the immune response. The activation mechanisms involved are known to be related to the carbohydrates, derived from the yeast cell wall. -glucans/whole yeast added to feed stimulates the phagocytic activity, respiratory burst activity and increase protection after challenge with pathogenic bacteria, similar finding ware reviewed by Robertson, 1999.

In this study oral administration of Baker's Yeast (S .cerevisiae) stimulate the non specific immunity level as measured through enhanced phagocytic activity, leucocrit level, respiratory burst activity and reduced A: G. Although enhanced leucocrit value does not necessarily relate to an immunostimulatory action of Baker's Yeast . The possible role of yeast as an imunostimulant due to its cell wall which composed of Lipopolysaccharide, such as glucan, which enhanced phagocytic activity of macrophages and globulin level as observed in the present experiment, the phagocytic indices seemed to be reliable indices of a hightened immunostimulatory response. Similar observation also observed by Swicki et al. [8]; Esteban et al. [31]; Orturo et al. [6]; Radriguezatal [7] in Rainbow trout and gilthread seabream. Albumin – Globulin ratio is a measurable humoral component at the non-specific defences. The reduction of A: G, might be due to the increase of Total serum globulin level with Significance protective mechanisms for fish [46]. Fish produce reactive oxygen species, which are considered to be toxic for fish bacterial pathogens [52,53] and are generated by phagocytes after stimulation by a variety of agents. In this experiment higher respiratory burst activity (O2 - production) was increased in all the experimental diets as compare to control diets, but highest number of NBT positive cells were observed in Exp-1 feed fed fishes, an effect also seen in rain bow trout (Siwicki at al,1994), Seabream (Ortuno et al., 2002) and turbot [54]. After challenge trial with Aeromonus hydrophila, highest survivability observed in Exp-1 fed fishes, the enhanced protection conferred by glucan is not surprising, similar report has been reported to induce resistance in carp (Cyprinus carpio) against E. tada [55]. Evidence suggests that glucan enhance disease resistance by stimulating non-specific components of fish immune system [56,57].

To conclude the present results provide evidences that Baker's yeast (S. cerevisiae) added in a common fish diet, exhibit better growth, better nutrient utilization and activate the innate immunity, as well as increase the survivability of L. rohita. Optimal doses and administration time have been established in an attempt to provide a useful approach for protecting culture fish against infectious diseases.

ACKNOWLEDGEMENT

The authors are thankful to the Department of Science and Technology, government of West Bengal for providing infrastructure facilities. The authors are also thankful to the Department of Zoology for extending all sorts of working facilities and the Research Scholars of the Aquaculture Research Unit, Department of Zoology, Vidyasagar University, for their valuable co-operation to prepare this manuscript.

REFERENCES

1. Gatesoupe FJ (1999) The use of probiotics in aquaculture. Aquaculture 180:147–65.
2. Masihi KN (2000) Immunomodulatory agents for prophylaxis and therapy of infections. Intern J of Antimicrobial Agents 14:181–91.
3. Fuller R (1989). Probiotics in man and animals. J of Appl. Bacteriology 66:365– 78.
4. Tannock GW (1997) Modification of the normal microbiota by diet, stress,antimicrobial agents, and probiotics. In: Mackie RI, With BA, Isaacson RE, editors. Gastrointestinal microbes and host interactions. New York: Intern. Thomson Pub 434–65.
5. Chen D, Anisworth AJ (1992) Glucan administration potentiates immune defense mechanisms of channel catfish. Ictahrus punctatus Rafineque. J Fish Dis 15:295-304.
6. Ortuno J, Cuesta A, Rodriguez A, Esteban MA, Meseguer J (2002) Oral administration of yeast, Saccharomyces cerevisiae, enhances the cellular innate immune response of gilthead seabream (Sparus aurata L.). Vet Immunol Immunopathol 85: 41-50.
7. Cuesta RA, Ortun˜o J, Esteban MA, Meseguer J (2003) Immunostimulant properties of a cell wall-modified whole Saccharomyces cerevisiae strain administered by diet to seabream (Sparus aurata L.). Veterinary Immunology and Immunopathology 96: 183-192.
8. Siwicki, AK, Anderson DP, Rumsey GL (1994) Dietary intake of immuno stimulants by rainbow trout affects non-specific immunity and protection against furunculosis. Veterinary Immunology and Immunopathology 4:125–139.
9. Bazaz MM, Keshavanath P (1993) Effect of feeding levels of sardine oil on growth, muscle composition and digestive enzyme activities of mahseer,Tor khudree. Aquaculture 115:111-119.
10. Keshavanath P, Renuka P (1998) Effect of dietary L-carnitine supplements on growth and body composition of fingerling rohu. Labeo rohita (Ham). Aqua Nutr 4:83-87.
11. Bandyopadhyay P, Swain SK , Mishra S (2005) Growth and dietary utilization in gold fish (Carrassius auratus Lin) fed diets formulated with various local agro-products. Bioresource Tech 96:731-740 .
12. Patra BC, Maity J, Debnath J, Patra S (2002) Making Aquatic weed useful II: Nymphoides cristatum (Roxb.) O .Kuntze as fish feed for an Indian major carp Labeo rohita (Ham.). Aquaculture Nutrition 8:33-42.
13. Bernfeld P (1955) In. Methods of Enzymology. S.P. Colowick and N.O. Kaplan (Eds), Vol.1, Academic Press, New York. 149.
14. Bramley TA (1975) Treatment of immature mice with gonadotropins. Effects on some enzymatic activities of unfractioned ovarian homogenates. J Biochem 140:451-460.
15. Rosauki SR (1993) In: Boehringer Mannheim Gmbh analysis protocol. Clin Chem 39: 648.
16. Wintrobe MM (1978) Clinical haematology, (Kipton, H), London, 85-89.
17. Kulow H (1967) Eine Schnellmethode zur Bestimmung der Serumproteine von Satzkarpfen (A rapid method of finding the serum proteins in young common carp). Dt. Fischerei-Ztg. 14: 241-149.
18. Blaxhall P, Daisely K (1973) Routine haematological methods for use in fish blood. J Fish Biol 5:771-781.
19. Park JH, Jeong HD (1996) Enhanced resistance against Edwardsiella trada infection in tilapia (Oreochromis niloticus) by administration of protein bound polysaccharide. Aquaculture 143:135-143.
20. Kaneko S (1971) Microbiological study of fresh fish (2). New Book Indus 13:76– 80.
21. Austin B, Stuckey LF, Robertson PAW, Effendi I, Griffith DRW (1995) A probiotic strain of Vibrio algilolyticus effective in reducing diseases caused by Aeromonas salmonicida, Vibrio anguillarum and Vibrio ordalii. J Fish Dis 18:93-96.
22. Duncan DB (1955) Multiple range and multiple F-tests. Biometrics 11:1-42.
23. SAS (1991) SAS Institute Inc. SASR system for regression (Second Edn), SAS Institute Inc, Cary, NC, 210.
24. Di Luzio NR (1979) Lysozyme, glucan-activated macrophages and neoplasia. J Reticuloendothel Soc 26:67–81.
25. Czop JK, Austen KF (1985) A beta-glucan inhabitable receptor on human monocytes: its identity with the phagocytic receptor for particulate activators of the alternative complement pathway. J Immunol 134:2588–2593.

26. Hardie LJ, Fletcher TC, Secombes CJ (1991) The effect of dietary vitamin C on the immune response of Atlantic salmon (Salmo salar). Aquaculture 95:201– 214.

27. Yano T, Matsuyma H, Mangindaan REP (1991) Polysaccharide-induced protection of carp. Cyprinus carpio L. against bacterial infection. J Fish Dis 14:577-582.

28. Sakai M (1999) Current research status of Fish Immunostimulants. Aquaculture 172: 63-92.

29. Thompson I, White A, Fletcher TC, Houlihan DF, Secombes CJ (1993) The effect of stress on the immune response of Atlantic salmon (Salmo salar L.) fed diets containing different amount of vitamin C. Aquaculture 114:1-18.

30. Esteban MA, Cuesta A, Ortuno J, Meseguer J (2001) Immunomodulatory effects of dietary intake of chitin in gilthead seabream (Sparus aurata L.) innate immune response. Fish Shellfish Immunol 11:305-315.

31. Esteban MA, Mulero V, Cuesta A., Ortun˜o J, Meseguer J (2000) Effects of injecting chitin particles on the innate immune responses of gilthead seabream (Sparus aurata L.). Fish Shellfish Immunol 10:543–554.

32. Ortun˜oJ, Cuesta A, Esteban MA, Meseguer J (2001) Effect of oral administration of high vitamin C and E dosages on the gilthead seabream (Sparus aurata L.) innate immune system. Vet Immunol Immunopathol 79:167–80.

33. Cuesta A, Esteban MA, Meseguer J (2001) Vitamin E increases natural cytotoxic activity in seabream (Sparus aurata L.). Fish Shellfish Immunol 11:293–302.

34. Cuesta A, Ortun OJ, Rodriguez A, Esteban MA, Meseguer J (2002) Changes in some innate defence parameters of seabream (Sparus aurata L.) induced by retinol acetate. Fish & Shellfish Immunol 13:279-291.

35. Charlon N, Bergot P (1984) Rearing system for feeding fish larvae on dry diets. Trial with carp (Cyprinus carpio) larvae. Aquaculture 54:83-88.

36. Alami- Durante H, Charlon N, Escafree AM, Bergot P (1991) Supplementation of artificial diets for common carp (Cyprinus carpio L.) larvae. Aquaculture 93:167-175.

37. Singh BN, Sinha VRP, Chakrobarty DP (1980) Feeding intake, absorption, conversion and growth of fry fingerlings of rohu, Labeo rohita. Indian J Fish 27:193-200.

38. Swain SK, Rangacharyulu PV, Sarkar S, Das KM (1996) Effect of a probiotic supplement on growth, nutrient utilization and carcass composition in mrigal fry. J Aqua 4: 29-35.

39. Fuller R, Turvy A (1971) Bacteria associated with the intestinal wall of the fowl (Galtus domesticus). J Appl Bacteriology 66:365-378.

40. Parker RB (1974) Probiotics-The other half of the antibiotics story. Anim Nutr Health. 29: 1416-1422.

41. Roach S, Tannock GW (1980) Indigenous bacteria that influence the number of Salmonella typhimurium in the spleen of intravenously challenged mice. Canadian J Microbiol 26:408-411.

42. Sugita H, Miyajima C, Deguchi H (1991) The vitamin B12 producing ability of the intestinal microflora of fresh water fish. Aquaculture 92:267-276.

43. Fuller R (1992) History of development of probiotics. In. Fuller,R. (Ed), Probiotics: the scientific basis, Chapman and hall, New York. 1-8.

44. Sugita H, Takahashi J, Deguchi H (1992) Production and consumption of biotin by the intestinal microflora of cultured fresh water fishes. Biosciences. Biotech and Biochem 56: 1678-1679.

45. Gibson GR., Saavendra J M, MacFarlane S, MacFarlane GT (1997) Probiotics and intestinal infection. In Probiotics 2, Application and Practical Aspects (ed. By R. Fuller). Chapman and Hall, London.

46. Sahoo PK, Mukherjee SC (2001) Immuonocompressive effect of aflatoxin B1 in Indian major carp (Labeo rohita). Comp immunol Micro Inf Dis 24:143-9.

47. Decie SIV, lewis SM (1991) Practical haematology (VII Edn). J and A Chuechill Ltd. , Livingston, London, Melbourne and NewYork.

48. Buckley LJ (1980) Changes in ribonucleic acid, deoxyribonuclc acid and protein content during ontogenesis in winter flounder Pseudopleuronectes amouricauses and effect of starvation . U.S. Fish Bull 77:703-708.

49. Khan MA, Jafri AK (1991) Protein and nucleic acid concentration in the muscle of catfish Clarias batrachus. Asian Fish Sci 4:75-84.

50. Banerjee SK, Patra BC, Bandyopadhyay P, Tewary A (2002) Changes of the blood parameters in an Indian major carps, Catla catla Ham. Due to Myxozoan parasites infection. Indian J of Aquatic Biol 17: 79-84.

51. Swicki RW, Barnstein PL, Strange JL (1965) The response of the isozymes of alanine aminotransferase to diet and hormones. J Biol Chem. 240:3341-3345.

52. Hardie LJ, Ellis AE, Secombs CJ (1996) In vitro activation of rainbow trout macrophages stimulates inhibition of Renibacterium salmoninarum growth concomitant with augmented generation of respiratory burst products. Disease of Aquatic Organism 25: 175-183.

53. Itou T, Iida T, Kawatsu H (1996) The importance of hydrogen peroxide in phagocytic bactericidal activity of Japanese eel neutrophils. Fish pathology 3:121-125.

54. Toranzo AE, Devesa S, Romalde JL, Lamas J, Riaza A, et al. (1995) Efficacy of intraperitoneal and immersion vaccination against Enterococcus sp. infection in turbot. Aquaculture 134:17-27.

55. Yano T Mangindaan REP, Matsuyama H (1989) Enhancement of the resistance of carp Cyprinus carpio to experimental Edwardsiella tarda infection, by some -1,3 -glucans, Nippon Suisan Gakkaishi 55:1815-1819.

56. Association of Officials Analytical Chemistry (AOAC), (1990) Official methods of analysis. 15th Edn. Edited by S. Williams, AOAC, Virginia, 1298 pp.

57. American Public Health Association (1998) Standard method for the Examination of water and wastewater, 20th edn. APHA, Washington, DC, USA

CHAPTER 9

USE OF PROBIOTICS IN DIETS OF TILAPIA

Thomas L. Welker and Chhorn Lim

INTRODUCTION

Tilapia, because of their enormous adaptability to a wide range of physical and environmental conditions, ability to reproduce in captivity, relative resistance to handling stress and disease-causing agents compared to other cultured finfish species, good flesh quality, feed on a low trophic level and excellent growth rate on a wide variety of natural and artificial diets, are the most abundantly cultured species worldwide. Although they are endemic to tropical freshwater in Africa, Jordan and Israel, their distribution has widened following introductions to other regions of the world in the early part and after the middle of the 20th century. They are presently cultured in virtually all types of production systems, in both fresh and salt water, and in tropical, subtropical and temperate climates [1]. Tilapia dominate both small- and large-scale aquaculture in many tropical and subtropical countries, both as low-price commodity for mass consumption as a staple protein source and as a high-value, upscale product for export markets. They are the second most cultured freshwater fish in the world (after carps). However, they are increasingly recognized as the species of choice for intensive aquaculture and are likely to become the most important cultured fish in the world [2]. According to the American Tilapia Association, global farm-raised tilapia production is expected to reach 3 million metric tons by 2010, compared to 2.6 million metric tons in 2007. Several species of tilapia are being cultured commercially, but Nile tilapia, Oreochromis niloticus and various hybrids are the predominant culture species worldwide.

A major problem associated with intensive fish culture operations is the increased susceptibility of fish to infectious diseases. The total losses from disease outbreaks in aquaculture worldwide have reached billions of dollars annually and have been identified as a major threat to the sustainability of aquaculture industry. Traditionally, antibiotics and chemicals have been used to treat diseases in aquatic animals [3]. However, given the fact that diseased fish eat poorly, a limited number of government-approved and efficacious drugs/chemicals are available, and the increasing problem of emerging drug-resistant pathogens and the resultant food and environmental contamination [4], disease prevention, rather than treatment of sick fish, is a better means of controlling infectious diseases. Moreover, the use of antibiotics in animal production, including aquaculture, is increasingly under public scrutiny and criticism in most developed countries. Consequently, there have been considerable interests in recent years to evaluate the feasibility of using non-nutrient dietary additives, particularly prebiotics and probiotics, to enhance growth, stimulate immune system function and/or improve the resistance of fish to infectious diseases. Numerous reviews on these subjects have been published in the past decade [5-12]. Although most research has focused on dietary supplementation of probiotics, other means of administration have also been used, e.g. by addition to rearing water [13].

This chapter provides an overview on the use of probiotics in diets and their effects on growth performance, feed utilization efficiency, gut microbiota, immune responses, and disease resistance of tilapia.

OVERVIEW AND DEFINITIONS

Considerable attention has been given to alteration of the gut microbiota to boost health in humans and other animals in recent years through the use of probiotics and prebiotics (termed biotics). The goal of these dietary supplements is similar, but the manner in which they alter the gut microbial community is varied. Furthermore, some of the probiotic and prebiotic products are similar in composition, containing inactivated microbes or microbial components, which has led to some confusion over what exactly constitutes a "probiotic" or a "prebiotic". The definitions of probiotic and prebiotic have changed to some degree since both terms were first introduced. The most widely used definition of probiotics is

Citation: Welker TL, Lim C (2011) Use of Probiotics in Diets of Tilapia. J Aquac Res Development S1:014. doi:10.4172/2155-9546.S1-014.

given by Fuller [14] as "a live microbial feed supplement which beneficially affects the host animal by improving intestinal balance". Current probiotic applications and scientific data on mechanisms of action indicate that non-viable microbial components act in a beneficial manner, and this benefit is not limited to the intestinal region [9], indicating that probiotics may include non-viable microbes and gut colonization is not necessary to produce benefits to the host. There is some debate as to whether the commonly accepted definition should be revised to include these changes. However, for the purposes of this review, living microorganisms, for the most part, will be considered as probionts. Prebiotics, on the other hand, are any non-digestible food ingredient that beneficially affects the host by selectively stimulating the growth and/or activity of one or a limited number of bacteria in the colon [12]. Synbiotics, a relatively recent concept, may prove more effective in controlling the microbiota of the gastrointestinal (GI) tract [8]. Synbiotics are supplements that contain both prebiotics and probiotics. An effective synbiotic pairing would allow alteration of the colonic environment by a prebiotic that would select for preferential growth conditions of known beneficial probionts. The benefits of this approach are obvious: fish culturists not only are able to control and provide favorable conditions in the colon but also ensure that a beneficial probiont is present in sufficient numbers.

As has been pointed out by numerous researchers [8,9,11,15-17], there is an increasing need to find safe alternatives to antibiotics. Intensive fish culture increases the prevalence of stress-related disease outbreaks and associated fish losses. Traditional use of antibiotics to prevent disease in fish has received criticism due to the potential for development of antibiotic resistant bacteria, the presence of antibiotic residues in fish tissue, negative impacts on microbial populations in the aquaculture environment, and suppression of the cultured species' immune system [12,15]. Probiotics have received the most attention as a viable alternative; however, their use poses a number of potential drawbacks: possible negative impact of untested probionts on the environment, regulatory constraints, food safety issues, and maintaining viable probionts through the feed manufacturing process (particularly extrusion) [18]. Use of bacterial species known to be safe for humans and the environment may be the most expedient approach for testing of prospective probionts and maintaining viable populations during storage [8]. Although there are a number of drawbacks to implementation of probiotics in fish diets, when used properly, they can be effective in improving growth performance and immunity in tilapia.

The study of the effects of probiotics supplemented in the diet of tilapia has not advanced as far as it has in other species, such as salmonids. Merrifield et al. [11] provide an excellent and detailed review of the status of biotics in salmonids. By comparison, the available research with tilapia is severely lacking, underscoring that more research is required. However, research on the supplementation of probiotics in the diets of tilapia is advancing rapidly with the majority of studies taking place since the mid-2000s, and further advances are likely to occur in the coming years.

Types of Probionts

There are generally two choices when selecting a probiont to supplement in diets of fish: 1) bacteria native to and isolated from the host fish species and 2) commercially available probionts typically of non-fish origin (the majority sold for use in terrestrial animals). The first choice may be the most sensible and potentially successful, since they are capable of colonizing the intestinal epithelial surface and grow within the intestinal mucus, are able to survive the pre-colon digestive tract environment of the host species, are not pathogenic to the host, and are likely environmentally safe. However, for most prospective probionts isolated from fish there are questions about potential safety issues and effectiveness as a probiotic, for example: it is unknown whether they are free of plasma-encoded antibiotic resistance genes, pose a potential health risk to humans and other animals, possess desirable growth characteristics, are easily cultured, and remain viable under normal storage conditions and after the feed making process [11]. The cost of undertaking the regulatory approval process for these indigenous species may prove to be a major obstacle for commercialization [8]. More familiar microorganisms, already tested in other animal species with a body of knowledge supporting their safe use presents a substantial advantage for approval, and many of these probiont species are already available in commercial preparations. Likewise, microorganisms isolated from the gut of fish but belonging to familiar groups (Bacillus spp., Lactobacillus spp., yeast spp, and others) may also be suitable [8,11].

GUT MICROBIOTA AND PROBIOTICS

Much more is known about the microbiota of the GI tract in mammals and its role in mammalian health compared to fish, but in recent years, substantial research has been conducted to characterize the gut microbiota of fish [17]. The GI microbiota perform a variety of functions that benefit the health of the host species by promoting nutrient supply, enhancing immune function, preventing colonization of pathogenic microbes, energy homeostasis, and maintenance of normal mucosal integrity and function. However, little is known about bacterial communities, their establishment and diversity, and role in fish nutrition and health. In tilapia, early work focused on describing the gut microbiota in naturally-occurring populations and cultured fish in Japan [19-25] and only later investigations have centered more on changes in the gut microbial community and health effects through the use of probiotics.

Microbial Community of the Tilapia GI Tract

Few studies have characterized the microbial composition of the GI tract of tilapia. The majority of those studies were conducted by Sugita and colleagues in the 1980s using traditional microbiological culture methods. Even more recent examinations of tilapia gut microbiota have continued to rely on traditional culture techniques, rather than more reliable and advanced molecular identification methods. Under normal conditions, the dominant microbial species in the intestine of fish are anaerobic (e.g., Bacillus, Bifidobacterium, Lactobacillus, Saccharomycetes, etc.) accounting for 99% of the microbial community; aerobic and facultative bacteria only account for 1% of the population [15]. Facultative anaerobic bacteria from the Vibrio-Aeromonas group and obligate anaerobic bacteria, Bacteroides type A and type B [19,21], Plesiomonas shigelloides, A. hydrophila [24], and Cetobacterium somerae

[26] are the primary bacteria comprising the GI tract of cultured Nile tilapia. Sugita et al. [22] found the microbiota in freshwater cultured tilapia to be rather stable. However, the rearing environment of tilapia can cause shifts in the gut microbial composition. Obligate anaerobes declined and were replaced by aerobic and facultatively anaerobic Gram negative rods (Vibrio-Aeromonas group) when tilapia were adapted from freshwater to seawater [21]. The authors suggest that dominance of the GI microbiota by Bacteroides type A and type B is restricted to fish cultured in freshwater. The effect of the salinity of the rearing environment on the intestinal microbiota of tilapia is also illustrated by examining the research of Al-Harbi and Uddin [27-29]. Gut bacteria in Nile tilapia cultured in brackish water were dominated by Vibrio spp. [29], while tilapia reared in freshwater, earthen ponds exhibited a greater diversity of gut bacteria, predominately comprised of Gram-negative rods (Aeromonas hydrophila, Shewanella putrefaciens, Corynebacterium urealyticum, E. coli, and Vibrio cholera) [27,28]. In these studies, the gut bacterial population correlated well with the predominant species found in the rearing water and pond sediment, showing that the rearing environment plays a large role in the gut microbial composition. Sugita et al. [23] also suggests that the predominant bacterial species in tilapia GI tract, including aerobes and anaerobes, originate from the culture environment. Molinari et al. [30] reported that the GI tract of Nile tilapia is dominated by eight bacterial species, Aeromonas veronii, A. hydrophila, Burkholderia cepacia, Chromobacterium violaceum, Citrobacter freundii, E. coli, Flavimonas zryzihabitans, and Plesiomonas shgelloides. Although some of these species are also dominant in work by Sugita et al. and Al-Harbi and Uddin, the differences, which are likely affected to some degree by isolation and identification methods (especially the lack of strict anaerobic isolation methods by Al-Harbi and Uddin and Molinari et al. [28]), also suggest the rearing environment should be taken into consideration when selecting probionts for supplementation in diets of tilapia.

Gut microbial composition can also vary seasonally and with changes in diet. Al-Harbi and Uddin [28] discovered that the bacterial composition of tilapia gut can fluctuate considerably depending on the time of year. Even with the most dominant bacterial species, numbers (total counts and as a percentage of the total population) change dramatically during the course of a year. Modifications of diet can also affect the microbiota composition. In rainbow trout, the gut microbiota make-up is altered when fish were switched from a fish meal to a plant meal based diet [31]. With continued emphasis on use of alternative feed ingredients in place of fish meal and fish oil, many currently used probiotics may prove ineffective with changes in diet formulation. However, probiotic applications proven successful in improving growth with traditional fish meal-based diets may also improve digestibility and utilization of alternative plant ingredients. Use of established probiotics may need to be re-evaluated with alterations in diet and rearing conditions, and selection of new probionts and/or development of new feeding regimes may be required.

Changes in Gut Microbiota

Although a number of studies have been conducted which examine the effect of dietary probiotics on tilapia health, only a few have determined the impacts of supplementation on the microbiota of the GI tract. The aim of probiotic use is to alter the gut microbial composition. The normal microbiota of the tilapia GI tract is established 20 to 60 days after hatching [22]. After the microbiota is established, modifying the microbial composition of the gut is complicated/complex, further complicated by several exogenous and endogenous factors that can also influence establishment or alteration of the GI tract microbiota of fish. Gut structure, rearing environment, and farming practices can all effect initial colonization and establishment [17]. The goal of probiotic supplementation (immune stimulation, disease resistance, growth performance, etc.) must also be taken into account [11]. Therefore, proper selection of probionts is critical to success, and as mentioned, use of more than one species or type of probiont in diet may be wise to ensure success with changing conditions and developmental stage, which has proven to be effective in mammals [32] but remains largely untested in fish.

Establishing the effects of probiotic dietary supplementation on the indigenous microbial population of fish has been difficult because researchers have predominantly enumerated the GI microbial population after probiotic supplementation with little attention paid to the composition of the indigenous microbiota [11]. This observation also holds true for tilapia, but a few studies have attempted to characterize the GI tract microbial make-up. Nile tilapia fed diets supplemented with viable Saccharomyces cerevisiae + Bacillus subtillus or non-viable S. cerevisiae produced different GI tract microbial compositions [33]. Tilapia fed non-viable S. cerevisiae showed no change in the intestinal microbial community, which was dominated by E. coli, Salmonella spp., Klebsiella spp., Morganella morganii, E. tarda, Aeromonas sobria,and P. fluorescens. Tilapia fed the diet supplemented with the viable yeast + B. subtillus mixture showed no presence of E. coli, Salmonella spp., Klebsiella spp., or P. fluorescens. Nile tilapia fed diets supplemented with Pediococcus acidilactici exhibited alteration of the indigenous gut bacterial population [34] (this is the only study which has examined the effects of probiotic supplementation on the gut microbiota in tilapia using a molecular, culture independent method). Although the total counts of aerobes and anaerobes did not change, lactic acid bacteria increased significantly in tilapia fed the probiont. P. acidilactici persisted in the gut of Nile tilapia for seventeen days after fish reverted to the control diet. P. acidilactici also provided antagonism against an unidentified bacterium. Knowledge on the impact of probiotic supplementation on the indigenous gut microbiota in tilapia and other species is lacking – this information is important to understanding which microbial species are best suited for dietary supplementation and how they may affect fish health.

PHYSIOLOGICAL CHANGES AND GROWTH PERFORMANCE

A number of specific modes of action by probiotic microorganisms has been attributed to physiological benefits in fish. Although gut colonization is often identified as the most important characteristic of effective probionts, the reality is that the benefits incurred by the host from probiotic supplementation are likely a synergistic product of multiple biological effects (some of which have nothing to do with gut colonization), including production of inhibitory compounds, competition for chemicals or available energy, competition for adhesion sites, inhibition of virulence gene expression or disruption of quorum sensing, enhancement of the immune response, source of macro and/or micronutrients, enzymatic contribution to digestion, and stimulation of local and systemic immune responses [11]. Identifying the specific effect(s) produced by probionts that benefit fish can prove difficult. However, such benefits have been documented in many fish species, including tilapia, some of which will be expanded on in this manuscript.

Changes in Gastrointestinal Morphology

The endogenous microbiota of the GI tract also affects GI morphology and function of fish. Gnotobiotic studies (animals cultured under axenic conditions) in fish have shown that the GI microbiota community is important in nutrient metabolism and absorption, regulation and energy balance, development and maturation of the mucosal immune system, and epithelial differentiation and maturation [17]. In zebrafish, the absence of GI microbiota results in arrestment of the gut epithelium from a lack of brushborder intestinal alkaline phosphatase activity [35]. The authors also found that proliferation of goblet cells, enteroendocrine cells, and other secretory cell lines are tied to gut microbiota. Maintenance of a healthy gut microbiota is a likely benefit to the development of the gut epithelial architecture, and because many fish pathogens can disrupt the integrity of the intestinal epithelium, a healthy GI microbial population may reduce mucosal damage, increase absorptive area, and prevent pathogenic disease [11]. Few studies have examined the effects of probionts on intestinal morphology of tilapia. Ferguson et al. [32] found no effect on intestinal morphology when tilapia were fed diets supplemented with P. acidilactici. Pirarat et al. [36] discovered that dietary supplementation of L. rhamnosus promoted the development of the intestinal structure of Nile tilapia. Fish fed the probiotic diet had greater villous height in the proximal and middle sections of the intestine. Similar results were found by Merrifield et al. [37] in rainbow trout, where dietary supplementation with P. acidilactici increase microvilli length but not density in the proximal intestine. More research is needed in tilapia and other species to determine the mechanisms by which probiotic microorganisms affect gastrointestinal morphology.

Nutrient Utilization, Digestion, and Growth

The gut microbial population is also important to the nutrition of fish by increasing nutrient uptake and utilization, production of enzymes, amino acids, short-chain fatty acids, and vitamins, and improved digestion [11,17]. Lovell and Limsuwan [38] found that the intestinal microbiota of Nile tilapia fed a B12 deficient diet were able to produce at least 11.2 ng vitamin B12/g body weight per day, almost ten times that of channel catfish. Sugita et al. [39,40] reported that obligate anaerobes, primarily Bacteroides type A and Clostridium spp., were likely responsible for vitamin B12 production in the gut of tilapia. Tsuchiya et al. [1] also found that 17 strains of Bacteroides type A isolated from the gut of tilapia were able to synthesize vitamin B12 at a rate of approximately 8 ng/mL culture in 48 hr. Tilapia do not require vitamin B12supplementation in diets to prevent deficiency because of the microbial vitamin B12 production capability of the gut microbial community. Other B vitamins (nicotinic acid and pantothenic acid) are also produced by intestinal bacteria of fish [41], although it is undocumented whether the same is true for tilapia. Production of other fat or water soluble vitamins by microorganisms in the GI tract of fish has not been reported.

Few studies have examined the impact of probiotics on nutrient uptake and utilization in fish, including tilapia. However, there are reports of improved nutrient utilization through probiotic use in tilapia and other species of fish. Rainbow trout exhibited relief from vertebral column compression syndrome due to improved bone formation/mineral utilization from improved mineral uptake in fish fed diets containing the probiotic P. acidilactici [37,42]. The authors hypothesize that improved mineral uptake may have been caused by acidification of the intestinal environment through short-chain fatty acid and lactic acid production by the supplemented probionts. No evidence has been reported supporting improved mineral uptake in tilapia through probiotic use, but probiont-assisted production and utilization of other nutrients has been observed. Premalac® or Biogen®, commercially available probiotic mixtures, supplemented in diets containing varying levels of protein, produced better growth performance in tilapia, suggesting improved protein utilization [43]. Tilapia fed the probiotic diet containing lower dietary protein (27.5%) had better growth than fish fed a 30% protein diet without probiotics. No explanation on the mechanism responsible for the improved protein utilization is provided, but bacteria, including Aeromonads commonly found in the gut of tilapia are known to produce proteases [17]. Gut microbes produce amino acids that are utilized by tilapia. Newsome et al. [44] found that tilapia appeared to be able to acquire their essential amino acid requirements directly from the GI microbiota when dietary sources were insufficient. Volatile, short-chain fatty acids can be produced from anaerobic microbes in the gut of tilapia by fermentation of dietary carbohydrates [45,46]. The evidence shows that symbiotic gut microbes in tilapia contribute to nutrient production, and this contribution may be fairly substantial, especially when nutrient requirements are not being fully met by the diet. However, the role of probiotics in the synthesis and/or utilization of most vitamins, minerals, and macronutrients in finfish, including tilapia, are yet unexplored, and future research should emphasize their contribution to physiological maintenance, homeostasis, and growth.

Probiotics may improve digestion by stimulating production of digestive enzymes or through other alterations in the gut environment, which could translate to improved growth performance. Enzymes involved in digestion (carbohydrases, phosphatases, esterases, lipases, peptidases, cellulases, and proteases) are produced by gut microbes in fish, including some species commonly used as probiotics [17]. Amylase, lipase, and protease production was enhanced in tilapia fed diets containing Bacillus subtilis and an unidentified "photosynthetic bacteria" [47]. The author attributes improved weight gain and feed efficiency to the increased enzyme production. Essa et al. [48] also reported improved growth performance of Nile tilapia fed either B. subtilis, Lactobacillus plantarum, a mixture of B. subtillis and L. plantarum, or S. cerevisiae. Fish fed the bacterial probionts, alone or in mixture, showed improved activity of amylase, protease, and lipase enzymes in the GI tract. Tilapia fed the diet containing S. cerevisiae exhibited enhanced amylase but not protease and lipase activities. Streptococcus faecium + Lactobacillus acidophilus or S. cerevisiae supplementation produced significantly higher weight gain and feed utilization efficiency in tilapia fed diets containing 27% or 40% crude protein compared to the control diet [49,50]. Disaccharidase but not peptidase activity increased in tilapia fed the probiotic supplemented diets. Fish fed the diet containing S. cerevisiae exhibited the greatest increase in growth performance, which was attributed to significantly higher activity of alkaline phosphatase. The authors suggest that the increase in alkaline phosphatase activity reflects the development of brush border membranes of enterocytes stimulated by the supplemented yeast. Activity of the brush border enzymes [51] and alkaline phosphatase [52] can indicate increased nutrient uptake, especially carbohydrate and lipid.

Improved growth performance in tilapia fed probiotic diets has been reported by many researchers (Table 1 included as supplementary data). Tilapia fed S. cerevisiae [49,50], B. subtilis + S. cerevisiae [33,49,50], Micrococcus luteus [53], Bacillus subtilus, Lactobacillus plantarum,B. subtilis + L. plantarum, [48], Bacillus pumilus [54], Lactobacillus acidophilus, Streptococcus faecium [49,50], the commercial probiotic mixtures Organic Green® [55], Biogen® [43,56,57], and Premalac® [43] have all shown to increase growth performance in tilapia. However, other researchers report no effect of some dietary probiotics on growth. Non-viable S. cerevisiae [33], Pseudomonas spp. [53], Pediococcus acidilactici [34], and Enterococcus faecium (Biomate SF-20®), B. subtilis + B. licheniformis (Bioplus 2B®), and P. acidilactici (Bactocell PA10 MD®), viable S. cerevisiae (Levucell SB 20®) [58] have shown to not affect growth of tilapia. Although improved growth has been linked to the production of digestive enzymes stimulated by probionts as reported earlier, metabolite production and improved nutrient utilization may also be responsible for improved feed efficiency and growth performance in tilapia fed probiotics. In most cases, the mechanism for improved growth performance is not known or reported. It is difficult to draw concrete conclusions and provide specific recommendations on the effects of dietary probiotics on growth performance of tilapia given that the studies vary widely with regard to fish age and size, stocking density, diet composition, dietary probiont concentration, feed allowances, feeding duration, and of course, type and source of probiont. However, early studies suggest dietary probiotic supplementation may have beneficial effects when used as growth promoters in tilapia. Furthermore, use of probiotics may allow for the improved utilization of difficult to digest dietary ingredients and nutrients in some fish species. For example, in ruminant and nonruminant mammals, use of a probiotic that produces high levels of cellulases allows for the increased use of high-cellulose carbohydrate sources [59]. This approach may prove useful in fish to overcome poor digestibility and anti-nutritional factors in plant-based alternative feed ingredients.

IMMUNITY AND PROBIOTICS

There is ample evidence to suggest that probiotics supplemented individually or in combination can enhance systemic and local gut immunity and disease resistance in tilapia (Table 2 included as supplementary data). In most studies, there appear to be some positive effect on either immune function, disease resistance, or both. Variation between studies is likely due to differences in the choice of probiont, dietary concentration, species strain and age/size of fish, feeding management and duration, dosage and virulence of challenge pathogen, and method of challenge. Other factors, such as environmental conditions, handling practices, and stocking densities, may also affect results. All these factors can influence the success or failure of dietary probiotics to affect immunity and/or disease resistance in fish. Furthermore, Merrifield et al. [11] suggests the success or potential of probiotics in many studies to prevent disease may be greater than the results show due to the use of intraperitoneal (IP) method of disease challenge. The IP method bypasses one of the most important ways probiotics can prevent infection: competitive exclusion in the GI tract. Intraperitoneal challenges do not reflect the effect of probiotics on resistance to infection – they demonstrate the effect of probiotics on infected fish [11]. In studies where disease resistance is improved, this suggests that the probiotic may be providing immune stimulation outside the GI tract. This is an important point to highlight in the application of probiotics to boost immunity of tilapia, because the vast majority of challenges performed in tilapia research studies are done by IP injection. Streptococcal disease, caused predominantly by Streptococcus iniae, is the biggest disease problem in tilapia culture [60]; however, it is difficult to reproduce reliably by bath immersion, so researchers have had to rely on IP injection to produce reliable, consistent infection generating the desired mortality rate. A degree of caution should be taken when attempting to apply the results of studies employing IP injection for disease challenge. However, as will be noted, the immune response outside the GI tract is enhanced in many probiotic studies, and probiont colonization of the gut may not be required to prevent infection.

Immune Function Enhancement

A large number of probiotic studies have been conducted in fish examining immune function and disease resistance, but their effects on immunity are hard to elucidate [11]. Certainly, it has been shown that one of the means by which probiotic microorganisms can boost immunity is by antagonistic colonization of the GI tract. However, the effect of the gut microbiota, and therefore probiont colonization, on the immune system is more far-reaching and complex than simple physical displacement of pathogenic microbes in the gut. Along with providing a physical barrier, the epithelium and gut-associated lymphoid tissue (GALT) interact with gut microbes to develop a complex immune response, involving the production of cytokines, chemokines, and different effector and regulatory T-cells [61]. In mammals, the gut microbiota also regulates immune gene expression of goblet cells, modifies glycosylation patterns (potentially affecting bacterial adhesion) [62], induces secretion of antimicrobial peptides [63], and ensures proper maturation of the GALT [61]. Although the level of GALT organization in fish is lower than in mammals, the GI microbiota is involved in epithelial differentiation and maturation in gnotobiotic studies in fish [17] and likely helps in GALT development. Nayak [16] also points out that probiotics interact with fish GALT to induce the immune response, increasing Ig+ cells, acidophilic granulocytes, and T-cell counts. Few studies in fish have focused on the effects of probiotic supplementation on gut immunity, and in tilapia, there are far fewer. Pirarat et al.[64] (in press) found an increase in the number of mucous cells in the distal portion of the intestine and a greater abundance of intraepithelial lymphocytes and acidophilic granulocytes in Nile tilapia fed diets containing Lactobacillus rhamnosus GG. No disease challenge was performed in the study. Although Ferguson et al. [34] did not find any changes in the number of leucocytes infiltrating the intestinal epithelium, blood leucocyte numbers and serum lysozyme activity were enhanced in Nile tilapia given the probiotic Pediococcus acidilactici.

Probiotic supplementation also improves the systemic immune response in fish. The process and mechanism in which probiotics in the gut stimulate the systemic response are not completely understood in mammals and are clearly undefined in fish. In mammals, the general consensus is that cells in the GALT process and recognize antigens via pattern recognition receptors, such as toll-like receptors (TLRs) [61]. The TLRs activate immune signaling pathways leading to the production of cytokines and other chemical signals which recruit immune cells (T-cells, natural killer or NK cells, neutrophils, etc.) and affect other immune functions. In tilapia, dietary supplementation of L. rhamnosusGG causes an increase in serum complement activity and enhanced phagocytosis and killing ability of head kidney leukocytes [36]. The authors suggest that L. rhamnosus GG in the gut affected these peripheral immune responses through an increase in the expression of tumor necrosis factor-α (TNF-α) and interleukin-1 (IL-1). A number of other systemic, non-specific immune functions have been shown to be enhanced by dietary probiotic supplementation, including lysozyme activity, peripheral blood immune cell counts, alternative complement activity, phagocytic ability of leucocytes, neutrophil migration and adherence, plasma bactericidal activity, respiratory burst, myeloperoxidase, and superoxide dismutase activities, and others (Table 2 included as supplementary data). However, not all studies result in enhanced immune function. Shelby et al. [58] did not find any affect on lysozyme activity, alternative complement, or total serum immunoglobulin in tilapia fed commercial probiotics containing B. subtilis + B. licheniformis P. acidilactici, and S. cerevisiae. There is far less evidence available suggesting that dietary probiotics influence the humoral immune response in tilapia. Shelby et al. [58] did not find an effect of dietary probiotic supplementation on the antibody response to S. iniae. Probiotic use can enhance the immune response of tilapia, and this has been linked to improved disease resistance.

Disease Prevention

Although it is important to understand how immune function is affected by probiotic supplementation, tilapia culturists are likely more interested in whether probiotics will prevent disease in fish. The success of probiotic use on disease prevention in fish has been mixed, but overall appears to be fairly effective in tilapia using a variety of probionts against a number of different bacterial pathogens (Table 2 included as supplementary data). Streptococcus iniae, Aeromonas hydrophila, and E. tarda are the primary bacterial pathogens that have been evaluated in tilapia probiotic studies. All but one study examining the effects of probiotics on disease resistance of tilapia used challenge by IP injection, which as outlined above, bypasses the gut and integument, and may not fully evaluate the potential of probiotics to prevent disease in tilapia. The effectiveness of probiotics in terms of protection against infection is often attributed to enhanced immunity. Pirarat et al. [64] has suggested that L. rhamnosus GG protection against E. tarda is accomplished by enhancing the alternative complement system thereby increasing phagocytic cell aggregation and phagocytic activity. As noted previously, the interaction of probiotics and the immune system is complex and identifying the exact immune mechanism(s) providing protection would be a difficult task. Probiotics can also be ineffective in preventing disease in tilapia. Shelby et al. [58] found that feeding commercial probiotics for 94 days did not prevent streptococcal disease. Probiotics have proven effective in preventing viral diseases in several fish species. Resistance of grouper (Epinephelus coiodes) to iridovirus is enhanced with supplementation of Lactobacillus plantarum in diet [65], and Pseudomonas sp., Vibrio sp., Aeromonas sp., and groups of coryneforms show antiviral activity to infectious hematopoietic necrosis virus (IHNV) [66]. Probiotic bacteria have also exhibited antiviral activity against poliovirus [67] and Oncorhynchus masou virus (OMV) [68] in fish. However, to our knowledge, the effect of probiotics has not been evaluated on any viral disease in tilapia.

Several factors affect the efficacy of probiotics on disease prevention in fish, especially the type of probiont and dietary dose concentration (dietary concentration + feeding duration). In tilapia, short-term feeding (2 weeks) and long-term (2 months or greater) have all proven to be effective in enhancing disease resistance in tilapia (Table 2 included as supplementary data). Few tilapia studies have explored the effects of dose concentration, although several have examined the effect of dietary concentration and feeding duration separately. Published information on immunostimulants suggests that, the larger the dose concentration, the less effective immunostimulants are in protecting fish against infection and can even result in immune suppression [5,11]. Little data exist to confirm whether this phenomenon also occurs with probiotic supplementation in fish diets, and no studies are known to have investigated the effects of pulsed-feeding (short-term, alternating feeding intervals of basal and probiotic diets) in tilapia, a method commonly used to prevent immune-suppression from excessive dose concentration when feeding immunostimulants to fish [11]. Aly et al. [54] found that supplementing Bacillus pumilus at 1012/g diet increased protection of Nile tilapia against A. hydrophila after 1 and 2 but not 8 months of feeding. The lower dietary concentration tested (106/g) did not provide any protection. In another study, Aly et al. [55] also found that dietary supplementation of L. acidophilus, B. subtilis, or a mixture of the two generally provided greater protection against A. hydrophila, P. fluorescens, and S. iniae after 2 months of feeding compared to 1 month. The form of probiotic administration can also impact effectiveness in affecting fish health. Improving viability of probionts during the feed making process and during feed storage can be improved by encapsulation in non-nutritive matrices, such as calcium alginate. Encapsulation of Shewanella putrefaciens in calcium alginate improved viability of the bacterium during feed storage, and its presence was found in the gastrointestinal track of Sengalese sole (Solea senegalensis) fed encapsulated but not non-encapsulated of S. putrefaciens [69]. Furthermore, the rout of probiotic administration can also affect the success of probiotic application. Addition of Lactococcus lactis RQ516 to rearing water increases the resistance of Nile tilapia to A. hydrophila [13]. Further research on the effects of dose dependency and form and route of probiont administration on disease resistance are needed for all fish species, including tilapia, in order to provide effective feeding and treatment regimens.

The combination of probiotics and prebiotics, also referred to as synbiotics, has shown promise in treating diseases in humans [70] and other mammals [71]. The prebiotic is thought to give a probiont a competitive advantage by providing a fermentable energy source enabling it to out-compete endogenous microbial populations [11,72]. The survival rate of probiotics is improved during their passage through the digestive tract, thereby contributing to the stabilization and/or enhancement of the probiotic effects. Few studies have examined the effects of a synbiotic approach on the health of fish. Japanese flounder fed a diet containing Bacillus clausii or in combination with prebiotics fructo- or mannan oligosaccharide showed improved non-specific immune function [73]. Although the diet containing either prebiotic with B. clausiiexhibited the highest immune function, activity was not significantly different compared to flounder fed B. clausii alone. No disease challenge was conducted in the study. Feeding a synbiotic combination of mannan oligosaccharide and Enterococcus faecalis improved survival of rainbow trout challenged with V. anguillarum than trout fed the individual prebiotic or probiotic [74]. Given the success observed with synbiotic approaches in mammals, a high priority should be given in future research in tilapia and other fish species.

CONCLUSIONS AND RECOMMENDATIONS

Research on the effect of probiotics on the health and growth of tilapia has increased significantly over the past few years, but many questions remain unanswered and more work needs to be conducted. Little effort has been undertaken to determine the microbial composition of the GI tract of tilapia. This basic information is vital to discovery and selection of potential probionts. Understanding the underlying mechanisms in which probionts are able to colonize the gut and alter the gut environment to influence digestion, nutrient absorption, growth performance, and immunity are also important for choosing potential microbes for use as probionts. Work in this area needs to continue and be expanded. Furthermore, insight as to how changes in the rearing environment and alterations of diet influence the gut microbial community will also help researchers and culturists select microbes with a better chance for success and to develop more effective feeding regimens and diet formulations. Use of probiotic mixtures containing several probionts may be required when this information is not available. Development of synbiotic mixtures may prove to be the most effective means of probiotic use, allowing fish culturists to control and provide favorable conditions in the gut and also ensure that a beneficial probiont is present and in sufficient numbers. The success of probiotic dietary supplementation to improve the health and growth performance of tilapia is encouraging given the small amount of research that has occurred. Their effectiveness will surely continue to improve as more questions and uncertainty surrounding probiotic use in tilapia diets are addressed. Future work must focus on applications of probiotics in order to achieve the maximum efficiency, and probiotic treatments must be used in concert with effective farm management and husbandry.

REFERENCES

1. Lim C, Webster CD (2006) Tilapia biology, culture, and nutrition. The Haworth Press, Inc., Binghamton, NY.
2. Fitzsimmons K (2006) Prospect and potential for global production. Pages 51- 72 in C. Lim and C.D. Webster, Editors. Tilapia: Biology, Culture and Nutrition. The Haworth Press, Inc., Binghamton, New York, NY.
3. Li P, Gatlin DM (2005) Evaluation of the prebiotic GroBiotic®-A and brewers yeast as dietary supplements for sub-adult hybrid striped bass (Morone chrysops x M. saxatilis) challenged in situ with Mycobacterium marinum. Aquaculture 248: 197-205.
4. FAO (Food and Agriculture Organization of the United Nations), 2002. Antibiotics residue in aquaculture products. Pages 74-82 in The State of World Fisheries and Aquaculture, Rome, Italy.
5. Sakai M (1999) Current research status of fish immunostimulants. Aquaculture 172: 63-92.
6. Gannam AL, Schrock RM (2001) Immunostimulants in fish diets. Pages 235- 266 in C. Lim and C.D. Webster, Editors. Nutrition and fish health. Food Product Press, New York, NY.
7. Irianto A, Austin B (2002) Probiotics in Aquaculture. J Fish Diseases 25: 633- 642.
8. Gatesoupe FJ (2005) Probiotics and prebiotics for fish culture, at the parting of the ways. Aqua Feeds: Formulation & Beyond 2: 3-5.
9. Kesarcodi-Watson A, Kasper H, Lategan MJ, Gibson L (2008) Probiotics in Aquaculture: The need, Principles and mechanisms of action and screening processes. Aquaculture 274: 1-14.
10. Wang YB, Li JR, Lin J (2008) Probiotics in aquaculture: Challenges and outlook. Aquaculture 281: 1-4.
11. Merrifield DL, Dimitroglou A, Foey A, Davies SJ, Baker RTM, et al. (2010a) The current status and future focus of probiotic and prebiotic applications for salmonids. Aquaculture 302: 1-18.
12. Ringø E, Oldsen RE, Gifstad TO, Dalmo RA, Amlund H, et al. (2010) Prebiotics in aquaculture: review. Aquaculture Nutr 16: 117-136.
13. Zhou X, Wang Y, Jiangtao Y, Li W (2010) Inhibition ability of probiotic, Lactococcus lactis, against A. hydrophila and study of its immunostimulatory effect in tilapia (Oreochromis niloticus). Int J Engin Sci Tech 2: 73-80.
14. Fuller R (1989) Probiotics in man and animals. J Appl Bacteriol 66: 365–378.
15. Zhou Q, Li K, Jun X, Bo L (2009b) Role and functions of beneficial microorganisms in sustainable aquaculture. Bioresour Technol 100: 3780- 3786.
16. Nayak SK (2010a) Probiotics and immunity: a fish perspective. Fish & Shellfish Immunol 29: 2-14.
17. Nayak SK (2010b) Role of gastrointestinal microbiota in fish. Aquac Res 41: 1553-1573.

18. Ringø E, Løvmo L, Kristiansen M, Bakken Y, Salinas I, et al. (2009) Lactic acid bacteria vs. pathogens in the gastrointestinal tract of fish: a review. Aquac Res 41: 451-467.

19. Sakata T, Sugita H, Mitsuoka T, Kakimoto D, Kadota H (1980a) Isolation of obligate anaerobes from the intestinal tracts of freshwater fish. Bull Jap Soc Sci Fish 46: 511.

20. Sakata T, Okabayashi J, Kakimoto D (1980b) Variations in the intestinal microflora of tilapia reared in fresh and sea water. Bull Jap Soc Sci Fish 46: 313-317.

21. Sugita H, IsidaY, Deguchi Y, Kadota H (1982a) Bacterial flora in the gastrointestine of Tilapia nilotica adapted in sea water. Bull Jap Soc Sci Fish 48: 987-991.

22. Sugita H, Enomoto A, Deguchi Y (1982b) Intestinal microflora in the fry of Tilapia mossambica. Bull Jap Soc Sci Fish 48: 875.

23. Sugita H, Oshima K, Tamura M, Deguchi Y (1983) Bacterial flora in the gastrointestine of freshwater fishes in the river. Bull Jap Soc Sci Fish 49: 987- 991.

24. Sugita H, Tokuyama K, Deguchi Y (1985) The intestinal microflora of carp Cyprinus carpio, grass carp Ctenopharyngodon idella and tilapia Sarotherodon niloticus. Bull Jap Soc Sci Fish 50: 1325-1329.

25. Sugita H, Tsunohara M, Fukumoto M, Deguchi Y (1987) Comparison of microflora between intestinal contents and fecal pellets of fresh water fishes. Nippon Suisan Gakkaishi 53: 287-290.

26. Tsuchiya C, Sakata T, Sugita H (2008) Novel ecological niche of Cetobacterium somerae, an anaerobic bacterium in the intestinal tracts of freshwater fish. Lett Appl Microbiol 46: 43-48.

27. Al-Harbi AH, Uddin N (2003) Quantitative and qualitative studies on bacterial flora of hybrid tilapia (Oreochromis niloticus x O. aureus) cultured in earthen ponds in Saudi Arabia. Aquac Res 34: 43-48.

28. Al-Harbi AH, Uddin MN (2004) Seasonal variation in the intestinal bacterial flora of hybrid tilapia (Oreochromis niloticus x Oreochromis aureus) cultured in earthen ponds in Saudi Arabia. Aquaculture 229: 37-44.

29. Al-Harbi AH, Uddin N (2005) Bacterial diversity of tilapia (Oreochromis niloticus) cultured in brackish water in Saudi Arabia. Aquaculture 250: 566-572.

30. Molinari LM, Scoaris D, de O, Pedroso RB, Bittencourt N, et al. (2003) Bacterial microflora in the gastrointestinal tract of Nile tilapia,Oreochromis niloticus, cultured in a semi-intensive system. Acta Scientiarum Biological Sciences Maringa 25: 267-271.

31. Heikkinen J, Vielma J, Kemilainen O, Tiirola M, Eskelinen P, et al. (2006) Effects of soybean meal based diet on growth performance, gut histopathology and intestinal microbiota of juvenile rainbow trout (Oncorhynchus mykiss). Aquaculture 261: 259-268.

32. Timmerman HM, Koning CJM, Mulder L, Rombout FM, Beynen AC (2004) Monostrain, multistrain and multispecies probiotics – a comparison of functionality and efficacy. Int J Food Microbiol 96: 219-233.

33. Marzouk MS, Moustafa MM, Mohamed NM (2008) The influence of some probiotics on the growth performance and intestinal microbial flora of Oreochromis niloticus. Proceedings of 8th International Symposium on Tilapia in Aquaculture, Cairo, Egypt, pp. 1059-1071.

34. Ferguson RMW, Merrifield DL, Harper GM, Rawling MD, Mustafa S, et al. (2010) The effect of Pediococcus acidilactici on the gut microbiota and immune status of on-growing red tilapia (Oreochromis niloticus). J Applied Microbiol 109: 851-862.

35. Bates JM, Mittge E, Kuhlman J, Baden KN, Cheesman SE, et al. (2006) Distinct signals from the microbiota promote different aspects of zebrafish gut differentiation. Dev Biol 297: 374-386.

36. Pirarat N, Pinpimai K, Endo M, Katagiri T, Ponpornpisit A, et al. (2011) Modulation of intestinal morphology and immunity in Nile tilapia (Oreochromis niloticus) by Lactobacillus rhamnosus GG. Res Vet Sci 91: e92-e97 (In press).

37. Merrifield DL, Harpr G, Baker RTM, Ringø E, Davies S J (2010b) Possible influence of probiotic adhesion to intestinal mucosa on the activity and morphology of rainbow trout (Oncorhynchus mykiss) enterocytes. Aquacult Res 41: 1268-1272.

38. Lovell T, Limsuwan T (1982) Intestinal synthesis and dietary non essentiality of vitamin B12 for Tilapia nilotica. Trans of Amer Fish Soc 11: 485-490.

39. Sugita H, Miyajima C, Deguchi Y (1990) The vitamin B12 producing ability of the intestinal bacteria isolated from tilapia and channel catfish. Nippon Suisan Gakkaishi 56: 701.

40. Sugita H, Miyajima C, Deguchi Y (1991) Vitamin B12 producing ability of the intestinal microflora of freshwater fish. Aquaculture 92: 267-276.

41. Teshima S, Kashiwada K (1967) Studies on the production of B vitamins by intestinal bacteria of fish. 3. Isolation of vitamin B12synthesizing bacteria and their bacteriological properties. Bull Jap Soc Sci Fish 33: 979-983.

42. Merrifield DL, Bradley G, Harper GM, Baker RTM, Munn CB, et al. (2011) Assessment of the effects of vegetative and lyophilised Pediococcus acidilactici on growth, feed utilisation, intestinal colonisation and health parameters of rainbow trout (Oncorhynchus mykiss Walbaum). Aquac Nutr 17: 73-79.

43. Ghazalah AA, Ali HM, Gehad EA, Hammouda YA, Abo-State HA (2010) Effect of probiotics on performance and nutrients digestibility of Nile tilapia (Oreochromis niloticus) fed low protein diets. Nature and Sci 8: 46-53.

44. Newsome SD, Fogel ML, Kelly L, Martinez del Rio C (2011) Contributions of direct incorporation from diet and microbial amino acids to protein synthesis in Nile tilapia. Funct Ecol 25: 1051-1062 (in press).

45. Kihara M, Sakata T (1997) Fermentation of dietary carbohydrates to shortchain fatty acids by gut microbes and its influence on intestinal morphology of a detritivorous teleost tilapia (Oreochromis niloticus). Com Biochem Physiol 118: 1201-1207.

46. Smith TB, Wahl DH, Mackie RI (2005) Volatile

fatty acids and anaerobic fermentation in temperate piscivorous and omnivorous freshwater fish. J Fish Biol 48: 829-841.

47. Honsheng M (2010) Effects of probiotics on activities of digestive enzyme for tilapia. J South China Normal Univ 2: 10-15.

48. Essa MA, El-Serafy SS, El-Ezabi MM, Daboor SM, Esmael NA, et al. (2010) Effect of different dietary probiotics on growth, feed utilization and digestive enzymes activities of Nile tilapia, Oreochromis niloticus. J Arabian Aquacult Soc 5: 143-161.

49. Lara-Flores M, Olvera-Novoa MA, Guzmán-Méndez BE, López-Madrid W (2003) Use of the bacteria Streptococcus faecium andLactobacillus acidophilus, and the yeast Saccharomyces cerevisiae as growth promoters in Nile tilapia (Oreochromis niloticus). Aquaculture 216: 193-201.

50. Lara-Flores M, Olivera-Castillo L, Olvera-Novoa MA (2010) Effect of the inclusion of a bacterial mix (Streptococcus faecium and Lactobacillus acidophilus), and the yeast (Saccharomyces cerevisiae) on growth, feed utilization and intestinal enzymatic activity of Nile tilapia (Oreochromis niloticus). Internat J Fish Aquaculture 2: 93-101.

51. Gawlicka A, Parent B, Horn MH, Ross N, Opstad I, et al. (2000) Activity of digestive enzymes in yolksac larvae of Atlantic halibut (Hippoglossus hippoglossus): Indication of readiness for first feeding. Aquaculture 184: 303- 314.

52. German DP, Horn MH, Gawlicka A (2004) Digestive enzyme activities in herbivorous and carnivorous prickleback fishes (Teleostei: Stichaeidae): Ontogenetic, Dietary, and Phylogenetic Effects. Physiol Biochem Zool 77: 789- 804.

53. El-Rhman AMA, Khattab YA, Shalaby AM (2009) Micrococcus luteus and Pseudomonas species as probiotics for promoting the growth performance and health of Nile tilapia, Oreochromis niloticus. Fish Shellfish Immuol 27: 175-180.

54. Aly SM, Abdel-Galil Ahmed Y, Abdel-Aziz Ghareeb A, Mohamed MF (2008a) Studies on Bacillus subtilis and Lactobacillus acidophilus, as potential probiotics, on the immune response and resistance of Tilapia nilotica (Oreochromis niloticus) to challenge infections. Fish Shellfish Immunol 25: 128–136.

55. Aly SM, Mohamed MF, John G (2008b) Effect of probiotics on the survival, growth and challenge infection in Tilapia nilotica (Oreochromis niloticus). Aquac Res 39: 647–656.

56. El-Haroun ER, Goda AMA-S, Kabir Chowdhury MA (2006) Effect of dietary probiotic biogens supplementation as a growth promoter on growth performance and feed utilization of Nile tilapia Oreochromis niloticus (L.). Aquac Res 37: 1473–1480.

57. Mehrim AI (2009) Effect of dietary supplementation of Biogen (commercial probiotic) on mono-sex Nile tilapia Oreochromis niloticus under different stocking densities. J Fish Aquat Sci 4: 261-273.

58. Shelby RA, Lim CE, Aksoy M, Klesius PH (2006) Effect of probiotic diet supplements on disease resistance and immune response of young Nile tilapia, Oreochromis niloticus. J Appl Aquac 18: 23-34.

59. Ushakova NA, Kotenkova EV, Kozlova AA, Nifatov AV (2006) A study of the mechanisms of probiotic effect of Bacillus subtilis Strain 8130. Appl Biochem Microbiol 42: 252-257.

60. Shoemaker C, Lim C, Yildirim-Aksoy M, Welker T, Klesius PH (2006) Growth response and acquired resistance of Nile tilapia,Oreochromis niloticus (L.) that survived Streptococcus iniae infection. Aquac Res 37: 1238-1245.

61. Sanz Y, De Palma G (2009) Gut microbiota and probiotics in modulation of epithelium and gut-associated lymphoid tissue function. Int Rev Immunol 28: 397-413.

62. Caballero-Franco C, Keller K, De Simona C, Chadee K (2007) The VSL#3 probiotic formula induces mucin gene expression and secretion in colonic epithelial cells. Am J Physiol Gastrointest Liver Physiol 292: G315-G322.

63. Vaishnava S, Behrendt CL, Ismail AS, Eckmann L, Hooper LV (2008) Paneth cells directly sense gut commensals and maintain homeostasis at the intestinal host-microbial interface. Proc Natl Acad Sci USA 105: 20858-20863.

64. Pirarat N, Kobayashi T, Katagiri T, Maita M, Endo M (2006) Protective effects and mechanisms of a probiotic bacterium Lactobacillus rhamnosus against experimental Edwardsiella tarda infection in tilapia (Oreochromis niloticus). Vet Immunol Immunopathol 113: 339-347.

65. Son VM, Chang CC, Wu MC, Guu YK, Chiu CH, et al. (2009) Dietary administration of the probiotic, Lactobacillus plantarum, enhanced the growth innate immune responses, and disease resistance of the grouper Epinephelus coioides. Fish Shellfish Immunol 26: 691-698.

66. Kamei Y, Yoshimizu M, Ezura Y, Kimura T (1988) Screening of bacteria with antiviral activity from fresh water salmonid hatcheries.Microbiol Immunol 32: 67–73.

67. Girones R, Jofre JT, Bosch A (1989) Isolation of marine bacteria with antiviral properties. Can J Microbiol 35: 1015–1021.

68. Direkbusarakom S, Yoshimizu M, Ezura Y, Ruangpan L, Danayadol Y (1998) Vibrio spp. the dominant flora in shrimp hatchery against some fish pathogenic viruses. J Mar Biotechnol 6: 266–267.

69. Rosas-Ledesma R, León-Rubio JM, Alarcón FJ, Moriñgo MA, Balebona MC (2011) Calcium alginate capsules for oral administration of fish probiotic bacteria: assessment of optimal conditions for encapsulation. Aquac Res. doi:10.1111/j.1365-2109.2011.02809.x

70. Haskey N, Dahl WJ (2006) Synbiotic therapy: a promising new adjunctive therapy for ulcerative colitis. Nutr Rev 64: 132-138.

71. Zimmerman B, Bauer E, Mosenthin R (2001) Pro- and prebiotics in pig nutrition – potential modulators of gut health? J Animal Feed Sci 10: 47-56.

72. Gibson GR, Roberfroid MB (1995) Dietary modulation of the human colonic microbiota: introducing the concept

of prebiotics. J Nutr 125: 1401–1412.

73. Ye J-D, Wang K, Li F-D, Sun Y-Z (2011) Single or combined effects of fructoand mannan oligosaccharide supplements and Bacillus clausii on the growth, feed utilization, body composition, digestive enzyme activity, innate immune response and lipid metabolism of the Japanese flounder Paralychthys olivaceus. Aquac Nutr 17: e902-e911.

74. Rodriguez-Estrada U, Satoh S, Haga Y, Fushimi H, Sweetman J (2009) Effects of single and combined supplementation of Enterococcus faecalis, mannan oligosaccharide and polyhydrobutyric acid on growth performance and immune response of rainbow troutOncorhynchus mykiss. Aquac Sci 57: 609–617.

75. Ali HM, Ghazalah AA, Gehad EA, Hammouda YA, Abo-State HA (2010) Practical aspects and immune response of probiotics preparations supplemented to Nile tilapia (Oreochromis niloticus) diets. Nature and Sci 8: 39-45.

76. Galindo AB, Hartemink R, Schrama JW, Verreth J, Bucio L, et al. (2009) Kinetics of Lactobacillus plantarum 44a in the faeces of tilapia (Oreochromis niloticus) after its intake in feed. J Appl Microbiol 107: 1967-1975.

77. Jatoba A, Vieira FdN, Buglione-Neto CC, Mouriño' JLP, Silva BC, et al. (2011) Diet supplemented with probiotic for Nile tilapia in polyculture system with marine shrimp. Fish Physiol Biochem 37: 725-732. (in press).

78. Vijayabaskar P, Somasundaram ST (2008) Isolation of bacteriocin producing lactic acid bacteria from fish gut and probiotic activity against common fresh water fish pathogen Aeromonas hydrophila. Biotechnol 7: 124-128.

79. El-Ashram AMM, Mohammed MF, Aly SM (2008) Effect of Biobuds as a commercial probiotic product in cultured tilapia. 8th International Symposium on Tilapia in Aquaculture. 1089-1096.

80. Aly SM, Abd-El-Rahman AM, John G, Mohamed MF (2008c) Characterization of some bacteria isolated from Oreochromis niloticus and their potential use as probiotics. Aquaculture 277: 1-6.

81. Taoka Y, Maeda H, Jo J-Y, Kim S-M, Park S-I, et al. (2006) Use of live and dead probiotic cells in tilapia Oreochromis niloticus. Fish Sci 72: 755-766.

82. Mohamed MH, Refat NAGA (2011) Pathological evaluation of probiotic, Bacillus subtilis, against Flavobacterium columnare in tilapia nilotica (Oreochromis niloticus) fish in Sharkia Governorate. Egypt J Amer Sci 7: 244-256.

CHAPTER 10

EVALUATION OF ESSENTIAL OIL OF FENNEL AND GARLIC SEPARATELY OR COMBINED WITH BACILLUS LICHENIFORMIS ON THE GROWTH, FEEDING BEHAVIOUR, HEMATO-BIOCHEMICAL INDICES OF OREOCHROMIS NILOTICUS (L.) FRY

Mohamed S Hassaan and Magdy A Soltan

Keywords: Essential oil, Fennel, Garlic, Growth, Feeding behavior, Hemato-biochemical, Tilapia

INTRODUCTION

Improving the bio-economic efficiency of aquaculture is dependent on advances in biology, nutrition and environmental management of the production cycle. Antimicrobials and other veterinary drugs are administered regularly as additives in fish food, are used as therapeutics or growth promoters [1]. Nevertheless, the use of veterinary drugs is becoming more restricted since they present numerous side-effects for the environment and health safety [2]. Use of antibacterial agents in aquaculture should be avoided as it leads to proliferation of resistant and creates a negative image for aquaculture industry. Use of essential oil extracted from herbs medicine, could be a valuable alternative instead of antibiotics because there are fewer chances for the development of resistant strains [3].

There are many aromatic plants such as oregano, rosemary, sage, peppermint, thymus, fennel and garlic in worldwide, especially in the Mediterranean area [4]. The extracted essential oil from aromatic plants are usually used for antioxidant [5], digestive stimulant [6].

Fennel, *Foeniculum vulgare* is a biennial medicinal plant belonging to the family Apiaceae (Umbelliferae). Essential oils of fennel havehepatoprotective effects [7], as well as anti-inflammatory, and antioxidant activities [8]. Many phytochemical studies have been conducted to study the chemical composition of the essential oil of fennel from different origins and have shown that the major components are phenylpropanoid derivatives and monoterpenoids [9].

Allium is the largest and most important genus of the Alliaceae family and comprises 450 species, which were widely distributed in the northern hemisphere. Among them, garlic (*Allium sativum* L.), is a well-known species being used in traditional medicine and food in many countries [10]. Garlic and its constituents have antimicrobial activity. It was showed that the antibacterial activity of garlic resulted from thio sulfinates, especially allicin, which is responsible for the antimicrobial activity, as well as its flavor and aroma [11]. Five types of garlic preparations are currently available on the market: garlic essential oils, garlic oil macerate, garlic powder, aged garlic extract and fresh garlic [12]. Among them, the antimicrobial activity of garlic oil is 200 times greater than garlic powder and 900 times the strength of fresh garlic [12].

Previous studies revealed that herbs medicine and essential extracted oil from herbs can be successfully used in fish nutrition such as Carum carvi seeds [13], *Foeniculum vulgare* seeds [14], Cuminum cyminum powder [15], ground garlic, *Allium sativum*

Citation: Hassaan MS, Soltan MA (2016) Evaluation of Essential Oil of Fennel and Garlic Separately or Combined with Bacillus licheniformis on the Growth, Feeding Behaviour, Hemato-biochemical Indices of Oreochromis niloticus (L.) Fry. J Aquac Res Development. 7: 422. doi:10.4172/2155-9546.1000422.

[16,17], oil of oregano,*Origanum heracleoticum* [18,19], oil of thyme and fennel [20]. Furthermore, some aromatic plants and their extracts have been reported to stimulate the growth of certain bacteria where used a prebiotic, but these studies are scarce [21]. As aforementioned, many authors have attempted to examine various essential oil components from herbal plants, but information regarding fry culture, especially the effects of essential oil of fennel and garlic supplements and their combined with probiotics on fish performance and hematobiochemical blood parameters, is little. Therefore, the present study was designed to investigate the effect of fennel oil, garlic oil and *Bacillus licheniformis* alone and combined as dietary feed additives on growth performance, feed utilization, hematological and biochemical blood parameters of Nile tilapia, *O. niloticus* fry.

MATERIALS AND METHODS

Fish, Facility and Feeding Trial

The feeding trial was performed at an indoor installation of the Aquaculture Laboratory, faculty of agriculture Benha University. Nile tilapia, *O. niloticus* fries were obtained from fish hatchery of central laboratory for aquaculture research, Abbassa, Abou-Hammad Sharkia, Egypt. Fry were acclimated for two weeks to the experimental conditions in a fiberglass tanks. During this period fish fed daily on the commercial diet to be adapted to pelleted feeds. At the beginning of the experiment, 18 (100-L) glass aquaria were stocked with 20 fish with an initial average weight (1.88 ± 0.12 g). Each aquarium was supplied with compressed air via air-stones from air pumps. Overhead fluorescent illuminating was set at 14:10 (light:dark). All fish were fed, initially, at a rate of 6% of total body weight daily for the first 6-week and then decreased to 4% of total body weight daily for the remaining 12-week [22]. The fish were fed four times a day at 10:00, 12:00, 14:00, and 16:00, 6 days a week, for 84 days. All aquaria were siphoned twice a day to remove faecal materials then replaced by aerated fresh water. Total fish weight in each aquarium was determined every 2 weeks to check their growth and to adjust the feeding rate. Feeding was stopped 24 h before sampling. Water samples for chemical analysis were monitored weekly during the experimental period. Dissolved oxygen and temperature were measured on site with an YSI model 58 oxygen meter (Yellow Spring Instrument Co., Yellow Springs, OH, USA). Total ammonia was measured using DREL/2 HACH kits (HACH Co., Loveland, Co. USA). pH was measured using a pH meter (Digital Mini-pH Meter, model 55, Fisher Scientific, USA). During the period of the feeding trial, the water-quality parameters were averaged (± SD): Water temperature was 26.25 ± 0.9°C: dissolved oxygen, 5.7 ± 0.5 mg L^{-1}: pH 8.33 ± 0.4 and total ammonia, 0.07 ± 0.01 mg L^{-1}. All tested water quality criteria were suitable and within the acceptable limits for rearing the Nile tilapia *O. niloticus* [23].

Extraction of Essential Oils and Analysis

Fresh fennel seeds were dried at 35°C and crushed to fine powders using a grinder. Essential oil was obtained by hydrodistillation of the powdered dry seeds using a Clevenger-type apparatus. The essential fennel oil (EFO) was separated, dried over anhydrous sodium sulfate. However, extraction of the essential garlic oils (EGO) from fresh cloves of garlic was performed with a hydrodistiller after mashed. The hydrodistillation lasted 3 hours. Essential oil with an unpleasant strong odor was obtained and dried by adding anhydrous sodium sulfate, filter sterilized through a 0.22-mmfilter. Essential fennel and garlic oil were stored in a dark glass bottle at 4°C until use. The analysis of EFO and EGO were analyzed using gas chromatography-mass spectrometry (GC-MS) to identify their chemical constituents according to Aazza and Douiri et al. [24,25], respectively.

Preparation Inoculum of (*B. licheniformis*)

Strain of *Bacillus licheniformis* used was obtained from Microbiological Resources Center (MIRCEN), faculty of Agriculture, Ain Shams University, Egypt. Inoclum was prepared by adding 5 g lyophilized cells of *B. licheniformis* to 100 ml of specific media which containing 5 g L^{-1}peptone, 3 g L^{-1} beef extract at pH 7.0 then was incubation at 37°C for 24 hours. After this period, cells were centrifugation for 15 min (2000 g) then washed with phosphate buffered saline (PBS) with pH 7.3. The washed cells was added dropwise into the ingredients of diet before to pellet [26] until to reach the level 7×10^7 (CFU g^{-1}) of $7{\times}10^7$ (CFU g^{-1}) of *B. licheniformis*. The control diet received the same volume PBS to maintain an equal volume of PBS in all experimental diets.

Experimental Diets

The basal practical diet was formulated to contain approximately (35% crude protein and 19.96MJ Kg^{-1} gross energy) (Table 1). The basal practical diet was divided into six parts. The first part prepared without feed additives as control diet (D1). Other five parts prepared to contain*B. licheniformis* at level $7{\times}10^7$ colony forming unit (CFU g^{-1} diet) diet (D2); 1ml EFO kg^{-1} diet (D3); 1ml EGO kg^{-1} diet (D4); 1ml EFO + *B. licheniformis* $7{\times}10^7$ CFU g^{-1} diet (D5); 1ml EGO + *B. licheniformis* $7{\times}10^7$ CFU g^{-1} diet (D6). Essential fennel and garlic oil were added separately to soybean oil. All dry ingredients were thoroughly mixed with soybean oil, and vitamins and minerals mixture using a homogenous mixture grinder, and then passing the mixed feed through a laboratory pellet mill (0.5-mm die) in National Institute of Oceanography and Fisheries, Cairo Governorate, Egypt (CMP California Pellet Mill, San Francisco, CA, USA).

Table 1: Composition and proximate analysis of basal diet (g/ kg diet)

Ingredients	g
Fish meal	280
Soybean meal	310
Yellow corn	240
Wheat bran	90
Soybean oil	40
Vitamins and Minerals[a]	35
Stay-C 35[b]	5
Proximate analysis	
Dry matter	952.30
Crude protein	351.40

Crude lipid	81.10
Ash	84.50
Total carbohydrate[b]	483.00
Gross energy (MJ/ kg)[c]	19.96

[a]Vitamin and mineral mixture kg^{-1} of mixture contains: 4800 I.U. Vit A, 2400 IU **cholecalciferol** (vit. D), 40 g Vit E, 8 g Vit K, 4.0 g Vit B_{12}, 4.0 g Vit B2, 6 g Vit B6, 4.0 g, Pantothenic acid, 8.0 g Nicotinic acid, 400 mg Folic acid, 20 mg Biotin, 200 gm Choline, 4 g Copper, 0.4 g Iodine, 12 g Iron, 22 g Manganese, 22 g Zinc, 0.04 g Selenium. folic acid, 1.2 mg; niacin, 12 mg; d-calcium pantothenate, 26 mg; pyridoxine. HCl, 6 mg; riboflavin, 7.2 mg; thiamin. HCl, 1.2 mg; sodium chloride (NaCl, 39% Na, 61% Cl), 3077 mg; ferrous sulfate ($FeSO_4.7H_2O$, 20% Fe), 65mg; manganese sulfate ($MnSO_4$, 36% Mn), 89 mg; zinc sulfate ($ZnSO_4.7H_2O$, 40% Zn), 150 mg; copper sulfate ($CuSO_4.5H_2O$, 25% Cu), 28 mg; potassium iodide (KI, 24% K, 76% I).

[b]L-Ascorbic acid monophosphate (35%).

[c]Total carbohydrate = 100 - (Crude protein + Crude lipid + Ash).

[d]Gross energy calculated using gross calorific values of 0.2363, 0.3952 and 0.1715 MJ/g for protein, lipid and carbohydrate, respectively according to (Brett, 1973).

Feeding Behaviour

The observation of fish behaviour was conducted for 6 days from June 20th until July 21st to study the effects of experimental diets on fish behavior according to Hussein et al. [27] to study the effects of experimental diets on fish behaviour. The types of fish behavior included were mouth wrestling and chasing. Fish were fed manually their daily amount of the experimental diets four times a day at 10:00, 12:00, 14:00 and 16:00. Ten minutes after the third meal (14:00) all aquarium were observed individually and each aquarium was given 3 min to quantify the behavioural interactions (i.e., as mouth wrestling and chasing separately) between fish.

Sample Collection and Analysis

At the end of the trial a random sample of five individual fish were sampled from each aquarium, then oven-dried 105°C for 24 h, ground, and stored at −20°C for subsequent analysis. Proximate analysis was conducted on diets and fish samples. Dry matter, crude lipids, crude protein and ash content were determined by the standard [28]. Dry matter was determined after drying the samples in an oven (105°C) for 3 h. Crude protein was determined by micro-Kjeldhal method, N × 6.25 (using Kjeltech auto analyzer, Model 1030, Tecator, Höganäs, Sweden) [28] (method number 984.13), crude lipid by soxhlet extraction with diethyl ether (40-60°C) [28] (method number 920.39) and ash by incineration at 550°C for 12 h [28] (method number 942.05). Nitrogenfree extract was computed by taking the sum of values for crude protein, crude lipid and ash then subtracting this sum from 100.

At the end of the feeding trial all fish were counted and weighed to calculate final body weight (FBW), weight gain (WG), specific growth rate (SGR%), feed conversion ratio (FCR) and protein efficiency ratio (PER) using the following equations:

WG (g/ fish) = FBW - IBW; SGR% = [ln FBW - ln IBW]/ t×100, where IBW is initial body weight (g); ln = natural logarithmic; t = time in days. FCR = FI /WG, where FI is feed intake (g); PER=WG/protein intake (g).

Blood Samples, Hematological and Biochemical Blood Parameters

At the end of experimental period, fish were fasted for 24 h prior to blood sampling. Fish were anaesthetized by tricaine methanesulfonate (MS222) at 250 mg L^{-1} in water. Blood was drawn near caudal peduncle from five fish from each treatment by a sterile syringe and transferred into a heparinized tube for hematological study. Blood was drawn from another five fish from each treatment kept outside without anticoagulant at room temperature (26°C) for few minutes then centrifuged at 5,000 g for 15 min for separation of serum (collected serum was stored at -20°C) for biochemical study. Hematocrit (Hct) was determined as described by Reitman and Frankel [29]. Hb was determined by the hemoglobin kits which is a standardized procedure of the cyanomet hemoglobin method. RBCs were counted under the light microscope using a Neubauer haemocytometer after blood dilution with phosphate-buffered saline (pH 7.2). Serum AST and ALT were determined colorimetrically using spectrophotometer using specific kits according to Reitman and Frankel [29]. Total serum protein and albumin and globulin were determined according to Doumas et al. [30].

Statistical Analysis

Data were statistically analyzed by ANOVA using SAS ANOVA procedure (Statistical Analysis System 2004). The data were submitted to one ways classification variance analysis. Duncan's multiple range test was used to compare differences between treatment means when significant F values were observed [31], at ($P < 0.05$) level. All percentage data were arc-sin transformed prior to analysis [32], however data are presented untransformed to facilitate comparisons.

RESULTS

Essential Oil Composition

The chromatographic analysis of hydro distilled oil revealed the presence of fifteen major compounds in EFO, representing 87.24% of the total fennel oil. The compounds present in the fennel oil at concentrations higher than 2% are: estragole (47.94); limonene (20.64); fenchone (7.22); trans-Anethole (4.9) α-Pinene (3.61) and eucalyptol (2.93). The chromatographic analysis of hydro distilled oil represented thirteen different compounds in EGO, representing 83.69% of the total garlic oil. The compounds present in the garlic oil at concentrations higher than 2% are: diallyl trisulfide (33.33%); diallyl disulfide (20.61%); methyl allyl trisulfide (14.11%); propenyl dithiopropanoate (10.21%); dimethyl trisulfide (3.20%) and diallyl tetrasulfide (2.23%). Diallyl tetrasulfide, which is generated from allicin is the main sulfurous constitute.

Growth and Feed Utilization

Survival at the end of the experiment was high (about 95%) and the effect of dietary treatments was significant (Figure 1). No significant differences in survival of fish ($P>0.05$) were observed among fish groups fed diet (D2, D3 and D4). The survival of fish fed diet D6 (EGO + *B. licheniformis*) was significantly higher that of other treatments. The differences in body weight were first observed after the 4 weeks of the feeding and increased as the feeding period continued (Figure 2). Moreover, a significant reduction in the final body weight of fish control was observed compared with fish fed other diets. Results of growth and feed utilization are presented in (Table 2) and indicated that the highest weight gain (WG) and specific growth rate (SGR) were recorded by fish fed D5 (EFO + *B. licheniformis*) and D6 (EGO + *B. licheniformis*). Furthermore, no significant differences were found in WG and SGR among D2 (*B. licheniformis*), D3 (EFO) and D4 (EGO). The highest FI was observed in fish fed D4 (EGO), D5 (EFO + *B. licheniformis*) and D6 (EGO + *B. licheniformis*). While, the best FCR and PER were observed in D3 (EFO), D4 (EGO), D5 (EFO + *B. licheniformis*) and D6 (EGO + *B. licheniformis*) in comparison to the other treatments groups.

Table 2: Fish performance and feed utilization of Nile tilapia fry fed diets containing *B. licheniformis*, essential oil of fennel and garlic and their combined for 84 days

Parameters	Experimental diets						P value
	D1	D2	D3	D4	D5	D6	
Initial body weight (g)	1.78±0.12	1.90±0.16	1.90±0.11	1.92±0.09	1.93±0.13	1.82±0.14	0.827
Weight gain (g)	12.37±1.55^{c}	17.66±1.23^{b}	18.86±1.43^{b}	18.80±1.52^{b}	20.02±1.47^{a}	19.58±1.61^{a}	0.010
Specific growth rate	2.34±0.04^{c}	2.59 ± 0.10^{b}	2.63 ± 0.03^{b}	2.64 ± 0.43^{b}	2.70 ± 0.05^{a}	2.70 ± 0.062^{a}	0.012
Feed intake(g feed/ fish)	20.69 ± 1.21^{c}	24.70 ± 1.25^{b}	24.64 ± 1.11^{b}	25.48 ± 1.40^{a}	25.95 ± 1.26^{a}	25.09 ± 1.32^{a}	0.001
Feed conversion ratio	1.67 ± 0.66^{a}	1.42 ± 0.68^{b}	1.32 ± 0.14^{c}	1.36 ± 0.14^{c}	1.30 ± 0.58^{c}	1.28 ± 0.66^{c}	0.001
Protein efficiency ratio	2.08 ± 0.11^{c}	2.38 ± 0.18^{b}	2.53 ± 0.14^{a}	2.47 ± 0.17^{a}	2.57 ± 0.13^{a}	2.60 ± 0.21^{a}	0.012

Results are presented as means ± SD of triplicate observations. Means in the same row with different superscripts were significantly different ($p < 0.05$). D1= Control; D2= *B. licheniformis*; D3 = Essential fennel oil (EFO); D4 = Essential garlic oil (EGO); D5 = (EFO + *B. licheniformis*); D6= (EGO + *B. licheniformis*).

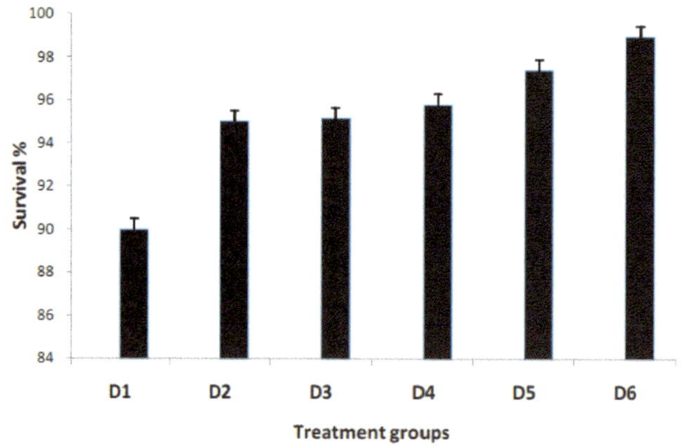

Figure 1: Survival of Nile tilapia fry fed diets containing *B. licheniformis*, fennel oil, garlic oiland their combinedfor 84 days. [D1= Control; D2 = *B. licheniformis*; D3 = Essential fennel oil (EFO); D4 = Essential garlic oil (EGO); D5 = (EFO +*B. licheniformis*); D6= (EGO + *B. licheniformis*)].

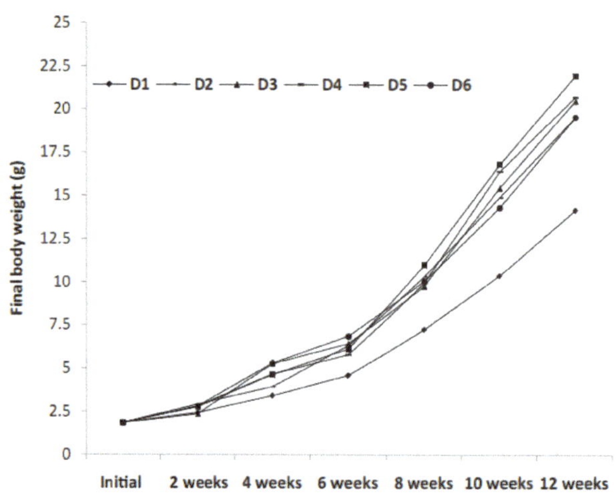

Figure 2: Final body weight of Nile tilapia fry fed diets containing *B. licheniformis*, fennel oil, garlic oiland their combinedfor 84 days. [D1= Control; D2 = *B. licheniformis*; D3 = Essential fennel oil (EFO); D4 = Essential garlic oil (EGO); D5 = (EFO +*B. licheniformis*); D6= (EGO + *B. licheniformis*)].

Feeding Behavior

The behavioural observation were recorded at week 7 (Figure 3) and the results showed significant differences among all treatments in terms of mouth wrestling and chasing behaviour. It showed that mouth wrestling and chasing behaviour were higher in fish feed D6 (EGO + *B. licheniformis*) resulting in more aggressive for pellets feed than all treatments diets. The opposite trend were observed in D1 (control diet).

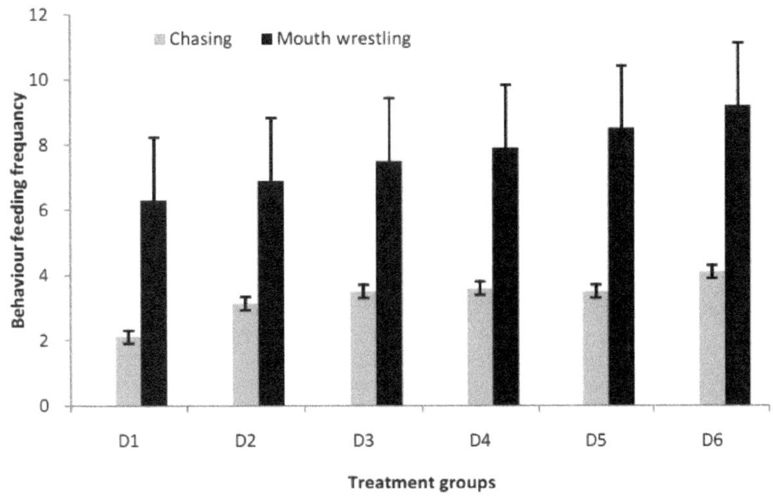

Figure 3: Behaviour feeding Final body weight of Nile tilapia fry fed diets containing *B. licheniformis*,fennel oil, garlic oiland their combined-for 84 days. [D1= Control; D2 = *B. licheniformis*; D3 = Essential fennel oil (EFO); D4 = Essential garlic oil (EGO); D5 = (EFO +*B. licheniformis*); D6= (EGO + *B. licheniformis*)].

Hematological Analysis and Biochemical Blood

The changes in hematological parameters of Nile tilapia fed treated diets are presented in Tables 3 and 4. The highest hematocrit (Hct), hemoglobin (Hb) and red blood cells (RBCs) values were recorded by fish of D5 (EFO + *B. licheniformis*) and D6 (EGO + *B. licheniformis*). No significant differences were found in Hct, Hb and RBCs among D1 (control diet), D2 (*B. licheniformis*), D3 (EFO) and D4 (EGO) treatments groups. Aspartate aminotransferase (AST) and alanine aminotransferase (ALT) were significantly lower ($P<0.05$) in fish D5 (EFO + *B. licheniformis*) and D6 (EGO + *B. licheniformis*) in comparison to the other treatments, but, no significant differences were found in ALT and AST among D1 (control diet), D2 (*B. licheniformis*), D3 (fennel oil) and D4 (EGO) treatments groups. The results showed the serum total protein and serum globulin was significantly higher ($P<0.05$) in fish fed D5 (EFO + *B. licheniformis*) and D6 (EGO + *B. licheniformis*). There was no significant ($P>0.05$) effect of different treated diets on serum albumin.

Table 3: Hematological indices of Nile tilapia fry fed diets containing B. licheniformis, essential oil of fennel and garlic and their combined for 84 days

Parameters	Experimental diets						P value
	D1	D2	D3	D4	D5	D6	
Hematocrit (%)	17.90 ± 0.52^b	17.45 ± 0.63^b	17.50 ± 0.75^b	17.05 ± 0.28^b	21.68 ± 0.85^a	23.00 ± 0.74^a	0.012
Hemoglobin(g/dl)	10.80 ± 0.24^b	10.90 ± 0.23^b	10.25 ± 0.41^b	11.00 ± .035^b	12.05 ± 0.35^a	13.80 ± 0.25^a	0.041
Red blood cells ($\times 10^6$ µL-1)	1.93 ± 0.02^b	1.91 ± 0.01^b	2.12 ± 0.05^b	2.12 ± 0.04^b	2.82 ± 0.02^a	2.78 ± 0.06^a	0.001

Results are presented as means ± SD of triplicate observations. Means in the same row with different superscripts were significantly different ($p < 0.05$). D1= Control; D2 = *B. licheniformis*; D3 = Essential fennel oil (EFO); D4 = Essential garlic oil (EGO); D5 = (EFO + *B. licheniformis*); D6= (EGO + *B. licheniformis*).

Table 4: Biochemical blood indices of Nile tilapia fry fed diets containing *B. licheniformis*, essential oil of fennel and garlic and their combined for 84 days

Parameters	Experimental diets						P value
	D1	D2	D3	D4	D5	D6	
Aspertataminotransferase (ul^{-1})	11.80 ± 0.47[a]	11.90 ± 0.65[a]	10.95 ± 0.68[a]	11.20 ± 0.23[a]	9.15 ± 0.39[b]	9.40 ± 0.62[b]	0.014
Alanine aminotransferase (ul^{-1})	10.35 ± 0.36[a]	10.05 ± 0.45[a]	9.75 ± 0.13[a]	9.90 ± 0.75[a]	8.55 ± 0.63[b]	8.25 ± 0.58[b]	0.001
Total protein (g dL^{-1})	2.70 ± 0.12[c]	2.90 ± 0.14[c]	3.00 ± 0.16[b]	3.10 ± 0.15[b]	3.69 ± 0.11[a]	3.67 ± 0.15[a]	0.042
Albumin (g dL^{-1})	0.95 ± 0.02	0.95 ± 0.01	0.99 ± 0.01	1.00 ± 0.04	1.05 ± 0.01	1.01 ± 0.2	0.562
Globulin (g dL^{-1})	1.75 ± 0.13[c]	1.95 ± 0.17[b]	2.01 ± 0.12[b]	2.10 ± 0.13[b]	2.64 ± 0.11[a]	2.66 ± 0.12[a]	0.013

Results are presented as means ± SD of triplicate observations. Means in the same row with different superscripts were significantly different ($p < 0.05$).D1= Control; D2 = *B. licheniformis*; D3 = Essential fennel oil (EFO); D4 = Essential garlic oil (EGO); D5 = (EFO + *B. licheniformis*); D6= (EGO + *B. licheniformis*).

Chemical Composition

Proximate analysis of whole body of *O. niloticus* fed the experimental diets was presented in Table 5. The chemical composition of whole body appeared that no significant differences were found in dry matter and ash content of fish fed control and treated diets.

Table 5: Chemical composition of Nile tilapia fry fed diets containing *B. licheniformis*, essential oil of fennel garlic and their combined for 84 days (g/kg wet basis)

Parameters	Experimental diets						P value
	D1	D2	D3	D4	D5	D6	
Dry matter	275.30 ± 1.23	277.20 ± 1.45	274.50 ± 2.32	275.01 ± 1.54	275.00 ± 1.25	278.52 ± 1.58	0.620
Crude protein	149.20 ± 1.23	152.13 ± 1.51	151.20 ± 1.45	152.35 ± 1.55	153.89 ± 1.63	153.69 ± 1.26	0.452
Crude lipid	64.23 ± 1.10	62.21 ± 0.98	61.13 ± 1.24	61.16 ± 1.44	60.73 ± 1.28	61.12 ± 1.35	0.521
Ash	33.12 ± 1.01	33.99 ± 1.20	34.13 ± 1.21	33.12 ± 1.00	32.95 ± 0.99	32.80 ± 1.01	0.232

Results are presented as means ± SD of triplicate observations. Means in the same row with different superscripts were significantly different ($p < 0.05$).D1= Control; D2 = *B. licheniformis*; D3 = Essential fennel oil (EFO); D4 = Essential garlic oil (EGO); D5 = (EFO + *B. licheniformis*); D6= (EGO + *B. licheniformis*).

DISCUSSION

The present study indicated that the highest growth and feed utilization were obtained at 1ml of EFO and EGO combined with *B. licheniformis*7 × 10^7 (CFU g^{-1}). This study of the application of EFO and EGO in diet showed the first report about the growth indices of Nile tilapia. Whereas, fennel seed as a powder was used as the growth promoter for Nile tilapia (*O. niloticus*) and the results are similar with the present study [14,33]. Besides, most of the authors have studied the effect of garlic as a powder form on growth and nutrient utilization [17,26,34,35]. The growth rates for the previous studies were significantly higher in fish fed diets supplemented with garlic than control groups. Garlic contains allicin, which promotes biogenic performance due to its positive effect on the intestinal flora, thereby improving digestion and enhancing nutrients utilization which influences the growth of fish [36]. Steam distillation of mashed garlic produces garlic oil containing methyl and allyl-sulphides of allicin and having the practical advantage of being more stable than allicin itself [37]. Due to the fact that the extracted essential oil from herbal plants has much higher contents of active substance than dried herbal plants35 resulting in the essential oil is usually used in low percentages in the diet compared to dried herbal plants.

Little information has been reported the effect of application of different essential oil in fish growth. These studies indicated that diets supplemented with essential oil of weet orange peel (*Citrus sinensis*) and *Aloysia triphylla* ([19,38] with different dose could enhance weight gain and specific growth rate. Supplementation of the essential oil of fennel or garlic in diet may improve digestibility, availability of nutrients and lead to a higher protein synthesis, which in turn could explain the better growth in this study. Meanwhile, stimulation of growth by *B. licheniformis* in the present study may attribute to improve digestibility by enhancing the synthesis of vitamins, enzymatic activity [39]. These results are in agreement with *Bacillus subtilis* [40]; carp [41], Tilapia O. nilotica.

On the other hand, the increase in growth for the combined application of EFO and/or EGO with *B. licheniformis* may be due to the EFO and EGO promoted the growth of exogenous probionts bacteria *B. licheniformis* in gastrointestinal of fish more than other endogenous populations bacteria, which improved the growth performance of Nile tilapia. The control of the microflora in intestinal fish by addition of oregano essential oil (*Origanum heracleoticum* L) can positively effect on growth performance [18]. The main way of the essential oil work seems to be through the regulation of intestinal microflora [42]. From aforementioned, the essential oil of fennel and garlic are good potential alternatives to the growth promoters.

The feeding behaviour of fish is complex [43], however, the acceptance or rejection of diet is physiologically dependent on inputs from chemoreception [44]. Fish have several chemosensory systems including gustation (taste), olfaction (smell), chemical sensory and chemoreceptor cells. The role of herbal plants for feeding behaviour represented in the flavor, thereby influences of eating patterns, the secretion of digestive fluids and total feed intake [45]. It seems that fish fed EGO combined with *B. licheniformis* could be stimulated the gustatory system is the most important in acquisition and ingestion of food. Allicin has an intense garlic flavor with a strong effect on olfaction in the most aquatic animals such as Pelodiscus sinensis, Ctenopharyngodon idellus, *Cyprinus carpio*, *Carassius auratus*, and *Oreochromis niloticus* [45]. Harada [46] indicated that garlic had a strong food calling effecting on loach (*Oriental weatherfish*) and Japanese amberjack,*Seriola quinqueradiata* Temminck et Schlegel.

Hematological variables are good predictors for explaining the health status of fish [47]. The highest Hct, Hb and RBCs count were obtained at D5 (EFO + *B. licheniformis*) and D6 (EGO + *Bacillus licheniformis*). Erythrocytes content in fish blood gives a guide to the health status of fish and can be helpful to determine any abnormalities arising from the use of immunostimulants. Accordingly, the elevate number of RBCs multiplies the concentration of hemoglobin ultimately resulting in a high capacity for oxygen carrying which improved the health of fish and consequently enhancing growth [48]. Furthermore, Fazllolahzadeh et al. [37] assumed that the increase of blood indices such as, Hct, Hb, and RBCs may be attributable to a defense reaction against garlic which occurs by stimulation of erythropoiesis. The number of RBCs was significantly ($P<0.05$) higher in Asian sea bass, *Lates calcarifer* [49], *Labeo rohita* [50] and Nile tilapia, *O. niloticus* fed diet supplemented with garlic as a powder form. The present study is inconsistent with Soltan and El-Laithy [51]. They reported that fennel seeds and garlic powder did not effect on Hct and Hb values for Nile tilapia, *O. niloticus*. Yılmaz and Ergün [52] reported that garlic oil did not produce any undesired effects on hematological characteristics for Sea Bass *Dicentrarchus labrax*. With regard to the different essential oil addition Acar et al. [19] showed that *O. mossambicus* fed a basal diet containing 1% thyme, rosemary or fenugreek for 45-day increased RBCs count and Hct value.

Alanine aminotransferase (ALT) and aspartate aminotransferase (AST) are liver enzymes and they have the function of transferring the amino group from alpha-amino acids to alpha-keto acids. A large amount of ALT and AST is released in the blood mostly during liver cell damage [53]. Thus, detection of the serum level of ALT and AST allows monitoring liver cell damage. The diets containing (EFO + *B. licheniformis*) and (EGO + *B. licheniformis*) could decrease the activity of ALT and AST compared to other treatment groups. Similar results were obtained by El-Dakar [33] who showed a significant lower ($P<0.05$) in levels of ALT and AST in Nile tilapia, *O. niloticus* fed fennel seed meal supplemented with diet. Also, Soltan and El-Laithy [51] found that ALT and AST levels were significantly decreased when Nile tilapia fed diets supplemented with 1% garlic powder or 1% fennel seed with compared to other diets. Application of garlic (*Allium sativum*) may cause stabilization cell membrane and protect the liver against deleterious agents and free radical-mediated toxic. This is reflected in the reduction of liver enzymes [37]. Also, Metwally [54] indicated that garlic-addition helps the liver to maintain its normal function by accelerating the generative capacity of its cells. The different active components found in essential oil of fennel and garlic may play an important role in enhancing the activity of liver enzymes but the exact mechanism for this need more advanced study. Globulin fractions are certainly important for maintaining a healthy immune system and immune function in the blood [55]. The increase in total protein and globulin contents is thought to reflect strong innate immunity [17]. In this study, diets supplemented with (EFO + *B. licheniformis*) and (EGO + *B. licheniformis*) showed a significant elevate in total protein and globulin in Nile tilapia. It can be inferred that the EFO and EGO which combined with *B. licheniformis* may enhance the non-specific immune response and more tolerant of stressful conditions of Nile tilapia. These results are similar with [20] who reported that the highest content of total protein was determined in fish fed diet supplemented with mixed oil of (thyme and fennel) with level 100 ml kg^{-1} diet, but serum albumin was not significantly different from the control group. The addition of garlic products, fresh oil and powdered forms increased the total protein in Nile tilapia *O. niloticus*, even at low level 0.25% of garlic oil [54]. Also, Shalaby et al. [16] reported that the highest total protein was determined in Nile tilapia *O. niloticus* when garlic powder was supplemented to the diet at 10 g kg^{-1} diet. Serum total protein and albumin contents decreased in sea bass exposed to garlic oil (0.005 ml L^{-1} water) via bath immersion for 96 h, but serum globulin was decreased at (0.02 ml of garlic oil L^{-1}) [52].

The present study indicated that no significant differences in chemical composition of body of fish fed different treated diets. In this regard, Abd El Hakim [14] found no significant differences in the chemical body composition of Nile tilapia fed diet supplemented with different level of fennel seed. Contrarily, Maniat et al. [56] showed that garlic supplemented in diets induce higher protein and lower fat content in body of benni fish (*Mesopotamichthys sharpeyi*). Ahmad and Abdel-Tawwab [13] reported that no significant differences (P>0.05) in moisture and crude protein in Nile tilapia, *Oreochromis niloticus* body meanwhile total lipids increased by increasing level of caraway seed (belonging to family Apiaceae as fennel). Shalaby et al. [16] reported the highest crude protein and the lowest crude lipid content were obtained by Nile tilapia fed diet supplemented with 30 g kg^{-1} garlic. Furthermore, Talpur and Ikhwanuddin [49] showed that sea bass fed diet supplemented with garlic was decreased in lipid content in body. Also, Metwally [54] showed that crude protein and ash was increased significantly in *O. niloticus* while crude lipid decreased significantly with diets containing different sources of *Allium sativum*.

In conclusion, addition of essential oil of fennel combined with *B. licheniformis* and/or essential oil of garlic combined with *B. licheniformis*resulting in increased growth performance and enhancement of hematological and biochemical blood parameters in Nile tilapia *O. niloticus*.

ACKNOWLEDGEMENTS

The authors are grateful to Animal production (Fish Nutrition Lab) staffs for their help in data collection and analysis. We thank National Institute of Oceanography and Fisheries (NIOF), Egypt, for their help in hematological study. The authors confirm that there is no conflict of interest.

REFERENCES

1. Rico A, Phu TM, Satapornvanit K, Min J, Shahabuddin AM, et al. (2013) Use of veterinary medicines, feed additives and probiotics in four major internationally traded aquaculture species farmed in Asia. Aquaculture 412: 231-243.
2. Reverter M, Bontemps N, Lecchini D, Banaigs B, Sasal P (2014) Use of plant extracts in fish aquaculture as an alternative to chemotherapy: current status and future perspectives. Aquaculture 433: 50-61.
3. Christaki E, Bonos E, Giannenas I, Florou-Paneri P (2012) Aromatic plants as a source of bioactive compounds. Agriculture 2: 228-243.
4. Kadri A,Zied Z, Ines BC, Ahmed B, Néji G, et al. (2011) Chemical constituents and antioxidant properties of Rosmarinusofficinalis L. essential oil cultivated from South-Western Tunisia. J Med Plants Res 5: 5999-6004.
5. Botsoglou N, Christaki E, Florou-Paneri P, Giannenas I, Papageorgiou G, et al. (2004) The effect of a mixture of herbal essential oils or á-tocopheryl acetate on performance parameters and oxidation of body lipid in broilers. South African Journal of Animal Science 34: 52-61.
6. Platel K,Srinivasan K (2004) Digestive stimulant action of spices: A myth or reality? Indian Journal of Medical Research119: 167-179.
7. Özbek H, Uğraş S, Dülger H, Bayram I, Tuncer I, et al. (2003) Hepatoprotective effect of Foeniculumvulgare essential oil. Fitoterapia 74: 317-319.
8. Choi EM, Hwang JK (2004)Antiinflammatory, analgesic and antioxidant activities of the fruit of Foeniculumvulgare. Fitoterapia 75: 557-565.
9. Özcan, U, Yilmaz E, Ozcan L, Furuhashi M, Vaillancourt E,et al. (2006) Chemical chaperones reduce ER stress and restore glucose homeostasis in a mouse model of type 2 diabetes. Science 313: 1137-1140.
10. Lanzotti V (2006)The analysis of onion and garlic. Journal of Chromatography A1112: 3-22.
11. Harris J, Cottrell SL, Plummer S, Lloyd D (2001) Antimicrobial properties of Allium sativum (garlic). Applied Microbiology and Biotechnology 57: 282-286.
12. Helou L, Harris IM (2007)Herbal products; toxicology and clinical pharmacology (2ndedn). Totowa, Human Press, New Jersey.
13. Ahmad MH, Abdel-Tawwab M (2011)The use of caraway seed meal as a feed additive in fish diets: Growth performance, feed utilization, and whole-body composition of Nile tilapia, Oreochromisniloticus (L.) fingerlings. Aquaculture 314: 110-114.
14. Abd El Hakim N (2010) Response of Nile tilapia, Oreochromisniloticus to diets supplemented with different levels of fennel seeds meal (Foeniculumvulgare). Abbassa International Journal of Aquaculture3: 215-230.
15. Yılmaz S, Ergün S, Soytaş N (2013) Dietary supplementation of cumin (Cuminumcyminum) preventing streptococcal disease during first-feeding of Mozambique tilapia (Oreochromismossambicus). J Bio Sci Biotech 2: 117-124.
16. Shalaby A, Khattab Y, Abdel RA (2006) Effects of garlic (Alliumsativum) and chloramphenicol on growth performance, physiological parameters and survival of Nile Tilapia (Oreochromisniloticus). JVenomAnimToxins incl Trop Dis 12: 172-201.
17. Nya EJ, Austin B (2009) Use of garlic, Allium sativum, to control Aeromonashydrophila infection in rainbow trout, Oncorhynchusmykiss (Walbaum). J Fish Dis 32: 963-970.
18. Zheng Z, Tan YWJ, Liu HY, Zhou XH, Xiang X, et al. (2009) Evaluation of oregano essential oil (Origanumheracleoticum L.) on growth, antioxidant effect and resistance against Aeromonashydrophila in channel catfish (Ictaluruspunctatus). Aquaculture 292: 214-218.
19. Acar Ü, Kesbic OS, Yilmaz S, Gultepe N, Turker A (20115) Evaluation of the effects of essential oil extracted from sweet orange peel (Citrus sinensis) on growth rate of tilapia (Oreochromismossambicus) and possible disease resistance against Streptococcus iniae. Aquaculture 437: 282-286.
20. Gulec AK, Danabas D, Ural M, Seker E, Arsian A, et al. (2013) Effect of mixed use of thyme and fennel oils on biochemical properties and electrolytes in rainbow trout as a response to Yersinia ruckeri infection. ActaVeterinaria Brno 82: 297-302.
21. Greathead H (2003) Plants and plant extracts for improving animal productivity. ProcNutrSoc 62: 279-290.

22. NRC (National Research Council) (1993) Nutrient requirements of fish. Committee onAnimal Nutrition. Board on Agriculture in National Research Council National AcademyPress, Washington DC, USA.

23. Boyd C (1979) Water quality in warmwater fish ponds. Auburn University Agricultural Experiment Station. Auburn, AL.

24. Aazza S, Lyoussi B, Miguel MG (2011) Antioxidant and antiacetylcholinesterase activities of some commercial essential oils and their major compounds. Molecules 16: 7672-7690

25. Douiri L, Boughdad A, Assobhei O, Moumni M (2013) Chemical composition and biological activity of Allium sativum essential oils against Callosobruchusmaculatus. J EnvSciToxicol Food Tech 3: 30-36.

26. Shelby RA, Lim C, Mediha Y, Mary AD (2006) Effects of probiotic diet supplements on disease resistance and immune response of young Nile tilapia, Oreochromisniloticus. Journal of Applied Aquaculture 18: 23-34.

27. Hussein EES, Konrad D, Deyab MSDE, Bong-Joo L (2013) Enhancing the growth of Nile tilapia larvae/ juveniles by replacing plant (gluten) protein with algae protein. Aquaculture Research 44: 937-949.

28. AOAC (1995) Official Methodsof Analysisofthe Association Official Analytical Chemists (16thedn). AOACInternational, Arlington, USA.

29. Reitman S, Frankel S (1957) A colorimetric method for the determination of serum glutamic oxalo-acetic and glutamic pyruvic transaminases. American journal of clinical pathology 28: 56-63.

30. Doumas BT, Bayse DD, Carter RJ, Peters T, Schaffer R (1981) A candidate reference method for determination of total protein in serum. I. Development and validation. Clinical Chemistry 27: 1642-1650.

31. Duncan DB (1955) Multiple range and multiple F tests. Biometrics 11:1-42.

32. Zar J (1984) Multiple comparisons. Biostatistical analysis 1: 185-205.

33. El-Dakar AY (2004) Use of medical and aromatic plants in fish diets: 2. Effect of dried basil leaves on performance of hybrid tilapia Oreochromisniloticus× O. aureus, Fingerlings., In: 3rd International Conference on Animal Production and Health in semi-Arid Areas, Suez Canal Univ.

34. Diab AS, Salah MA, George J, Yasser A, Mohamed FM (2008) Effect of garlic, black seed and Biogen as immunostimulants on the growth and survival of Nile tilapia, Oreochromisniloticus (Teleostei: Cichlidae), and their response to artificial infection with Pseudomonas fluorescens. African Journal of Aquatic Science33: 63-68.

35. Amin F, Milad K,Mohammad S,Mohsen SI, Morteza D,et al. (2010) Effect of garlic (Allium sativum) on growth factors, some hematological parameters and body compositions in rainbow trout (Oncorhynchusmykiss) 3: 317-323.

36. Khalil R, Nadia B, Soliman M (2001) Effects of Biogen and LevamisolHcl on the immune response of cultured Oreochromisniloticus to Aeromonashydrophila vaccine. Beni-Suef Vet Med J Egypt 11: 381-392.

37. Fazllolahzadeh F, Keramati K, Saeed N, Shirian S, Samira S (2011) Effect of garlic (Allium sativum) on hematological parameters and plasma activities of ALT and AST of rainbow trout in temperature stress. Australian Journal of Basic and Applied Sciences 5: 84-90.

38. Zeppenfeld C, Hernández DR, Santinón JJ, Heinzmann BM, Schmidt D, et al. (2015) Essential oil of Aloysiatriphylla as feed additive promotes growth of silver catfish (Rhamdiaquelen). Aquaculture Nutrition.

39. Gatesoupe F (1999)The use of probiotics in aquaculture. Aquaculture 180: 147-165.

40. Zhang Q, Ma H, Mai K, Zhang W, Liufu Z, et al. (2010) Interaction of dietary Bacillus subtilis and fructooligosaccharide on the growth performance, non-specific immunity of sea cucumber, Apostichopusjaponicus. Fish Shellfish Immunol 29: 204-211.

41. Aly SM, Abdel-Galil Ahmed Y, Abdel-Aziz Ghareeb A, Mohamed MF (2008) Studies on Bacillus subtilis and Lactobacillus acidophilus, as potential probiotics, on the immune response and resistance of Tilapia nilotica (Oreochromisniloticus) to challenge infections. Fish Shellfish Immunol 25: 128-136.

42. Hernández A, Benjamin GG, Maria JC, Maria DH (2015)The inclusion of thyme essential oil in the feed of gilthead seabream (Sparusaurata) promotes changes in the frequency of lymphocyte aggregates in gut-associated lymphoid tissue. Aquaculture Research.

43. Houlihan D, Boujard T, Jobling M (2008) Food intake in fish, John Wiley & Sons.

44. Hara T (1994) Olfaction and gustation in fish: an overview. ActaPhysiolScand 152: 207-217.

45. Lee JY,Gao Y (2012) Review of the application of garlic, Allium sativum, in aquaculture. Journal of the World Aquaculture Society 43: 447-458.

46. Harada K (1990) Attraction activities of spices for oriental weatherfish and yellowtail. Nippon Suisan Gakkaishi 56: 2029-2033.

47. Hrubec TC, Cardinale JL, Smith SA (2000) Hematology and plasma chemistry reference intervals for cultured tilapia (Oreochromis hybrid). Vet ClinPathol 29: 7-12.

48. Hassaan M, Soltan M, Ghonemy M (2014) Effect of synbiotics between Bacillus licheniformis and yeast extract on growth, hematological and biochemical indices of the Nile tilapia (Oreochromisniloticus). The Egyptian Journal of Aquatic Research 40: 199-208.

49. Talpur AD,Ikhwanuddin M (2012) Dietary effects of garlic (Allium sativum) on haemato-immunological parameters, survival, growth, and disease resistance against Vibrio harveyi infection in Asian sea bass, Latescalcarifer (Bloch). Aquaculture 364: 6-12.

50. Sahu S, Das BK, Mishra BK, Pradhan J, Sarangi N (2007) Effect of Allium sativum on the immunity and survival of Labeorohita infected with Aeromonashydrophila.

Journal of Applied Ichthyology 23: 80-86.

51. Soltan MA, El-Laithy S (2008) Effect of probiotics and some spices as feed additives on the performance and behaviour of the Nile tilapia, Oreochromisniloticus. J AquatBiolFish 12: 63-80.

52. Yılmaz S,Ergün S (2012)Effects of garlic and ginger oils on hematological and biochemical variables of Sea Bass Dicentrarchuslabrax. JAquatAnimHealth 24: 219-224.

53. Hassaan MS, Goda AM, Soaad AM, Tayel SI (2014)Protective effect of dietary vitamin E against fungicide copperoxychloride stress on Nile tilapia, Oreochromisniloticus (L.), Fingerlings.

54. Metwally M (2009)Effects of garlic (Allium sativum) on some antioxidant activities in tilapia nilotica (Oreochromisniloticus). World Journal of Fish and Marine Sciences 1: 56-64.

55. Jha AK, Pal AK, Sahu NP, Kumar S, Mukherjee SC et al. (2007)Haemato-immunological responses to dietary yeast RNA, Omega-3 fatty acid and β-carotene in Catlacatla juveniles. Fish Shellfish Immunol 23: 917-927.

56. Maniat M, Ghotbeddin N, Ghatrami ER (2014) Effect of garlic on growth performance and body composition of benni fish (Mesopotamichthyssharpeyi). International Journal of Biosciences (IJB) 5: 269-277.

CHAPTER 11

PROBIOTICS AS ANTIVIRAL AGENTS IN SHRIMP AQUACULTURE

Bestha Lakshmi, Buddolla Viswanath, and D. V. R. Sai Gopal

INTRODUCTION

Aquaculture is a worldwide activity and considered as a major economic and food production sector as it is an increasingly important source of protein available for human consumption. According to FAO, the supplies of fish, crustaceans, and molluscs from aquaculture increased from 3.9% of total production by weight in 1970 to 27.3% in 2000, and aquaculture is growing more rapidly than all other animal-food-producing sectors [1]. In Europe itself, it is estimated that aquaculture production will exceed 2.5 million tonnes by 2015 and reach 4 million tonnes by 2030 [1]. Aquaculture production is heavily dominated by China and other developing countries in the Asia Pacific region, which accounts for 89% by volume [2]. Intensification of aquaculture had led to the promotion of conditions that favor the development of a number of diseases and problems related to biofouling. Shrimp farming is an aquaculture business; that is, it exists in either a marine or freshwater environment, producing shrimp or prawns. Over the past five years, there have been major developments in shrimp farming.

Diseases are primary constraint to the growth of many shrimp species, which are exposed to stressful conditions, adverse environmental conditions. Consequently, wide ranges of chemicals particularly antimicrobial agents are used in shrimp farming to prevent and to treat diseases. The usage of these antimicrobial agents has increased enormously and tonnes of antibiotics are distributed in the biosphere during an antibiotic era of only about 60-year duration [3]. In the United States alone, 18,000 tonnes of antibiotics are produced each year for medical and agriculture purposes; 12,600 tonnes are used for the nontherapeutic treatments of livestock in order to promote growth [3]. In the last 20 years, there has been a fourfold growth in aquaculture industries worldwide [4–6]. This impressive industrial development has been accompanied by some practices that are potentially damaging to human and animal health [4, 7] which include passing of large amount of veterinary drugs into the environment [8, 9]. For instance, aquaculture of shrimp and salmon has been accompanied by an important increase in the use of prophylactic antibiotics in the aquatic environments [10–12]. The emergence of antibiotic-resistance among shrimp pathogens undermines the effectiveness of the prophylactic use of antibiotics in aquaculture [13, 14] and increases the possibilities for passage not only of these antibiotic-resistant bacteria but also of their antibiotic resistance determinants to bacteria of terrestrial animals and human beings, including pathogens. Another problem created by the excessive use of antibiotics in industrial aquaculture is the presence of residual antibiotics in commercialized fish, shrimp, and shellfish products [10, 14–19]. In shrimp farming, the passage and permanent existence of large amounts of antibiotics in the environment of water and sediments also have the potential to affect the presence of the normal flora and plankton in those niches, resulting in shifts in the diversity of the microbiota [20–22].

In this view, the development of nonantibiotic agents for health management in shrimp farming is necessary. Probiotics were known to control pathogens through a variety of mechanisms; hence, they are exploited as an alternative to antibiotic treatment. The review mainly focuses on the benefits of probiotics in shrimp farming especially against viral diseases. It also illustrates different methods to study the antiviral activity of probiotics with a brief explanation of shrimp immune system and its antiviral immune response.

VIRAL DISEASES OF SHRIMP

Shrimp aquaculture is the most valuable marine aquaculture industry. Despite the explosive growth in world production of

Citation: Bestha Lakshmi, Buddolla Viswanath, and D. V. R. Sai Gopal, "Probiotics as Antiviral Agents in Shrimp Aquaculture," Journal of Pathogens, vol. 2013, Article ID 424123, 13 pages, 2013. doi:10.1155/2013/424123.

cultivated shrimp, there have also been staggering, periodic losses due to diseases. Hence, diseases are now considered as one of the critical limiting factors in the shrimp culture. Serious viral disease outbreaks of shrimp challenge the shrimp industry to be better prepared in the view of a broadened knowledge about shrimps and their pathogens so that disease prevention methods could be improved. This requires shifted awareness to biosecurity, that is, possible methods of cultivating shrimp in restricted systems designed to prevent the entry of potential pathogens. The industry also realized that a good number of disease outbreaks originated from careless transboundary movement of contaminated but grossly normal aquaculture stocks. Estimates of economic losses suggest that developing countries in Asia lost at least US$1.4 thousand million due to diseases in 1990 alone. Since then, losses have increased. Reports from China suggest that losses in 1993 one of US$1 thousand million due to shrimp viral disease outbreaks (ADB/NACA, in press). A 1995 estimate suggests that aquatic animal disease and environment-related problems may cause annual losses to aquaculture production in Asian countries of more than US$3 thousand million per year (ADB/NACA, in press). According to a recent World Bank report, global losses due to shrimp disease are around US$3 thousand million, and the Bank recommends the investment of US$275 million in shrimp disease research over the next 15 years [23]. Some of major viral diseases that affect various species of shrimp are shown in the Table 1.

Table 1: Viral diseases of shrimp

Disease	Virus	Abbreviation	Genome	Family	Genus	Species affected
White spot syndrome	White spot syndrome virus	WSSV	dsDNA	Nimaviridae	Whispovirus	All farmed marine (Penaeid) shrimp species
Taura syndrome	Taura syndrome virus	TSV	(+) ssRNA	Dicistroviridae	Cripavirus	*Penaeus vannamei*
Yellow head disease	Yellow head disease virus	YHV	(+) ssRNA	Roniviridae	Okavirus	*P. monodon, P. vannamei, P. stylirostris*
Infectious hypodermal and hematopoietic necrosis	Infectious hypodermal and hematopoietic necrosis virus	IHHNV	ssDNA	Parvoviridae	Brevidensovirus	*P. monodon, P. vannamei, P. stylirostris*
Infectious myonecrosis	Infectious myonecrosis virus	IMNV	dsRNA	Totiviridae	Giardiavirus	*P. vannamei*
White tail disease	Macrobrachium rosenbergii nodavirus	MrNV	(+) ssRNA	Nodaviridae	Related to Alphanodavirus Betanodavirus	*Macrobrachium rosenbergii, P. vannamei*

TRADITIONAL METHODS OF DISEASE CONTROL IN SHRIMP AQUACULTURE

Currently, there are advanced molecular techniques available for detection and control of diseases in shrimp aquaculture. Here are some traditional methods followed from long back, that is, before the molecular techniques are available. Intensive and superintensive culture systems will become more common and will compete well with traditional methods.

Use of Postlarvae (PL)

The postlarvae (PL) were extensively used by the first shrimp cultivation systems. These were collected from the tidal flow or hand collection from nearby geographical areas. Primarily, the stocking densities were low; hence the disease problems were also low, but the production was also relatively low. As the demand for shrimp increased after 1980 the stocking density gradually increased which caused an increase in production volume and relative increase in disease problems. The solution for this is to select the PL, which are labeled as specific pathogen free (SPF). The seed before farming is diagnosed for the presence of disease and certified as pure, and then they are supplied to farmers.

Disease Management

Regarding the disease management, the essential factors to be considered are treatment of wastewater, sludge, disposal of diseased dead shrimps from ponds, and postharvest processes. Usually farmers in Asian countries release wastewater without treatment, that is, direct disposal of dead or diseased shrimp which is not a good practice. The farmer must be sure that the wastewater and effluent are free of pathogens. The direct disposal of the diseased dead shrimp causes the transmission and spread of the disease. Horizontal transmission of white spot syndrome virus (WSSV) through water and feeding of infected shrimps and movement of infected live animals have been known to be a probable route for the spread of the disease [61].

Effluent Management

Different procedures were applied for treatment of wastewater and effluent. Fishes like tilapias and milkfish were reared in the settlement ponds as biological filters, and the discharge water was released into the settlement ponds for some time before releasing into the open water. To reduce the negative impact of effluents, the use of effective microorganism product was suggested where the microorganisms do the recycling of sludge and used it as fertilizer. Usually the farmers dry the pond for a period of one to two months, then plough and turnover the sludge at the pond bottom, and carry out the disinfecting, drying, and flushing methods to ensure that the dark smelly pond bottom is cleaned and made suitable for shrimp cultivation [62]. Recycling of sludge and providing settlement ponds are some of the approaches recommended to mitigate shrimp pond effluents [63].

Phage Therapy

The use of bacteriophage in the control of bacterial population is not a new science, but this application in shrimp hatcheries is recently much focused. Phages are obligate intracellular parasites, which have no intrinsic metabolism and require the metabolic machinery of the host cell to support their reproduction. They subsist on the bacterial cells, lead lytic and lysogenic life cycle, and make the survival of host extremely difficult. They are species specific, self-perpetuating, and genetically flexible in nature. Bacteriophages are highly abundant in the aquatic environment ranging from $10^4 mL^{-1}$ to in excess of $10^8 mL^{-1}$. Numbers are typical 3–10 times greater than the bacterial counts although there is substantial variation between ecosystems. Bacteriophages specific to Vibrio harveyi isolates were isolated from oyster tissue and shrimp hatchery water lysed 70% of the V. harveyiisolates tested. In hatchery trials bacteriophage treatment at 2×10^6 pfu/mL showed a 85% survival of Penaeus monodon compared to the control and antibiotic treated ponds [64]. Good numbers of phages have also been isolated against important bacterial fish pathogens. Thus, these bacteriophages could be used as biocontrol agents in the shrimp hatcheries.

Use of Chemicals

Sodium hypochlorite, EDTA, ortho-toluidine, sodium thiosulfate, iodine-PVP formalin, caustic soda (NaOH) and chlorine liquid, Treflan, and muriatic acid are some of the chemicals that are commonly used in various steps of shrimp cultivation. The pesticides that are frequently used in shrimp farming are organochlorines (endosulfan), organophosphates (azinphosethyl, chlorpyrifos, diazinon, dichlorvos, malathion, monocrotophos, parathion, and trichlorfon), carbamates (carbaryl), and others including paraquat, rotenone, nicotine, copper sulphate, formalin, trifluralin, and butachlor.

Use of Antimicrobial

Oxytetracycline (mixed in feed) is the most commonly used antimicrobial and used in combination with chloramphenicol, oxolinic acid, and formalin. Other antibiotics used in shrimp farming are sulfonamides, fluoroquinolones, nonfluorinated quinolones, tetracyclines, chloramphenicol, gentamicin, trimethoprim, and so forth.

PROBIOTICS IN SHRIMP AQUACULTURE

The use of probiotics in humans is a success. The conventional definition of probiotics is "live microorganisms when added to food help to reconstruct a balanced indigenous microflora in the GIT of host" [65–68]. An expert with the Joint Food and Agriculture Organization (FAO/WHO) stated that probiotics are live microorganisms, which, when consumed in adequate amounts, confer a health benefit for the host [69]. While coming to define the probiotics for aquaculture, some specific factors are to be considered. It can be assumed in aquaculture that the intestinal microbiota does not exist as entity by itself but that there is a constant interaction with the environment and the host functions. Considering all these, Verschuere et al. [70] proposed a modified definition of a probiotic as "a live microbial adjunct which has a beneficial effect on the host by modifying the host-associated or ambient microbial community, by ensuring improved use of the feed or enhancing its nutritional value, by enhancing the host response towards disease, or by improving the quality of its ambient environment." Generally, probiotic strains have been isolated from indigenous and exogenous microbiota of aquatic animals. Gram-negative facultative anaerobic bacteria such as Vibrio and Pseudomonasconstitute the predominant indigenous microbiota of a variety of species of marine shrimp [71].

In contrast to saltwater fish and shrimp, the indigenous microbiota of freshwater species tends to be dominated by members of the genera Aeromonas, Plesiomonas, representatives of the family Enterobacteriaceae, and obligate anaerobic bacteria of the genera Bacteroides, Fusobacterium, and Eubacterium [72]. Lacticacidproducingbacteriaaregenerallysub-dominantinGUT and represented essentially by the genus Carnobacterium [73]. In aquaculture, however, Vibrio spp., Bacillus spp., lactic acid bacteria, and microalgae are mainly utilized as probiotics for growth and survival enhancement and reduction of pathogen [30, 74–79]. The significance of these microbiota is listed in Table 2. They have gained acceptance as being more effective than administering antibiotics or chemical substances. More recently, beneficial microbes for aquaculture have been isolated from seawater, sediment, and gastrointestinal tract of aquatic animals that have the ability to produce substances that inhibit pathogens [80–84].

Table 2: Indigenous microbiota of shrimp and their significance

Name of the organism	Significance	References
Aeromonas media A199	Decrease of mortality and suppression of the pathogen of Pacific oyster larvae when challenged with a pathogenic *Vibrio tubiashii*	[24]
Plesiomonas spps.	Show inhibitory activity against primary pathogens *Pseudomonas aeruginosa, P. fluorescens,* secondary pathogens *Salmonella, Shigella, E. coli, Streptococcus*	[25]
Pseudomonas spp.	Potential antagonistic bacterium against pathogenic vibrios in penaeid shrimp. Produces extracellular antivibrio component	[26]
Enterobacteriaceae	Indicators of hygienic quality of foods and water. Their presence in prawn may be attributed to the feed or animal manure commonly used to fertilize ponds	[27]
Vibrio P62, *Vibrio* P63	Immunostimulatory effect in *Penaeus vannamei*	[28]
V. harveyi	Cause vibriosis in shrimp	[29]
Bacillus strain S11	Better yield and good control of disease and immunity enhancement in *Penaeus monodon*	[30]
Lactic acid bacteria *Lactobacillus plantarum*	Stimulate nonspecific immune response in *Litopenaeus vannamei* when challenged with *V. harveyi*	[31]

Criteria for Selection of Probiotics for Shrimp Aquaculture

For the consideration of a microbial strain as a probiotic and for its application in shrimp farming it should be evaluated in a systematic scientific method. Stepwise procedure for the evaluation of the probiotic potential of a microbial strain and its application in shrimp farming is shown in Figure 1. Once the strain has been successfully evaluated by the above procedure, it can be assured as a potential probiotic strain and can be safely applied in the shrimp aquaculture.

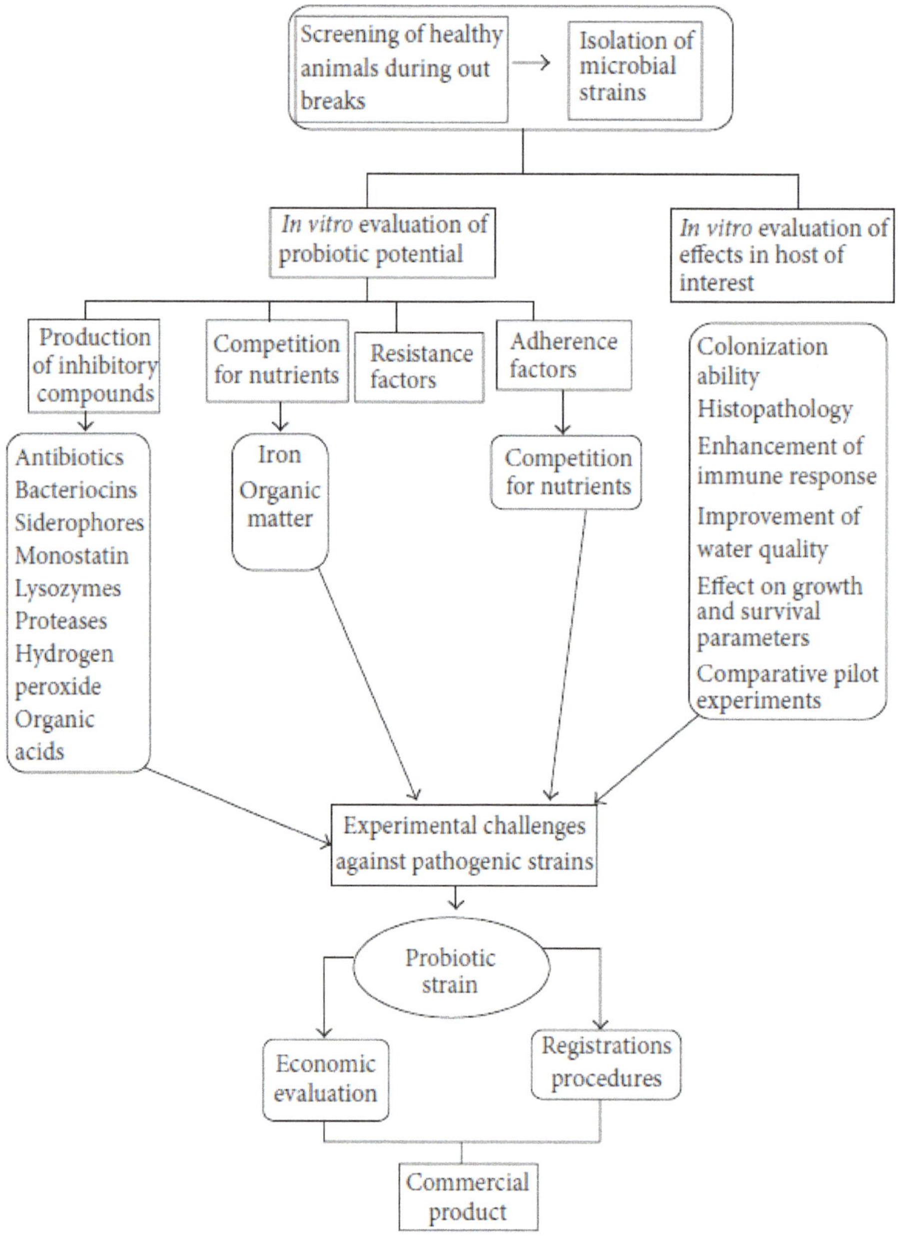

Figure 1: Procedure for evaluation of probiotic potential of microbial strain for shrimp aquaculture.

Evaluation of Probiotic Potential of Microbial Strains Other Than Animal Origin

Some of the probiotic strains are isolated from fermented foods, pond sediments, soil, water, and so forth. The procedure for evaluation of probiotic potential of microbial strains other than animal origin is illustrated in Figure 2. The experimental conditions of the probiotic potential tests vary according to the target host and the further application of probiotics. After the above evaluation process, the strain is further tested for economic evaluation.

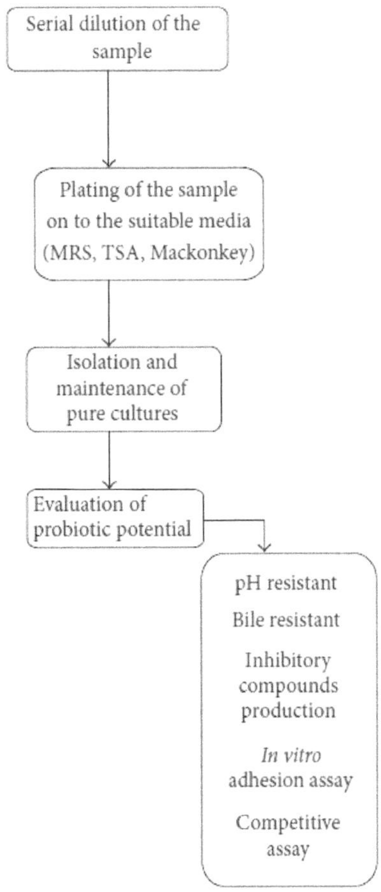

Figure 2: Microbiological procedure for evaluation of probiotic potential of microbial strain.

Application of Probiotics in Shrimp Aquaculture

Probiotic activity is mediated by a variety of effects that are dependent on the probiotic itself, the dosage employed, treatment duration and route, and frequency of delivery. Some probiotics exert their beneficial effects by elaborating antibacterial molecules such as bacteriocins that directly inhibit other bacteria or viruses, actively participating in the fight against infections, whereas others inhibit bacterial movement across the gut wall (translocation), enhance the mucosal barrier function by increasing the production of innate immune molecules, or modulate the inflammatory/immune response. Several studies have demonstrated that pattern recognition receptors (PRPs), such as toll-like receptors (TLRs) signaling pathways, immune responses, and the secretion of antimicrobial peptides such as defensins and chemokines by the epithelium play important roles in these mechanisms [85, 86].

Here are some reports, which stand as evidence for the beneficial effects of Probiotics. Bacillus S11, previously isolated from the GIT of P. monodon brood stock caught in the gulf of Thailand, demonstrated effective probiotic protection with P. monodon [30]. After a 100-day feeding trial with probiotic supplemented and nonsupplemented (control) feeds, P. monodon (from PL30 onwards) exhibited significant differences in growth, survival, and external appearance between the two groups. After challenging shrimps with a shrimp pathogen, Vibrio harveyi by immersion for 10 days, all probiotic treated groups had 100% survival, whereas the control group had only 26% survival which suggested competitive exclusion by probiotic Bacillus S11. Probiotic Bacillus subtilis UTM 126 was known to produce antimicrobial activity against vibriosis in juvenile shrimp, Litopenaeus vannamei [32].

Consideration of bacterial strains selected as probiotics should be safe to use as biological control. The extent of safe use of the microbes that have been used traditionally in probiotics can be confirmed through a long period of experience [87]. Yasuds and Taga [88] suggested that probiotic bacteria would be found to be useful not only as food but also as biological controllers of fish disease and activators of nutrient regeneration. The biological control in aquaculture emerges and since then the research effort has continuously increased. Generally, bacteria play two major roles as beneficial bacteria and pathogenic forms; beneficial bacteria are helpful in nutrient recycling and organic matter degradation and thus clear the environment [89]. Pathogenic bacteria are the causative agents of bad water quality, stress, and diseases as they act as primary and secondary pathogens [90].

The inhibitory effects of Bacillus sp. may be due to the production of antibiotics, bacteriocins, lysozymes, proteases, and hydrogen peroxide and the alteration of pH values by the production of organic acids [70]. Probiotics also influence the immune system of the fish, shrimp, and other aquatic species. Streptomyces has been applied as a probiotic in the laboratory culture of Penaeus monodon, which showed better water quality parameters than the control tank and increased length and weight in terms of growth [33]. Some probiotic products like Super-biotic, Super Ps, Zymetin, and Mutagen [91] were reported to play a vital role in postlarvae of P. monodon by maintaining good water quality parameters throughout the culture period. It was reported that Bacillus subtilis E20, isolated from the human health food, was used for white shrimp Litopenaeus vannamei larvae where it showed a significant decrease in the cumulative mortality and also increased gene expression of prophenoloxidase I, prophenoloxidase II, and lysozyme of larvae [34].

The above reports suggest that for the immune power and disease resistance enhancement the probiotics need to be given at regular intervals throughout the culture period and any halt in the probiotic supplementation leads to the untreated conditions and the animal become susceptible to the diseases. The benefits of probiotics were summarized in Table 3.

Table 3: Benefits of Probiotics in aquaculture

Probiotic strain	Used on	Effect of probiotic strain	Reference
Bacillus S11	*Penaeus monodon*	Protection against *Vibrio harveyi* by stimulation of cellular and humoral immune defenses	[30]
Bacillus subtilis UTM 126	*Litopenaeus vannamei*	Control vibriosis by producing bacitracin, gramicidin, polymyxin, tyrotricidin, and competitive exclusion	[32]
Streptomyces	*Penaeus monodon*	Better water quality parameters, increased length and weight of the animal	[33]
Bacillus subtilis E20	*Litopenaeus monodon*	Enhance humoral immune response	[34]

Probiotics in Activation of Shrimp Immune Defences

Probiotics were successfully reported for their beneficial effects in warm-blooded animals. Experiments indicate that probiotic bacteria administered orally may induce increased resistance to enteric infections [92]. As mentioned earlier, shrimp has a poorly developed immune system and probiotics were known to play an important role in the enhancement of immune response in shrimp. The probiotic bacteria Lactobacillus plantarum was reported to enhance the immune responses and gene expression in white shrimp, Litopenaeus vannamei, when given in diet. The bacteria influenced both the cellular and humoral immune defences in the shrimp. L. plantarum was known to enhance the phenoloxidase (PO) activity, prophenoloxidase (ProPO) activity, respiratory bursts, superoxide dismutase (SOD) activity and clearance efficiency of Vibrio alginolyticus, peroxinectin mRNA transcription, and survival rate after challenge with V. alginolyticus. These effects were observed when the bacteria was given in the diet at 10^{10} cfu (kg diet)$^{-1}$ for 168 hrs [93].

The Lactobacillus plantarum was also effective on Vibrio harveyi. On experimental challenge with V. harveyi, Litopenaeus vannamei showed increased resistance when compared to the control group. This was because the probiotic strain showed an immune reactive effect on the host. The probiotic bacteria are known to produce extracellular compounds that can stimulate the nonspecific immune response. The L. plantarum was responsible for the increase in total haemocyte count and phenoloxidase activity. In this report, the most effective elimination of the haemolymph and the hepatopancreas by the shrimp fed with the probiotic-supplemented diet could be related to the elevated agglutinating activity [31].

Probiotic Pediococcus acidilactici showed an effect on antioxidant defenses and oxidative stress of Litopenaeus vannamei when challenged with Vibrio nigripulchritudo. It was effective on the antioxidant defenses like SOD, catalase (CAT), glutathione peroxidase, total antioxidant status (TAS), glutathiones, and induced tissue damage. The probiotic strain was efficient in maintaining the antioxidant defence levels for a longer period than the control and uninfected groups [94]. This suggests that the probiotic bacteria besides enhancing the immune defenses also maintain the defence levels in the shrimp offering a prolonged protection. Probiotics strains Vibrio P62, Vibrio P63, and Bacillus P64 were isolated from hepatopancreas of healthy wild shrimp Penaeus vannamei, and their immunostimulatory effect was studied. Among the three, P64 showed a significantly higher immunity index and showed immune response similar to that of V. alginolyticus whereas the other two only showed good probiotic properties. Here, the P64 gave the immune alert with a significant increase in the hyaline cell population [28].

Some Vibrio spp. were assessed for their probiotic potential for L. vannamei. Among the species tested V. alginolyticus (NCIMB 1339) and V. gazogenes (NCIMB 2250) showed antagonistic activity towards shrimp pathogens vibriosis. When the juvenile shrimps were fed with chitin and V. gazogenes, they caused a significant decline in the number of vibrio-like bacteria in the fore and hind gut. In this study, the Vibrio and chitin mixture caused significant changes in haemocyte numbers. This change in haemocyte number probably reflects the immunological status of shrimp because these are involved in both cellular and humoral defences of the shrimp [95].

PROBIOTICS FOR VIRAL DISEASES

As mentioned earlier, probiotics have the capability of enhancing the immune response in fish and shrimp. For the protection from viral diseases, no specific drug was designed; besides the use of antibiotic is giving rise to a new type of resistant strains. Enhancement of the disease resistance in animal and the development of the immune power are the best option to prevent and resist the viral infections. For this purpose, one needs to have a proper understanding of the immune system and the type of immune response in the animal.

Outlines of the Immune System of Shrimp

Fish and shrimp differ significantly in their ability and the degree to which they respond to immune challenge. Shrimps have a poorly evolved defense mechanism and the capacity to recognize, expand the specific recognition, express specific recognition, and coordinate defense is much lower in shrimp when compared to fish. They do not have the ability to produce immunoglobulins; that is, adaptive memory is completely absent, and so they totally depend only on innate immune system [96]. The innate immune system includes both the humoral and cellular components which work in co-ordination with each other for the detection and elimination of all foreign organisms potentially hazardous for the host [97]. The cellular and humoral components are illustrated in Figure 3.

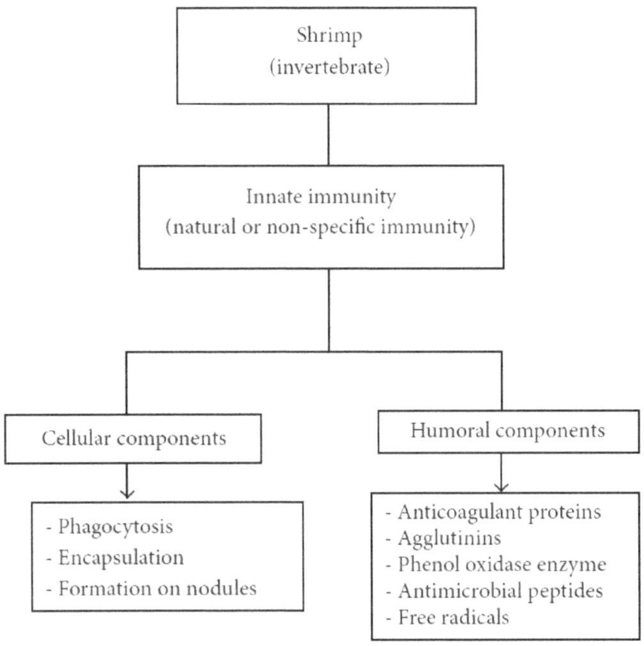

Figure 3: Outline of immune defense system in shrimp.

There was a report on the passive immunization of the tiger prawn, Penaeus monodon, using rabbit antisera to Vibrio harveyi [98]. In this method, the extracellular product of bacterium, V. harveyi strain 820514, has been isolated and it was carried on for purification process and injected into the rabbit for production of antiserum. The antiserum was collected from the rabbit and was injected intramuscularly into the prawn with an interval of 10 days and 17 days. When compared to the control ponds, the experimental ponds showed the postponement of the mortality for 2 weeks after postbacterial challenge. It was concluded that the protection given by this passive immunization is a relatively short-lived one. As mentioned earlier, shrimps or prawns lack a specific defense system, but the later research and reports suggest that specific immunity can be induced in the shrimp. Oral vaccination of the shrimp, P. monodon,towards WSSV showed a significant decrease in the mortality rate when compared to the control group [99]. In this experiment, inactivated bacteria overexpressing the WSSV envelope proteins VP19 and VP28 coated on food pellets were selected for delivery of the WSSV proteins. After vaccination, it was observed that vaccination with either VP28 alone or a mixture of VP28 and VP19 resulted in lower mortalities of 30% and 50% compared to the group vaccinated with the empty vectors. It was also reported that the DNA vaccines encoding viral envelope proteins confer protective immunity against WSSV in black tiger shrimp [100]. But further studies like comparison of immune response between the subunit vaccines and DNA vaccines, immune response towards enveloped, and nonenveloped virus are necessary.

Antiviral Immune Response in Shrimp

Antiviral immune response in shrimp is mediated through the pattern recognition receptors (PRRs). When a virus enters into the shrimp, the infected cells contain viral components like genomic DNA and RNA or dsRNA, and these viral components are sensed by means of PRRs. These PRRs are known to trigger effective and appropriate antiviral responses, including production of various cytokines and induction of inflammatory and adaptive immune reactions [101]. Antiviral immune responses include many antiviral-related proteins/ genes, antibody or immune stimulants mediated antiviral activity, cytokine-activated mediated antiviral response, apoptosis [102], phagocytosis, and prophenoloxidase system (Table 4).

In a comparative study of the efficiency of antiviral immune responses including phagocytosis, apoptosis, and ProPO system in the shrimp Marsupenaeus japonicas, it was reported that the phagocytosis and apoptosis play a major role in the antiviral immune response whereas the ProPO system is of only minor part [103]. The viral envelope proteins play very important role in the mechanism of infection of virus. It was documented that the viral proteins VP68, VP281, and VP466 have a very significant role in the infection of WSSV [104–106]. The VP466 is known to enhance the innate host immune response in shrimp. It acts as a GTPase-activating protein and forms a complex with the host shrimp Rab and increases its GTPase activity in vivo and in vitro. The complex induced rearrangements on the actin cytoskeleton, resulting in the formation of actin stress fibers which promoted the phagocytosis against virus [107]. This idea paves the way for the application of viral proteins as the immune enhancers in the host (shrimp) and can also be employed in the development of subunit vaccines.

Table 4: Proteins/genes involved in WSSV antiviral defense in shrimp

Proteins/genes involved in WSSV antiviral defense in shrimp	Species	Role in immune response	Reference
Actin	*Litopenaeus vannamei*	WSSV major structural protein VP26 binds to actin VP26-actin interaction is seen in the early stages of viral infection	[35]
ALF	*Pacifastacus leniusculus*	Interferes with WSSV replication by RNAi mechanism	[36]
β-Integrin	*Marsupenaeus japonicus*	Integrin mediates signal transduction and activates focal adhesion kinase (FAK). Enhance immune cell adhesion	[37]
Calreticulin	*Fenneropenaeus chinensis*	Modulate cell adhesion, phagocytosis, and integrin-dependent Ca^{+2} signalling	[38]
Caspase-3-like gene	*Penaeus monodon*	Upregulated during WSSV infection and cause increased apoptosis	[39]
C-type lectin (CTL) (pattern recognition protein)	*Litopenaeus vannamei*	Participate in nonself-innate immune defense in invertebrates	[40]
C-type lectin (LvCTL) (mannose binding CTL)	*Litopenaeus vannamei*	Binds to WSSV envelope proteins and exert antiviral activity	[41]
Fc lectin	*Fenneropenaeus chinensis*	Enhance innate immunity, that is, immune recognition, phagocytosis	[42]
Fortilin	*Penaeus monodon*	Antiapoptotic protein, found in high levels during onset of viral infections	[43]
Hemocyanin	*Marsupenaeus japonicus*	Expression of Pj Hc, Pj HcL hemocyanin subunit genes could delay the infection to WSSV	[44]
	Penaeus monodon	Nonspecific antiviral properties and no cytotoxicity to host cells	[45]
LGBP	*Penaeus stylirostris*	Enhance innate immune responses, activates prophenoloxidase (proPO) case	[46]
Manganese superoxide dismutase	*Fenneropenaeus chinensis*	Enhance immune defense reactions by eliminating oxidative stress	[47]
PmAV (First antiviral gene)	*Penaeus monodon*	Inhibit virus-induced cytopathic effect	[48]
Pm CBP (chitin-binding protein)	*Penaeus monodon*	Upregulated in late stages of WSSV infection and interacts with WSSV O67C (ORF)	[49]
Pm Rab7	*Penaeus monodon*	Binds to WSSV and VP28, Inhibits WSSV-induced histopathology	[50]
Rab GTPase	*Marsupenaeus japonicas*	Upregulated in WSSV-resistant shrimp	[51]
Ran protein	*Marsupenaeus japonicus*	Upregulated in WSSV-resistant and infected shrimp	[52]
Syntenin	*Penaeus monodon*	Upregulated in acute phase of a WSSV infection	[53]
Syntenin-like protein gene	*Penaeus monodon*	Involved in signaling pathway of antiviral shrimp immune response	[54]

Another mechanism in the antiviral immune response is the RNA interference (RNAi) method, which has been applied to silence viral genes in eukaryotic organisms. It was reported that a specific 21 bp short interfering RNA (VP28-SiRNA) was designed targeting a major envelope gene VP28 of WSSV. It readily silenced the transcription and translation of the viral gene and significantly reduced the mortality rate of the shrimp [108]. Thus, the RNAi mechanisms stand as an effective therapeutic strategy for viral infections in shrimp.

Study of Antiviral Activity of Probiotics

The exact and accurate methods for the study of antiviral activity of probiotics were not mentioned. From the earlier reports of Kamei et al. [109] and Direkbusarakom et al. [110], the antiviral activity was studied by plaque assays. In this plaque assay the bacterial cultures and the virus were mixed in equal volumes at regular intervals. From this mixture, the aliquots were withdrawn for plaque assay. This mixture was added to the cell cultures, and the antiviral activity was investigated from the rate of plaque reduction. The rate of plaque reduction was figured out by the following formula:

$$\frac{\text{P.F.U.}\ +0\ \text{time} - \text{P.F.U. at each reaction time}}{\text{P.F.U. at 0 time}} \times 100\%.$$

From the plaque reduction assay the bacterial strains Pseudomonas, Aeromonas/Vibrio, and Coryneforms were reported for their antiviral substances [109]. A novel eukaryotic cell culture model was proposed to study the antiviral activity of potential probiotic bacteria [111]. This model was successfully applied for human virus. This method is based on the mechanism of probiotic bacteria by which they fight infections for exclusion of pathogens by means of competition for attachment and stimulation of host cell immune defenses [112]. The approved model of this study has not been established yet in aquaculture. With some modifications, this cell culture model can be applied for aquaculture. The antiviral assays in this cell culture model include the following. (i)Pretreatment of cells with bacteria, where the monolayers/cell lines were first incubated with viable probiotic bacteria and the unbound bacteria were washed off. Now monolayers/cell lines were challenged immediately with the virus and incubated. The cell survival was considered for the antiviral activity. (ii) Coincubation of bacterial and virus (competition assay). In this method, the viable probiotic bacteria and virus were added simultaneously to the cell lines and further incubated under prescribed conditions. Dose dependence between the titer of the virus and number of bacteria was assessed for antiviral activity.(iii)Virus adsorption to the bacteria includes the coincubation of the viable probiotic bacteria with the virus. Then, the bacteria were removed by centrifugation, and the pellet was prepared for immunofluorescence. Residual viral infectivity in supernatants was assayed on cell lines for comparing the TCID50 to the inoculum titer. The virus alone was treated in the same way as the control.(iv)Antiviral effect of bacterial supernatants: Here the supernatants of the viable probiotic bacteria (after inoculation and incubation) were collected and were added on the cell lines/monolayer as incubation medium followed by immediate virus challenge. For the antiviral activity, the CPE was determined after incubation. For the application of the above, the method of aquaculture is important, and it is also important to consider the physical and biochemical factors of the probiotic bacteria and the virus chosen. It is also necessary to have a thorough knowledge of the probiotic bacteria and virus that are being tested.

Generally the virus in the laboratory is cultured in the suitable cell lines, and the virus growth was confirmed from the CPE studies. For example, Taura syndrome virus is cultivated in primary hemocyte culture of shrimp (Penaeus vannamei) [113]. The virus growth was confirmed and the titer was considered by studying CPE of the shrimp primary hemocyte culture at regular intervals of 6 hrs, 12 hrs, and 48 hrs. In the same way, the viable probiotic bacteria can be added to these cell lines in any way as mentioned in the cell culture model and CPE can be studied for the antiviral activity of probiotic bacteria.

Real-time PCR was conducted to measure WSSV copies in shrimp [103]. In this method, the results (Ct value) of viral quantity would be converted into copy number of virus based on the standard curve. Attempts are being made for the application of quantitative real-time PCR in the study of antiviral activity of probiotic bacteria, and the process is under progress. Quantitative real-time PCR is the advanced molecular technique, which is applied to the study of the virus titer. Shrimp mortality assay [103] is a physical method to study the antiviral activity of probiotic bacteria where the survival rate of the shrimp is considered as the parameter for the antiviral activity. This survival rate was considered by comparing with the control ponds and the experimental ponds.

Disease problems and environmental issues in shrimp farming have caused worries about the sustainability of traditional farming practices [114]. Certified, disease-free postlarvae and pond preparation were recognized as two of the most important steps in disease prevention. Several measures had been applied in health management to reduce disease, which included postlarvae selection, specific pathogen free brooders, closed systems, recirculation systems, probiotic application, and some form of biosecurity. Farmers reported that the use of probiotic base products and vitamins is helpful for health management and for reducing disease risk by fortifying natural defenses of the stock. Evidence that shows the effectiveness of probiotics in inhibiting a wide range of fish and shrimp pathogens and diseases problems in shrimp farming was also mentioned earlier in this review. Some microorganisms have been authorized for use as probiotics in feeding stuffs in the European Union. In addition, other probiotics are commercialized on the market. The last authorized list of feed additives published by the Commission [115] is tabulated in Table 5.

Table 5: List of microbial strains authorized as probiotics under Council Directive 70/524/EEC.

Probiotic strain	Host and applied host	Beneficial effect	Method of application	Reference
Bacillus cereus	*Farfantepenaeus brasiliensis*	Control vibrio concentration as well as a commercial probiotic. Mean final weight and specific growth rate of shrimp were significantly higher	Addition to culture water	[55]
Bacillus licheniformis	*Penaeus monodon*	Occur naturally in the intestinal tracts of prawns Compete with other bacteria in ponding and clearing the organic matter	Addition to culture water	[56]
Bacillus subtilis	*Penaeus monodon*	Occur naturally in the GIT. Higher shrimp growth, FCR, increased immunity to *Vibrio harveyi* 639 infection	Addition to culture water	[57]
Saccharomyces cerevisiae	*Litopenaeus vannamei,*	Immunostimulation and protection to *Vibrio harveyi*	Addition to diet	[58]
Streptococcus sp.	*Fenneropenaeus indicus*	Antagonism to *Vibrio alginolyticus*	Enrichment to live food, addition to diet	[59]
Pediococcus acidilactici	Artemia culture	Control *Vibrio alginolyticus* infection	Addition to diet	[60]

FUTURE PROSPECTIVE AND CONCLUDING REMARKS

The incidence of infectious diseases in shrimp aquaculture is a serious problem due to the overuse or misuse of antibiotics and antibiotic resistance genes among opportunistic pathogens such as Vibrio species. The application of probiotics against viruses in shrimp cultivation is a novel and safe approach. Probiotics for bacterial diseases like vibriosis is well reported but for viral diseases the authentic strains still need research. The current research is focused on the molecular interactions between the virus-probiotics and virus-host as the information available is very mere. Experiments are also under process in order to increase the potentiality of probiotic strains towards viral diseases. From the available literature and reports till date, the probiotic supplements showed better results when they are given from the starting of the culture than after the outbreak of disease. Thus, it is the best suggestion to include probiotics in the regular diet of the animal in order to prevent it from different infections and keep the animal healthy, which increases its economic value.

ACKNOWLEDGMENTS

The authors want to thank Dr. Emmanuel Vijay Paul Pandeeti, The University of TX, Health Science Center, Tyler, Texas, USA, and anonymous reviewers for their valuable suggestions to improve this review.

REFERENCES

1. FAO, "The State of world fisheries and aquaculture," 2002, http://www.fao.org/docrep/005/y7300e/y7300e00.htm.
2. FAO, "State of World Fisheries and Aquaculture, Food and Agricultural Organization of the United Nations," 2008, http://www.fao.org/docrep/011/i0250e/i0250e00.htm.
3. SCAN, "Opinion of the Scientific Committee on Animal Nutrition on the criteria for assessing the safety of microorganisms resistant to antibiotics of human clinical and veterinary importance," European Commission Health and Consumer Protection Directorate- General, 2003, http://ec.europa.eu/food/fs/sc/scan/out108_en.pdf.
4. R. Naylor and M. Burke, "Aquaculture and ocean resources: raising tigers of the sea," Annual Review of Environment and Resources, vol. 30, pp. 185–218, 2005.
5. R. L. Naylor, J. Eagle, and W. L. Smith, "Salmon aquaculture in the Pacific Northwest: a global industry with local impacts," Environment, vol. 45, no. 8, pp. 18–39, 2003.
6. R. L. Naylor, R. J. Goldburg, J. H. Primavera et al., "Effect of aquaculture on world fish supplies," Nature, vol. 405, no. 6790, pp. 1017–1024, 2000.
7. R. Goldburg and R. Naylor, "Future seascapes, fishing, and fish farming," Frontiers in Ecology and the Environment, vol. 3, no. 1, pp. 21–28, 2005.
8. A. B. Boxall, L. A. Fogg, P. A. Blackwell, P. Kay, E. J. Pemberton, and A. Croxfor, "Veterinary medicines in the environment," Reviews of Environmental Contaminationand Toxicology, vol. 180, pp. 1–91, 2004.
9. K. Haya, L. E. Burridge, and B. D. Chang, "Environmental impact of chemical wastes produced by the salmon aquaculture industry," ICES Journal of Marine Science, vol. 58, no. 2, pp. 492–496, 2000.
10. K. Grave, E. Lingaas, M. Bangen, and M. Rønning, "Surveillance of the overall consumption of antibacterial drugs in humans, domestic animals and farmed fish in Norway in 1992 and 1996," Journal of Antimicrobial Chemotherapy, vol. 43, no. 2, pp. 243–252, 1999.
11. T. X. Le, Y. Munekage, and S. I. Kato, "Antibiotic resistance in bacteria from shrimp farming in mangrove areas," Science of the Total Environment, vol. 349, no. 1–3, pp. 95–105, 2005.
12. T. X. Le and Y. Munekage, "Residues of selected antibiotics in water and mud from shrimp ponds in mangrove areas in Viet Nam," Marine Pollution Bulletin, vol. 49, no. 11-12, pp. 922–929, 2004.
13. T.M.L'abée-LundandH.Sørum,"Class1integronsmediate antibiotic resistance in the fish pathogen Aeromonas salmonicida worldwide," Microbial Drug Resistance, vol. 7, no. 3, pp. 263–272, 2001.
14. H. Sorum, F. M. Aarestrup, and D. C. Washington, "Antimicrobial drug resistance in fish pathogen," in Antimicrobial Resistance in Bacteria of Animal Origin, pp. 213–223, American Society for Microbiology press, Washington, DC, USA, 2006.
15. F. J. Angulo, V. N. Nargund, and T. C. Chiller, "Evidence of an association between use of anti-microbial agents in food animals and anti-microbial resistance among bacteria isolated from humans and the human health consequences of such resistance," Journal of Veterinary Medicine Series B, vol. 51, no. 8-9, pp. 374–379, 2004.
16. F. C. Cabello, "Antibiotics and aquaculture in Chile: implications for human and animal health," Revista Medica de Chile, vol. 132, no. 8, pp. 1001–1006, 2004.
17. F. C. Cabello, "An analysis of their potential impact upon the environment, human and animal health in Chile," Antibiotics and Aquaculture, vol. 17, pp. 1–16, 2003.
18. R. J. Goldburg, M. S. Elliot, and R. L. Naylor, "Marine Aquaculture in the United States: environmental impacts and policy options," Tech. Rep., PEW Oceans Commission, Stanford, Calif, USA, 2001.
19. K. Grave, A. Markestad, and M. Bangen, "Comparison in prescribing patterns of antibacterial drugs in salmonid farming in Norway during the periods 1980–1988 and 1989–1994," Journal of Veterinary Pharmacology and Therapeutics, vol. 19, no. 3, pp. 184–191, 1996.
20. J. E. Davies, M. C. Roberts, S. B. Levy, G. H. Miller, T. F. Brien, and F. C. Tenover, "Antimicrobial resistance: an ecological perspective," A Report from the American Academy of Microbiology, American Academy of Microbiology, Washington, DC, USA, 1999.
21. C. J. Hunter, D. Karl, and M. Buckley, "Marine microbial diversity: the key to earth's habitability," A Report from the American Academy of Microbiology, Marine Microbial Diversity, San Francisco, Calif, USA; American Academy of Microbiology, Washington, DC, USA, 2005.
22. C. D. Miranda and R. Zemelman, "Antibiotic resistant bacteria in fish from the Concepción Bay, Chile," Marine Pollution Bulletin, vol. 42, no. 11, pp. 1096–1102, 2001.
23. C. G. Lundin, "Global attempt to address shrimp disease," Marine Environmental Paper 4, The World Bank, 1996.
24. L. F. Gibson, J. Woodworth, and A. M. George, "Probiotic activity of Aeromonas media on the Pacific oyster, Crassostrea gigas, when challenged with Vibrio tubiashii," Aquaculture, vol. 169, no. 1-2, pp. 111–120, 1998.
25. S. K. Nayak and S. C. Mukherjee, "Screening of gastrointestinal bacteria of Indian major carps for a candidate probiotic species for aquaculture

practices," Aquaculture Research, vol. 42, no. 7, pp. 1034–1041, 2011.

26. S. Rahman, S. N. Khan, M. N. Naser, and M. M. Karim, "Safety issues of isolated probiotic natured bacteria from Bangladesh coastal waters for controlling shrimp diseases," Journal of Scientific Research, vol. 3, no. 3, pp. 659–668, 2011.

27. K. V. Lalitha and P. K. Surendran, "Bacterial microflora associated with farmed freshwater prawn Macrobrachium rosenbergii (de Man) and the aquaculture environment," Aquaculture Research, vol. 35, no. 7, pp. 629–635, 2004.

28. M. Gullian, F. Thompson, and J. Rodriguez, "Selection of probiotic bacteria and study of their immunostimulatory effect in Penaeus vannamei," Aquaculture, vol. 233, no. 1–4, pp. 1–14, 2004.

29. T. Defoirdt, N. Boon, P. Sorgeloos, W. Verstraete, and P. Bossier, "Alternatives to antibiotics to control bacterial infections: luminescent vibriosis in aquaculture as an example," Trends in Biotechnology, vol. 25, no. 10, pp. 472–479, 2007.

30. S. Rengpipat, W. Phianphak, S. Piyatiratitivorakul, and P. Menasveta, "Effects of a probiotic bacterium on black tiger shrimp Penaeus monodon survival and growth," Aquaculture, vol. 167, no. 3-4, pp. 301–313, 1998.

31. F. N. Vieira, C. C. Buglione, J. P. L. Mouriño et al., "Effect of probiotic supplemented diet on marine shrimp survival after challenge with Vibrio harveyi," Arquivo Brasileiro de Medicina Veterinaria e Zootecnia, vol. 62, no. 3, pp. 631–638, 2010.

32. J. L. Balcázar and T. Rojas-Luna, "Inhibitory activity of probiotic Bacillus subtilis UTM 126 against Vibriospecies confers protection against vibriosis in juvenile shrimp (Litopenaeus vannamei)," Current Microbiology, vol. 55, no. 5, pp. 409–412, 2007.

33. S. Das, P. S. Lyla, and S. Ajmal Khan, "Application of Streptomyces as a probiotic in the laboratory culture of Penaeus monodon (Fabricius)," Israeli Journal of Aquaculture, vol. 58, no. 3, pp. 198–204, 2006.

34. K. F. Liu, C. H. Chiu, Y. L. Shiu, W. Cheng, and C. H. Liu, "Effects of the probiotic, Bacillus subtilis E20, on the survival, development, stress tolerance, and immune status of white shrimp, Litopenaeus vannamei larvae," Fish and Shellfish Immunology, vol. 28, no. 5-6, pp. 837–844, 2010.·

35. X. Xie and F. Yang, "Interaction of white spot syndrome virus VP26 protein with actin," Virology, vol. 336, no. 1, pp. 93–99, 2005.

36. H. Liu, P. Jiravanichpaisal, I. Söderhäll, L. Cerenius, and K. Söderhäll, "Antilipopolysaccharide factor interferes with white spot syndrome virus replication in vitro and in vivo in the crayfish Pacifastacus leniusculus," Journal of Virology, vol. 80, no. 21, pp. 10365–10371, 2006.

37. D. F. Li, M. C. Zhang, H. J. Yang, Y. B. Zhu, and X. Xu, "Beta-intergrin mediates WSSV infection," Virology, vol. 368, no. 1, pp. 122–132, 2007.

38. W. Luana, F. Li, B. Wang, X. Zhang, Y. Liu, and J. Xiang, "Molecular characteristics and expression analysis of calreticulin in Chinese shrimp Fenneropenaeus chinensis," Comparative Biochemistry and Physiology B, vol. 147, no. 3, pp. 482–491, 2007.

39. K. Wongprasert, P. Sangsuriya, A. Phongdara, and S. Senapin, "Cloning and characterization of a caspase gene from black tiger shrimp (Penaeus monodon)-infected with white spot syndrome virus (WSSV)," Journal of Biotechnology, vol. 131, no. 1, pp. 9–19, 2007. ·

40. T. H. T. Ma, S. H. K. Tiu, J. G. He, and S. M. Chan, "Molecular cloning of a C-type lectin (LvLT) from the shrimp Litopenaeus vannamei: early gene down-regulation after WSSV infection," Fish and Shellfish Immunology, vol. 23, no. 2, pp. 430–437, 2007.

41. Z. Y. Zhao, Z. X. Yin, X. P. Xu, S. P. Weng, X. Y. Rao, and Z. X. Dai, "A novel C-type lectin from the shrimp Litopenaeus vannamei possesses anti-WSSV activity," Journal of Virology, vol. 83, no. 1, pp. 347–356, 2009.

42. Y. C. Liu, F. H. Li, B. Dong et al., "Molecular cloning, characterization and expression analysis of a putative C-type lectin (Fclectin) gene in Chinese shrimp Fenneropenaeus chinensis," Molecular Immunology, vol. 44, no. 4, pp. 598–607, 2007.

43. M. Tonganunt, B. Nupan, M. Saengsakda et al., "The role of Pm-fortilin in protecting shrimp from white spot syndrome virus (WSSV) infection," Fish and Shellfish Immunology, vol. 25, no. 5, pp. 633–637, 2008. ·

44. K. Lei, F. Li, M. Zhang, H. Yang, T. Luo, and X. Xu, "Difference between hemocyanin subunits from shrimp Penaeus japonicus in anti-WSSV defense," Developmental and Comparative Immunology, vol. 32, no. 7, pp. 808–813, 2008.

45. X. Zhang, C. Huang, and Q. Qin, "Antiviral properties of hemocyanin isolated from shrimp Penaeus monodon," Antiviral Research, vol. 61, no. 2, pp. 93–99, 2004. ·

46. M. M. Roux, A. Pain, K. R. Klimpel, and A. K. Dhar, "The lipopolysaccharide and β-1,3-glucan binding protein gene is upregulated in white spot virus-infected shrimp (Penaeus stylirostris)," Journal of Virology, vol. 76, no. 14, pp. 7140–7149, 2002.

47. Q. Zhang, F. Li, B. Wang et al., "The mitochondrial manganese superoxide dismutase gene in Chinese shrimp Fenneropenaeus chinensis: Cloning, distribution and expression," Developmental and Comparative Immunology, vol. 31, no. 5, pp. 429–440, 2007.

48. T. Luo, X. Zhang, Z. Shao, and X. Xu, "PmAV, a novel gene involved in virus resistance of shrimp Penaeus monodon," FEBS Letters, vol. 551, no. 1-3, pp. 53–57, 2003. ·

49. L. L. Chen, L. C. Lu, W. J. Wu, C. F. Lo, and W. P. Huang, "White spot syndrome virus envelope protein VP53A interacts with Penaeus monodon chitin-binding protein (PmCBP)," Diseases of Aquatic Organisms, vol. 74, no. 3, pp. 171–178, 2007.

50. K. Sritunyalucksana, W. Wannapapho, C. F. Lo, and T. W. Flegel, "PmRab7 is a VP28-binding protein involved in white spot syndrome virus infection in shrimp," Journal

of Virology, vol. 80, no. 21, pp. 10734–10742, 2006.

51. W. Wu and X. Zhang, "Characterization of a Rab GTPase up-regulated in the shrimp Peneaus japonicusby virus infection," Fish and Shellfish Immunology, vol. 23, no. 2, pp. 438–445, 2007.·

52. F. Han and X. Zhang, "Characterization of a ras-related nuclear protein (Ran protein) up-regulated in shrimp antiviral immunity," Fish and Shellfish Immunology, vol. 23, no. 5, pp. 937–944, 2007.

53. M. Tonganunt, A. Phongdara, W. Chotigeat, and K. Fujise, "Identification and characterization of syntenin binding protein in the black tiger shrimp Penaeus monodon," Journal of Biotechnology, vol. 120, no. 2, pp. 135–145, 2005.

54. P. Bangrak, P. Graidist, W. Chotigeat, K. Supamattaya, and A. Phongdara, "A syntenin-like protein with postsynaptic density protein (PDZ) domains produced by black tiger shrimp Penaeus monodon in response to white spot syndrome virus infection," Diseases of Aquatic Organisms, vol. 49, no. 1, pp. 19–25, 2002.

55. D. Moreira, M. Sabrina, F. P. Leivas, L. A. Romano, E. Luis, and E. Ballester, New Bacillus Probiotic Tested for Shrimp, Global Aquaculture Advocate, 2011.

56. D. J. W. Moriarty and O. P. Decamp, Probiotics in Aquaculture, AQUA Culture Asia Pacific Magazine, 2005.

57. P. Utiswannakul, S. Sangchai, and S. Rengpipat, "Enhanced growth of black tiger shrimp Penaeus monodon by dietary supplementation with Bacillus (BP11) as a probiotic," Journal of Aquatic Research and Development, vol. 3, no. 4, pp. 2155–9546, 2011.

58. U. Scholz, G. Garcia Diaz, D. Ricque, L. E. Cruz Suarez, F. Vargas Albores, and J. Latchford, "Enhancement of vibriosis resistance in juvenile Penaeus vannamei by supplementation of diets with different yeast products," Aquaculture, vol. 176, no. 3-4, pp. 271–283, 1999.

59. S. Ajitha, M. Sridhar, N. Sridhar, I. S. B. Singh, and V. Varghese, "Probiotic effects of lactic acid bacteria against Vibrio alginolyticus in Penaeus (Fenneropenaeus) indicus," Asian Journal of Fishery Sciences, vol. 17, no. 1, pp. 71–80, 2004.

60. L. Villamil, A. Figueras, M. Planas, and B. Novoa, "Control of Vibrio alginolyticus in Artemia culture by treatment with bacterial probiotics," Aquaculture, vol. 219, no. 1–4, pp. 43–56, 2003.·

61. M. G. Bondad-Reantaso, R. P. Subasinghe, J. R. Arthur et al., "Disease and health management in Asian aquaculture," Veterinary Parasitology, vol. 132, no. 3-4, pp. 249–272, 2005.

62. D. Norasma and M. Saleem, Effluent and Disease Management in Traditional Practices of Shrimp Farming: A Case Study on the West Coast of Sabah, Malaysia, Research and Farming Techniques, 2008.

63. F. PÁez-Osuna, "The environmental impact of shrimp aquaculture: causes, effects, and mitigating alternatives," Environmental Management, vol. 28, no. 1, pp. 131–140, 2001.

64. I. Karunasagar, M. M. Shivu, S. K. Girisha, G. Krohne, and I. Karunasagar, "Biocontrol of pathogens in shrimp hatcheries using bacteriophages," Aquaculture, vol. 268, no. 1–4, pp. 288–292, 2007.

65. R. Fuller, Probiotics 2: Applications and Practical Aspects, Chapman and Hall, London, UK, 1st edition, 1997.

66. R. Fuller, Probiotics: The Scientific Basis, Chapman and Hall, London, UK, 1st edition, 1992.

67. R. Fuller, "Probiotics in man and animals," Journal of Applied Bacteriology, vol. 66, no. 5, pp. 365–378, 1989.

68. G. W. Tannock, R. Fuller, and K. Pedersen, "Lactobacillus succession in the piglet digestive tract demonstrated by plasmid profiling," Applied and Environmental Microbiology, vol. 56, no. 5, pp. 1310–1316, 1999.

69. FAO/WHO, "Report of a Joint FAO/WHO expert consultation on evaluation of health and nutritional propeties of probiotics in food including powder milk with live lactic acid bacteria," in Health and Nutritional Properties of Probiotics in Food Including Powder Milk with Live Lactic Acid Bacteria, FAO/WHO, Cordoba, Argentina, 2001.

70. L. Verschuere, G. Rombaut, P. Sorgeloos, and W. Verstraete, "Probiotic bacteria as biological control agents in aquaculture," Microbiology and Molecular Biology Reviews, vol. 64, no. 4, pp. 655–671, 2000.

71. A. M. Onarheim, R. Wiik, J. Burghardt, and E. Stagkebrandt, "Characterization and identification of two Vibrio species indigenous to the intestine of fish in cold sea water; description of Vibrio iliopiscarius sp. nov.," Systematic and Applied Microbiology, vol. 17, no. 3, pp. 370–379, 1994. ·

72. T. Sakata, "Microflora in the digestive tract of fish and shell-fish," in Microbiology in Poecilotherms, R. Lesel, Ed., pp. 171–176, Elsevier, 1990.

73. E. Ringø and O. Vadstein, "Colonization of Vibrio pelagius and Aeromonas caviae in early developing turbot (Scophthalmus maximus L.) larvae," Journal of Applied Microbiology, vol. 84, no. 2, pp. 227–233, 1998.

74. B. Austin, E. Baudet, and M. Stobie, "Inhibition of bacterial fish pathogens by Tetraselmis suecica," Journal of Fish Diseases, vol. 15, no. 1, pp. 55–61, 1992.

75. B. Austin, L. F. Stuckey, P. A. W. Robertson, I. Effendi, and D. R. W. Griffith, "A probiotic strain of Vibrio alginolyticus effective in reducing diseases caused by Aeromonas salmonicida, Vibrio anguillarum and Vibrio ordalii," Journal of Fish Diseases, vol. 18, no. 1, pp. 93–96, 1995.

76. P. A. Douillet and C. J. Langdon, "Use of a probiotic for the culture of larvae of the Pacific oyster (Crassostrea gigas Thunberg)," Aquaculture, vol. 119, no. 1, pp. 25–40, 1994.

77. A. Gildberg, A. Johansen, and J. Bogwald, "Growth and survival of Atlantic salmon (Salmo salar) fry given diets supplemented with fish protein hydrolysate and lactic acid bacteria during a challenge trial with Aeromonas salmonicida," Aquaculture, vol. 138, no. 1–4, pp. 23–34,

1995.

78. A. Gildberg, H. Mikkelsen, E. Sandaker, and E. Ringø, "Probiotic effect of lactic acid bacteria in the feed on growth and survival of fry of Atlantic cod (Gadus morhua)," Hydrobiologia, vol. 352, no. 1–3, pp. 279–285, 1997.

79. W. Phianphak, S. Rengpipat, S. Piyantiratitivorakul, and P. Menasveta, "Probiotic use of Lactobacillus spp. for black tiger shrimp. Penaeus monodon," Journal of Scientific Research, Chulanokorn University, vol. 24, pp. 41–51, 1999.

80. B. Austin and J. G. Day, "Inhibition of prawn pathogenic Vibrio spp. by a commercial spray-dried preparation of Tetraselmis suecica," Aquaculture, vol. 90, no. 3-4, pp. 389–392, 1990.

81. B. Austin and A. C. Billaud, "Inhibition of fish pathogen, Serratia liquefaciens, By an antibiotic producing isolate of Plamococcus recovered from sea water," Journal of Fish Diseases, vol. 13, no. 6, pp. 553–556, 1990.

82. C. P. Dopazo, M. L. Lemos, C. Lodeiros, J. Bolinches, J. L. Barja, and A. E. Toranzo, "Inhibitory activity of antibiotic-producing marine bacteria against fish pathogens," Journal of Applied Bacteriology, vol. 65, no. 2, pp. 97–101, 1988.

83. P. D. Munro, H. A. McLean, A. Barbour, and T. H. Birkbeck, "Stimulation or inhibition of growth of the unicellular alga Pavlovalutheri by bacteria isolated from larval turbot culture systems," Journal of Applied Bacteriology, vol. 79, no. 5, pp. 519–524, 1995.

84. A. Westerdahl, J. C. Olsson, S. Kjelleberg, and P. L. Conway, "Isolation and characterization of turbot (Scophtalmus maximus)-associated bacteria with inhibitory effects against Vibrio anguillarum," Applied and Environmental Microbiology, vol. 57, no. 8, pp. 2223–2228, 1991.

85. E. M. M. Quigley, "Prebiotics and probiotics; modifying and mining the microbiota," Pharmacological Research, vol. 61, no. 3, pp. 213–218, 2010.

86. P. M. Sherman, J. C. Ossa, and H. K. Johnson, "Unravelling mechanisms of action of probiotics," Nutrition Clinical Practices, vol. 24, no. 1, pp. 10–14, 2009.

87. M. A. Mayra, M. Bigret, S. V. Salminen, A. Wright, and D. Marcel, "Industrial use and production of lactic acid bacteria," in Lactic Acid Bacteria, p. 65, InTech, New York, NY, USA, 1993.

88. K. Yasuds and N. A. Taga, "Mass culture method for Artemia salina using bacteria as food," Medical Education Resources, vol. 18, p. 53, 1980.

89. D. J. W. Moriarty, "The role of microorganisms in aquaculture ponds," Aquaculture, vol. 151, no. 1–4, pp. 333–349, 1997.

90. I. Karunasagar, S. K. Otta, I. Karunasagar, and K. Joshua, "Applications of Vibrio vaccine in shrimp culture," Fishing Chimes, vol. 16, p. 49, 1996.

91. P. Soundarapandian, V. Ramanan, and G. K. Dinakaran, "Effect of probiotics on the growth and survival of Penaeus monodon (Fabricius)," Current Research Journal of Social Sciences, vol. 2, no. 2, pp. 51–57, 2010.

92. W. H. Holzapfel, P. Haberer, J. Snel, U. Schillinger, and J. H. J. Huisss, "Overview of gut flora and probiotics," International Journal of Food Microbiology, vol. 41, no. 2, pp. 85–101, 1998. ·

93. C. H. Chiu, Y. K. Guu, C. H. Liu, T. M. Pan, and W. Cheng, "Immune responses and gene expression in white shrimp, Litopenaeus vannamei, induced by Lactobacillus plantarum," Fish and Shellfish Immunology, vol. 23, no. 2, pp. 364–377, 2007.

94. M. Castex, P. Lemaire, N. Wabete, and L. Chim, "Effect of probiotic Pediococcus acidilactici on antioxidant defences and oxidative stress of Litopenaeus stylirostris under Vibrio nigripulchritudochallenge," Fish and Shellfish Immunology, vol. 28, no. 4, pp. 622–631, 2010.

95. J. Thompson, S. Gregory, S. Plummer, R. J. Shields, and A. F. Rowley, "An in vitro and in vivo assessment of the potential of Vibrio spp. as probiotics for the Pacific White shrimp, Litopenaeus vannamei," Journal of Applied Microbiology, vol. 109, no. 4, pp. 1177–1187, 2010. ·

96. P. Roch, "Defense mechanisms and disease prevention in farmed marine invertebrates," Aquaculture, vol. 172, no. 1-2, pp. 125–145, 1999.

97. P. Jiravanichpaisal, B. L. Lee, and K. Söderhäll, "Cell-mediated immunity in arthropods: hematopoiesis, coagulation, melanization and opsonization," Immunobiology, vol. 211, no. 4, pp. 213–236, 2006.

98. K. K. Lee, P. C. Liu, G. H. Kou, and S. N. Chen, "Passive immunization of the tiger prawn, Penaeus monodon, using rabbit antisera to Vibrio harveyi," Letters in Applied Microbiology, vol. 25, no. 1, pp. 34–37, 1997.

99. J. Witteveldt, C. C. Cifuentes, J. M. Vlak, and M. C. W. Van Hulten, "Protection of Penaeus monodonagainst white spot syndrome virus by oral vaccination," Journal of Virology, vol. 78, no. 4, pp. 2057–2061, 2004.

100. N. Rout, S. Kumar, S. Jaganmohan, and V. Murugan, "DNA vaccines encoding viral envelope proteins confer protective immunity against WSSV in black tiger shrimp," Vaccine, vol. 25, no. 15, pp. 2778–2786, 2007.

101. T. Kawai and S. Akira, "Innate immune recognition of viral infection," Nature Immunology, vol. 7, pp. 131–137, 2006.

102. L. Haipeng, K. Soderhall, and P. Jiranvanichpaisal, "Antiviral immunity in Crustaceans," Fish and Shellfish Immunology, vol. 27, no. 2, pp. 79–88, 2009.

103. W. Wang and X. Zhang, "Comparison of antiviral efficiency of immune responses in shrimp," Fish and Shellfish Immunology, vol. 25, no. 5, pp. 522–527, 2008.

104. W. Liu, F. Han, and X. Zhang, "Ran GTPase regulates hemocytic phagocytosis of shrimp by interaction with myosin," Journal of Proteome Research, vol. 8, no. 3, pp. 1198–1206, 2009.

105. W. Wu, L. Wang, and X. Zhang, "Identification of white spot syndrome virus (WSSV) envelope proteins involved in shrimp infection," Virology, vol. 332, no. 2, pp. 578–583, 2005.

106. J. Xu, F. Han, and X. Zhang, "Silencing shrimp white spot syndrome virus (WSSV) genes by siRNA," Antiviral Research, vol. 73, no. 2, pp. 126–131, 2007.

107. Y. Ting, Z. Rangrong, and Z. Xiaobo, "The role of White spot syndrome virus (WSSV) VP466 protein in shrimp antiviral phagocytosis," Fish and Shellfish Immunology, vol. 33, no. 2, pp. 350–358, 2012.

108. J. Xu, F. Han, and X. Zhang, "Silencing shrimp white spot syndrome virus (WSSV) genes by siRNA," Antiviral Research, vol. 73, no. 2, pp. 126–131, 2007.

109. Y. Kamei, M. Yoshimizu, Y. Ezura, and T. Kimura, "Screening of bacteria with antiviral activity from fresh water salmonid hatcheries," Microbiology and Immunology, vol. 32, no. 1, pp. 67–73, 1988.

110. S. Direkbusarakom, M. Yoshimizu, Y. Ezura, L. Ruangpan, and Y. Danayadol, "Vibrio spp., the dominant flora in shrimp hatchery against some fish pathogenic viruses," Journal of Marine Biotechnology, vol. 6, no. 4, pp. 266–267, 1998.

111. T. Botić, T. D. Klingberg, H. Weingartl, and A. Cencič, "A novel eukaryotic cell culture model to study antiviral activity of potential probiotic bacteria," International Journal of Food Microbiology, vol. 115, no. 2, pp. 227–234, 2007.

112. E. Isolauri, "Probiotics for infectious diarrhoea," Gut, vol. 52, no. 3, pp. 436–437, 2003. ·

113. S. K. George, K. N. Kaizer, Y. M. Betz, and A. K. Dhar, "Multiplication of Taura syndrome virus in primary hemocyte culture of shrimp (Penaeus vannamei)," Journal of Virological Methods, vol. 172, no. 1-2, pp. 54–59, 2011.

114. C. A. Otoshi, S. M. Arce, and S. M. Moss, "Growth and reproductive performance of broodstock shrimp reared in a biosecure recirculating aquaculture system versus a flow-through pond," Aquacultural Engineering, vol. 29, no. 3-4, pp. 93–107, 2003.

115. Council Directive 70/524/EEC, "List of the authorized additives in feeding stuffs published in application of Article 9t of Council Directive 70/554/EEC concerning additives in feedingstuffs," Official Journal of European Union, pp. C50–C144, 2004.

CHAPTER 12

ENHANCED GROWTH OF BLACK TIGER SHRIMP PENAEUS MONODON BY DIETARY SUPPLEMENTATION WITH BACILLUS (BP11) AS A PROBIOTIC

Ponpisit Utiswannakul, Siripen Sangchai and Sirirat Rengpipat

Keywords: *Bacillus subtilis*, BP11, *Penaeus monodon*, Black tiger shrimp, Hemocyte count, Antibacterial activity, *Vibrio harveyi*

INTRODUCTION

Thailand became the world's largest farmed-shrimp exporter in 1994, exporting a total of around 30% of the world market share [1], and the value of the exported frozen and chilled shrimp from Thailand has remained high (~$ 2,500 million per year) during the last ten years (2001-2010) [2]. However, the yield per crop in most black tiger shrimp, Penaeus monodon, farms in Thailand began to seriously decline since 2005 with some complete losses, as seen elsewhere in SE Asia, due to the low quality of brood stocks and outbreaks of viral and bacterial infections. The resulting stunted growth and high mortality rates induced by these infections have led to a change in the major farmed shrimp species in Thailand and elsewhere away from P. monodon to the white shrimp, Litopenaeus vannamei, which now accounts for some 90% of shrimp aquaculture in Thailand and SE Asia [3,4]. Currently, the addition of substantial amounts of antibiotics or chemotherapeutics remains the current method of choice for disease control throughout the shrimp aquaculture industry. However, in addition to the imposed increased costs of such antibiotic administration, and some concern of the health and environmental affects, there is the increasing concern about the predictable spread and increase in antibiotic-resistant bacteria [5-9]. Moreover, it is not clear if the release of virulent pathogen infected aquaculture media back into the sea may lead to transient, but important, mortality in wild shrimp including short term future brood stocks for aquaculture. This has led to alternative suggestions for more sustainable and environmental friendly disease prevention, including the use of non-pathogenic probiotic bacteria [10-13]. This alternative direction of biocontrol aims to make shrimp farming more sustainable. Selection for non-pathogenic probiotic bacteria with a high specificity to the cultured shrimp host and that can provide a healthy balance of indigenous organisms in the hosts' intestines, or other benefits to the host, is the initial starting point of such an approach. However, the degree of regional (population specific) host-probiont-pathogen coevolution is unknown, and so screening of probiotic candidates should include the same brood stock populations as used for the shrimp aquaculture. In addition, the bioactive component(s) or the entire probiotic organism(s), of benefit to the health or safety of commercially farmed shrimp needs to be able to be prepared at commercial scale easily and be relatively stable to at least short-term storage (shelf life) to be of practical applicable use [12,14-20].

Although several probiotics have recently been developed for aquaculture crops, they are still not widely used. There are, however, substantial advantages for the use of probiotics in shrimp and fish aquaculture [10,11,21-26]. Among the potential bacterial probiotics for shrimp, some Bacillus spp. have been extensively evaluated as an additive in the feed [22-25,27-29] and in the culture water [30-32]. Several Bacillus spp. naturally constitute a part of the bacterial flora of the intestinal tracts of several marine taxa, including for example B. subtilis in fish [33-35], and B. subtilis and B. pumilus in shrimp [22,25,36]. Concurrently, Bacillus spp. can be isolated from shrimp culture ponds [23-25] and marine environments [37]. Bacillus spp. have rarely been reported as shrimp pathogens, whereas several studies have demonstrated that culturing shrimp with feed supplemented with Bacillus sp. reduced the subsequent mortality induced by vibriosis, in particular, some strains of V. harveyi [12,22,25,27]. To further study the potential of

Citation: Utiswannakul P, Sangchai S, Rengpipat S (2011) Enhanced growth of black tiger shrimp Penaeus monodon by dietary supplementation with Bacillus (BP11) as a probiotic. J Aquac Res Development S1:006. doi:10.4172/2155-9546. S1-006.

Bacillus spp. as a probiotic biocontrol agent in shrimp farming, and to sustain commercial black tiger shrimp aquaculture, we isolated Bacillus spp. from the intestines of healthy black tiger shrimp brood stocks derived from the Andaman sea, which is a different area from that previously reported [22]. Thus, this offers the potential of either new Bacillus isolates or those probiotic bacteria that are better adapted to the local shrimp-pathogen isolates of this region. Therefore, the objectives of this research were to isolate potential probiotic Bacillus candidates from the intestines of P. monodon from the Andaman sea and to evaluate their effects on the growth, survival and disease resistance of post larva shrimp P. monodon from the same region but reared in concrete tanks and in earthen pond, respectively.

MATERIALS AND METHODS

Bacterial strains

P. monodon brood stocks (~150-200 g of ~9-12 inches each) from the Andaman sea along the West coast of Thailand were collected and their intestines were aseptically dissected out. The removed intestine was longitudinally cut and rinsed gently with sterile 0.85% (w/v) NaCl (NSS) three times and homogenized. Serial dilutions (10^{-1} to 10^{-8}) in NSS of the intestine homogenate were then plated in triplicate on tryptic soy agar (TSA) plus 1% (w/v) NaCl plates and cultured at 37o C for 24hr, and the number of colonies were counted. Isolated colonies were selected and checked for antibacterial properties against Vibrio harveyi 639 and Escherichia coli ATCC 25922 using an agar diffusion technique, as described in detail by Naclerio et al. [38].

V. harveyi 639 isolated from Penaeus monodon dying of luminescent disease was kindly provided by the Shrimp Culture Research Center, Charoen Pokphan Feedmill, Samutsakorn, Thailand. It was cultured in tryptic soy broth (TSB) or TSA containing 2% NaCl (w/v) at 30o C and identity was confirmed as described previously [39], Presumptive concentrations or colony-forming units (CFU ml^{-1}) of Vibrio spp. were determined using spread plates of thiosulfate citrate bile sucrose agar (TCBS). Escherichia coli ATCC 25922 was kindly provided from Department of Microbiology, Faculty of Science, Chulalongkorn University. E.coli was cultured on TSA and TSB at 30o C. All media were obtained from Difco, Sparks, MD, USA.

BP11 Identification

The culture purity and identity of the selected isolate (BP11), which produced the greatest antibacterial performance, was routinely checked during preparation using both conventional methods and the api- 20E® and api-50 CHB® Medium test kits (BioMérieux, Marcy-l'Etoile, France). The results were analyzed with the APILAB Plus software. Gram staining, spore staining, oxidase and catalase tests were performed following conventional methods and the bacterial morphology was examined as cells / spores under the microscope and as colonies on agar plates. Physiological characteristics were determined in tryptic soy broth (TSB) with various concentrations of NaCl (1-4% (w/v)) and at different pH (6-8) and temperatures (35-50o C).

Determination of the 16S rRNA nucleotide sequence

Genomic DNA of the selected isolate (BP11) was extracted following the procedure of the Geneclean II kit (Qiagen, Germany), as previously described [40]. The PCR amplification was carried out in a DNA Thermal Cycler (Perkin Elmer, USA) using the universal bacterial 16S rRNA primers: 10F 5' -AGTTTGATCCTGGCTC- 3' and 1540R 5' -AAGGAGGTGATCCAGCC- 3'. PCR reactions were carried in a total volume of 50 µl, comprised of ~1 µg of DNA template, 1µM of each primer, 1.5 mM Mg^{2+}, 200 µM dNTPs, 1.5 U Taq DNA polymerase in 1 x Taq DNA polymerase buffer. All PCR reagents were purchased from Promega, USA. The PCR conditions consisted of 35 cycles of 95°C for 1 min, 46°C for 1 min and 72°C for 1 min, followed by a final 72°C for 7 min [41]. After that, the PCR products were sequenced commercially at the BioService Unit (NSTDA, Bangkok, Thailand) using the same F and R primers for sequencing the leading and lagging strand, as used for the PCR. The consensus sequence was compared with those 16S rDNA sequences available in the NCBI GenBank database using the BLASTn search algorithm. The highest sequence identities were aligned using Clustal, checked by eye and used to compute the pair-wise sequence identity and also construct a phylogenic tree using the Neighbor Joining distance method in the PHYLIP Version 3.5. on line software, (http:// evolution.genetics. washington.edu/phylip.html) (June 12, 2010).

Determination of Antimicrobial Substances

The evaluation of the production of antimicrobial substances by the BP11 isolate was preliminarily determined using an antimicrobial residue screening test kit (AM-Test), developed by researchers at the Center for Antimicrobial Resistance Monitoring in Food-borne Pathogens (in collaborating with WHO), Faculty of Veterinary Science, Chulalongkorn University, Bangkok, Thailand. The principle is based on the tube diffusion method and the threshold for detection limit for 18 known standard antibiotics which are in the range of > 0.001-0.1 ppm, depending upon the antibiotic. The BP11 culture broth after a 24 h (late log phase growth) and 48 h (spore forming stage) incubation at 30o C was harvested by centrifugation (at 6,230x g, 10 min at 4o C) to pellet the bacteria, and the supernatant passed through a 0.22 µm sterile filter and transferred into a tube containing semi-solid agar with spores of the Gram positive Geobacillus stearothermophilus and pH indicator. Milk containing antibiotic provided in test kit was used as positive control. Tubes were incubated at 65 ± 1o C for 2-2.5 h. During this incubation period any relatively small antimicrobial substances in the sample would diffuse into the agar and inhibit the growth of G. stearothermophilus, allowing the purple color to remain, whereas in the absence of any inhibitor a yellow color would be detected.

Determination of the Enterotoxin Produced by BP11 Culture

Determination of the enterotoxin from filtrates of the BP11-cultured broth, were performed following the procedures according to manufacturer's instructions (TECRA® test kit; for the detection of Bacillus diarrheal enterotoxin (BDE);

TECRA International Pty Ltd, Frenchs Forest NSW 2086, Australia). The principle of this enterotoxin test is based on the "sandwich" configuration of an enzyme-linked immunosorbent assay (ELISA).

Shrimp Feed Preparation

Black tiger shrimp (P. monodon) were fed a commercially formulated feed purchased from Grobest and Phokaphan Aquatech Corporation Co, Ltd., Thailand. The nutrition in feed by weight were as follows: protein (>35 %), lipid (>5%), fiber (>4%), moisture (<10%), and a trace of ash. For the BP11-supplemented feed, 1 kg wet weight (~100 g dry weight) of BP11 (~10-10 CFU g^{-1}) was thoroughly mixed with 4 kg of feed (i.e., approximately 2.5% (w/w) BP11). For this purpose, a single BP11 colony from a TSA plate was cultured in TSB at 30o C with shaking at 200 rpm for 24 h and then harvested by centrifugation at 6,230x g, 4o C for 10 min and washed three times with NSS before being weighed and re-suspended in fresh NSS. Fresh BP11 cells were then thoroughly mixed with the feed at a 1:4 (BP11: feed), and the mixture was spread out and dried in an oven for 1–2 h at 37o C. Feed was then stored in clean plastic bags at 4o C until use. Shrimp feed was prepared twice weekly, and each batch was analyzed for the level of viable BP11 cells as CFU g-1 by the total plate count method. This BP11-supplemented feed typically contained viable BP11 at ~3.9 x 108 - ~ 3.6 x 109 CFU g^{-1}. Aliquots from each feed preparation were stored at room temperature (~ 28-32o C) and at 4o C and assayed for viable BP11 levels (as CFU g^{-1} feed) up to six months in order to monitor their survival upon storage.

Experiment I: PL-30 shrimp Culture in Concrete Tanks

Shrimp age of ~ post larva -15 (PL-15) were purchased from a reputable hatchery at Chachoengsao province in Thailand, and acclimatized to the PL-30 stage (0.13±0.01 g) in concrete tanks with flat bottoms (each measuring 80 x 74 x 87 cm) with a closed recirculating water system of 400 L [42]. The culture water salinity level was initially set at 20%. These PL-15 shrimp were bred from broodstock caught from the Andaman sea. After acclimatization to the PL-30 stage, healthy active shrimp were randomly assigned into the two groups: a control and probiotic treatment. Shrimp in the control group were fed with shrimp feed (CF), whilst those in the probiotic treatment group were fed with the BP11-supplemented regular shrimp feed (PF). Each treatment was comprised of three replicates, each containing 70 or 60 shrimp/tank for the two trials, respectively. Shrimp were fed three times daily at 10% body weight and after 90 days of culture the survival (%) and total live weights (g) were determined.

Experiment II: PL-50 Shrimp Culture in net cages in Earthen Pond

Hatchery reared P. monodon from PL-50 (0.23±0.06 g), obtained from the same shrimp hatchery and broodstock as the PL-15 above, were first acclimatized in the cement tank. Shrimp were then preselected for healthy active shrimp and then randomly assigned to the CF- and PF- treatments, as detailed above, and stocked in ten 1.5 m^2 net cages (1.0 mm mesh) at 100 shrimp/cage. The net cages were all placed at a 1.3 m water depth in a single 1000 m^2 surface area earthen pond (typically 1.5 m deep) along an access walkway. Cages were suspended 20 cm above the pond bottom and extended 40 cm above the water surface. Five cages contained the control (CF) shrimp, while the PF shrimp were housed in the remaining five cages. Individual cages were separated by a distance of approximately 1 m. A row of five cages were separated by a distance 1.5 m on opposite sides of the walkway for two rows. Two rows of walkways were separated by a distance of one meter. Pond water salinity was ~6-7‰ during the dry season of Thailand and was aerated with a closed-system operation. All shrimp were fed three times daily at 10% body weight in the net cages for 80 days.

Intestinal Microbial Investigation of CF- and PF Shrimp Intestines

The CF- and PF shrimp of 80 days culture in an earthen pond were randomly selected and kept on ice. Their gastrointestinal tracts were dissected out. The intestines were longitudinally cut and rinsed vigorously with sterile NSS three times before fixation and a portion was fixed in 2.5% (w/v) glutaraldehyde in 0.1 M phosphate buffer pH 7.2 for 2 hr. After fixation these samples were then processed following conventional procedures, as reported previously [43], and examined by scanning electron microscopy (SEM) using a Jeol JSM-5410LV (Japan) electron microscope.

Water samples collection: Water samples were collected from the center of each tank (20 ml) and net cage (200 ml) from each treatment for determination of the viable BP11 cells every 15 or 20 days. Water quality measurements included the dissolved oxygen, ammonium, nitrite and phosphate levels plus the temperature, pH and salinity, and were evaluated as described by Strickland and Parsons [44].

Vibrio harveyi challenge tests: After 90 and 80 days, shrimp cultured in concrete tanks and in an earthen pond, respectively were randomly taken and challenged with Vibrio harveyi 639. The PF- and CF shrimp were subjected to static, disease challenge tests using the luminescent bacteria V. harveyi 639 at ~107 CFU ml^{-1}, according to Austin et al. [10] and modified from Rengpipat et al. [22]. Each challenge included 12 tanks with a 2 x 2 factorial design. Treatments were the PFand CF shrimp challenged with V. harveyi, and unchallenged PF- and CF shrimp. Treatments were performed in triplicate containing 30 and 25 shrimp per tank for shrimp cultured from concrete tanks of Trial I and II, respectively. Triplicate replications of each treatment, using 20 shrimp per tank, for shrimp cultured from the net cages in an earthen pond.

During V. harveyi challenge tests, no water was exchanged and the water quality parameters and shrimp survival were measured every 2 d. The cumulative number of dead shrimp was recorded each day. Shrimp that died following V. harveyi 639 exposure were dissected and the hepatopancreas-intestine and heart muscle were removed, cut and fixed in Davidson's fixative solution for 24 h and then processed for paraffin sectioning by standard techniques. Tissue sections (8 μm thick) were prepared and processed for haematoxylin and eosin staining (H & E) followed by indirect immunoperoxidase staining using first the anti-Vibrio VH3-3H murine monoclonal antibody [45] and secondly the horseradish peroxidase conjugated goat-anti-mouse (GAM-HRP) secondary antibody, diluted to 1:1000 in

10% (v/v) normal calf serum in PBS, as reported previously [45]. A positive immunoreactivity was visualized as a deep brown coloration against the pink and purple background of the H & E staining. In addition, the bacteria isolated from the hepatopancreas-intestine and heart muscle of each shrimp, were purified and identified using Gram staining, oxidase and motility tests as described previously [39], and were compared with the original V. harveyi 639 culture to confirm the similarity to the V. harveyi strain.

Haemolymph Collection

For determination of the total haemocyte count and antibacterial activity before and after challenge with V. harveyi, the haemolymph from shrimp that were cultured in the net cages in the earthen pond was collected from the ventral-sinus cavity of each shrimp using a 26-gauge needle and a 1-ml syringe containing modified KC-199 medium (K-199 medium plus Hepes 2.38 gl^{-1}, supplemented with 5% (w/v) L-cysteine as an anticoagulant [46]).

Total Haemocyte Count

A 100μl aliquot of the haemolymph, collected from three randomly selected shrimp per tank, was diluted into 0.4 ml of modified KC-199 medium and mixed gently, and then the haemocytes were counted using a haemocytometer under a light microscope at 400x magnification, and calculated as cells ml^{-1} of haemolymph.

Antibacterial Activity

One hundred microlitre of the shrimp haemolymph was added to 1.4 ml of ice cold, sterile Van Harrevald's salt solution (VHS) [47], mixed and then centrifuged at 600x g for 10 min at 4o C. The plasma supernatant was collected and sterilized by filtration (millipore membrane filter, 0.45 mm pore size) for subsequent assays for antibacterial activity. V. harveyi 639 was cultured in TSB supplemented with 2% (w/v) NaCl overnight at 37o C. The concentration was adjusted to ~104 CFU ml^{-1}, collected and washed in 2% (w/v) NSS by centrifugation at 92,006x g for 15 min at 4o C. The V. harveyi 639 suspension and shrimp plasma (100 μl each) were mixed in a sterile test tube and incubated for 1 hr at 37o C. Three 50 ml aliquots were taken from each tube and spread on thiosulfate-citrate-bile-sucrose (TCBS) agar plates to estimate the viable bacterial numbers as CFU ml^{-1} after incubating at 37o C for 24 hr. For the positive control, V. harveyi 639 were incubated with VHS. The percentage inhibition was calculated as reported elsewhere [48], where; %inhibition = 100 - (mean CFU ml-1 sample/ mean CFU ml^{-1} positive control) x 100.

Statistical Analysis

The effect of BP11 on the shrimp growth and survival were evaluated using student's t-test [49] with p

RESULTS

BP11 Identification and its Characteristics

Of the 245 bacterial isolates obtained from screening the intestines of freshly caught P. monodon broodstocks from the Andaman sea, BP11 isolate produced the greatest antimicrobial activity against V. harveyi 639 and E. coli (data not shown), and so was selected for further evaluation. BP11, a Gram-positive rod bacteria with a cell size of ~0.55-0.75 x 2.5-3.5 μm (width x length), can form a central spore (Figure 1), and gave good growth in TSB containing 1 - 4% (w/v) NaCl; at temperature of 35-50o C and pH of 6-8. The best growth was found in TSB with 2% (w/v) NaCl at 45o C and a pH of 6.5 (data not shown). The BP11 isolate was identified using the api-20E® and api-50 CHB® Medium test kits, where, after analysis with the APILAB Plus software, the percentage similarity of BP11 to Bacillus subtilis was found to be 99.9%. Moreover, sequence analysis of the 1491 bp partial 16S rRNA gene fragment from BP11 (GenBank accession code HM585370) exhibited the highest sequence identity to those sequences annotated as B. subtilis or B. amyloliquefaciens (>99% sequence identity), supported by a NJ distance based phylogenetic tree which grouped BP11 within the B. subtilis/B. amyloliquefaciens clade (data not shown). Thus, the biochemical and morphological analysis combined with the 16S r-RNA based molecular analysis place BP11 as a potential isolate of B. subtilis within the Firmicutes bacterial division.

Antimicrobial and Enterotoxin Detection

In contrast to the positive control (milk containing antibiotics) that showed a purple color, a yellow color was detected in the AM-test for filtrates of BP11 broth. Thus, no antimicrobial residue was detected in the BP11 broth after 24 and 48 hr of culture. In addition no BDE was detected in the BP11 broth culture by the TECRA test kit.

PL-30 and PL-50 Shrimp Culture in Concrete tanks and in Earthen Pond

Water Quality

The water quality values, during the two experiments each, for the shrimp cultures in the concrete tanks and in the earthen pond were similar for dissolved oxygen (~10 and 6.1-7.7 mg l^{-1}), ammonium (0.0-0.5 and 0.0 mg l^{-1}), nitrite (0.0-0.5 and 0.1mg l^{-1}) and phosphate (0.1-0.2 and <0.1-0.25 mg l^{-1}), but slightly different for the pH (6.5-7.0 and 7.7-8.4), temperature (26-27 and 28.4-31.4o C) and salinity (20-23 and 6.1^{-7}.2%). However, all of these water quality parameters were considered safe for shrimp culture [42,50,51]. BP11 counts of ~10^2-10^3 CFU ml^{-1} were found in all culture water taken from the concrete tanks or net cages of shrimp fed with BP11-supplemented feed (data not shown).

Experiment I: PL-30 Shrimp Growth in Concrete Tanks

The average live weights of the CF shrimp in the concrete tanks after 90 days of culture for the two separate trials (6.17 ± 0.61 g and 6.99 ± 0.19 g) were significant different ($p < 0.05$) from those of the PF shrimp (7.48 ± 0.23 g and 8.94 ± 0.43 g) (Table 1).

Experiment II: PL-50 shrimp Growth in the net cages in the Earthen Pond

The influence of BP11 on PL-50 shrimp growth and survival over 80 days was evaluated by simulating to a certain extent the conditions used in commercial shrimp culture using 1.5 m^2 surface area net cages, located in a 1,000 m^2 surface area earthen pond. The average live weights of the PF shrimp (13.74±0.77 g) in the net cages after 80 days culture were significantly different from that of the CF shrimp (12.86±0.38 g) (Table 2). In addition, the survival of PF-group shrimp (76.77 ± 3.78 %) was higher than that in the CF shrimp (65.2 ± 7.60) (Table 2). The food conversion ratio and productive index in the PF-group shrimp were better than those in the CF shrimp (Table 2).

Intestinal Microbial Investigation of CF- and PF shrimp Intestines

After culture for 80 days in the earthen pond, both the CF- and PF shrimp were randomly selected. The intestines were dissected out for sample preparation and examination by SEM to detect adherent bacteria on the intestinal wall. No bacteria conforming to the morphology of the BP11 was detected on the surface of the intestine of the CF shrimp (Figure 2a and b), whereas many rod-shaped bacteria were observed on certain surface area of some intestine portions from PF shrimp (Figure 2c and d). Concurrently, BP11 counts from the total bacteria counts (~10^6 -10^7 CFU g-1 intestine of either CF-or PF shrimp, data not shown) were < 10^1 -10^2 and ~10^5 -10^6 CFU g^{-1} intestine for the CF- and PF shrimp, respectively (data not shown), which supports the possible presence of BP11 on the surface of the PF shrimp's intestines.

Longevity of BP11 in dried feed Viable BP11 counts in the BP11 supplemented feed were found to be ~ 109 and ~108 CFU g^{-1} of feed after storage for 6 months at 4o C and room temperature (28-32o C), respectively. No changes in the physical morphology or the biochemical properties of BP11 were detected upon storage at either temperature for 6 months. Moreover, a clear probiotic affect of BP11, including the growth and vibriosis resistance enhancement in P. monodon (in vivo), was evident after storage at -18o C for 2 years, where for example the data of Table 2 is derived from the use of a 2 year old storage culture of BP11,

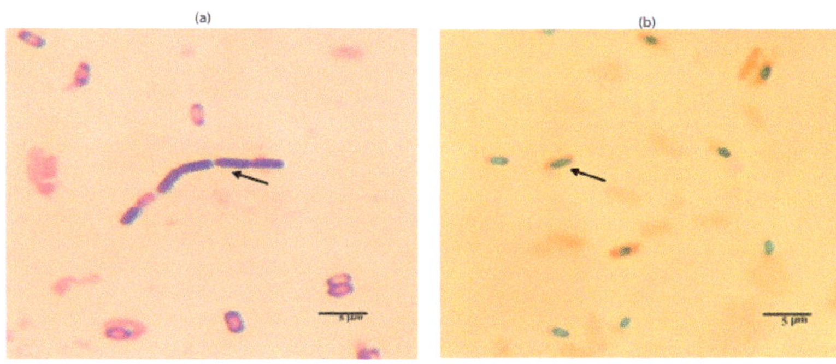

Figure 1: Light microscope images (x 1,000 magnification) of the Gram-positive rods and spores of BP11. (a) The cells, sized ~0.55-0.75 x 2.5-3.5 μm (width x length) and (b) the central spore of BP11 as pointed by arrow.

Table 1: Survival and average live weight of shrimp after 90 days culture in concrete tanks fed with (PF) or without (CF) BP11 supplemented feed, and their cumulative mortality after challenge by Vibrio harveyi 639

Experiment I	Control (CF)	Probiotic (PF)
Trial I (n = 210)		
Average shrimp live weight (g) on day 90 of culture (n = 30)	6.17 ± 0.61	7.48 ± 0.23*
Average daily gain (mg)	68.13 ± 6.73	82.68 ± 2.52*
Survival (%)	67.14 ± 6.54	84.29 ± 3.78*
After challenge by *V. harveyi*:		
Cumulative mortality (%) on day 4	55.0 ± 5.29	28.33 ± 3.06*
Trial II (n = 180)		
Average individual live weight (g) on day 90 of culture (n = 30)	6.99 ± 0.19	8.94 ± 0.43*
Average daily gain (mg)	77.21 ± 2.07	98.89 ± 4.74*
Survival (%)	61.11 ± 2.48	63.89 ± 1.47
After challenge by *V. harveyi*:		
Cumulative mortality (%) on day 5	56.0 ±5.33	17.78± 2.04*
Cumulative mortality (%) on day 9	100	44.0 ±3.53*

*Indicates a significant difference ($p < 0.05$) between the control (CF) and probiotic (PF) treatment groups.
Average daily gain; Total weight per days of culture

Table 2: Survival and total live weight of shrimp after 80 days of culture in net cages in an earthen pond fed with (PF) or without (CF) BP11 probiotic supplement in the feed (Experiment II)

Parameters	Control(CF)	Probiotic(PF)
Total live weight (kg)	4.56	5.33
Average individual live weight (g) on day 80 of the culture (n = 100)	12.86 ±0.38	13.74 ± 0.77*
FCR(1)	2.02	1.78
Survival (%)	65.2±7.6	76.8±3.78
PI(2)	184	287.5

*Indicates a significant difference ($p < 0.05$) between the control (CF) and probiotic (PF) shrimp
(1)Food conversion ratio; dried weight of ingested feed per live weight of produced shrimp
(2)Productive index; Shrimp weight x %survival x 100 per FCR x age

Longevity of BP11 in Dried Feed

Viable BP11 counts in the BP11 supplemented feed were found to be ~ 10^9 and ~108 CFU g^{-1} of feed after storage for 6 months at 4o C and room temperature (28-32°C), respectively. No changes in the physical morphology or the biochemical properties of BP11 were detected upon storage at either temperature for 6 months. Moreover, a clear probiotic affect of BP11, including the growth and vibriosis resistance enhancement in P. monodon (in vivo), was evident after storage at -18°C for 2 years, where for example the data of Table 2 is derived from the use of a 2 year old storage culture of BP11,

V. harveyi Challenge Test

The shrimp cultured in the concrete tanks for 90 days were collected and tested for disease resistance to a V. harveyi challenge by external exposure. It was found that the cumulative shrimp mortality was significantly lower in the PF shrimp than in the CF shrimp after external challenge with V. harveyi 639 for 4 or 5 days. After 9 days, 100% cumulative mortality was found in the CF shrimp compared to only 44% for the PF shrimp (Table 1). Thus, the BP11 supplemented feed appeared to afford (prophylactic) protection to the P. monodon shrimp against V. harveyi 639 induced mortality.

Likewise, when PL-50 shrimp collected from the net cage culture after 80 days in the earthen pond were challenged by V. harveyi 639 for 5 days, significantly lower cumulative mortality was observed in the PF shrimp than in the CF shrimp (Table 3). Survival was 100% in unchallenged control shrimp.

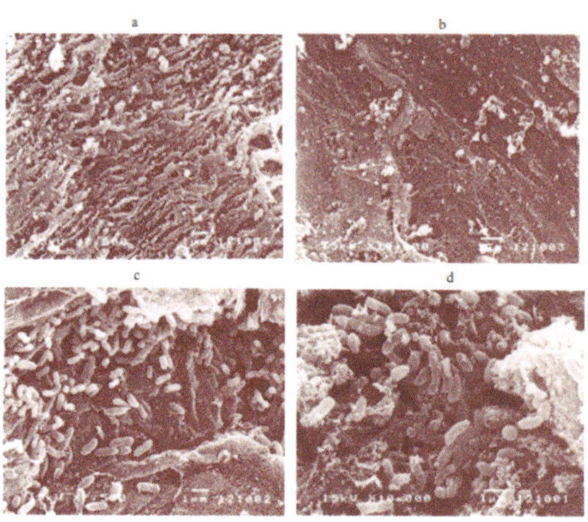

Figure 2: Representative SEM images of intestinal bacteria of 80-day earthen pond cultured shrimps' intestines (a, b) without (CF) or (c, d) with BP11 supplementation and shown at (a, c) x 7,500 or (b, d) x 10,000 magnification.

Table 3: Mean immunity indices values and cumulative mortality (%) of shrimp fed with (PF) or without (CF) BP11 supplement in the feed in net-cage culture in an earthen pond, before and after challenge with Vibrio harveyi 639 for 5 days

Parameters	Before *V. harveyi* challenge		After *V. harveyi* challenge	
	Control (CF)	Probiotic (PF)	Control (CF)	Probiotic (PF)
Total haemocytes (x10^7 cells ml^{-1})	1.32 ± 0.69	1.62 ± 0.19	0.35 ± 0.24	0.11 ± 0.10
Antibacterial activities (% inhibition)	17.3 ± 4.0^a	39.0 ± 3.9^b	63.7 ± 8.1^c	71.1 ± 5.6^c
Cumulative mortality (%); day 5	–	–	81.7 ± 16.1	67.7 ± 10.4

Data are shown as the mean + 1 SD, and are derived from nine shrimp per treatment. Means within a row followed by a different superscript letter are significantly different ($p < 0.05$).

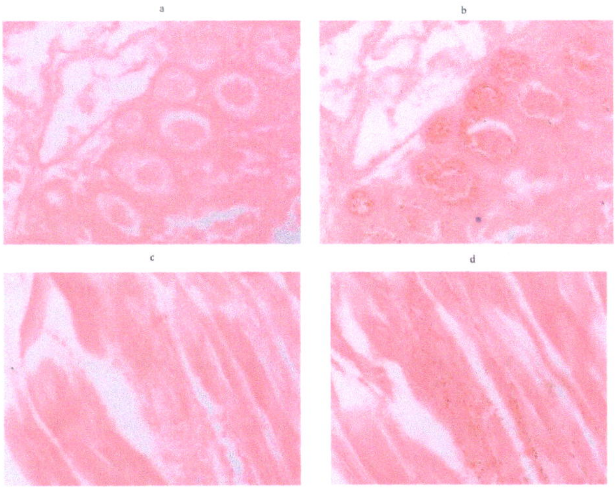

Figure 3: Representative immunohistochemical detection of V. harveyi infected tissues from dead P. monodon following exposure to V. harveyi 639. The (a,b) hepatopancreas and (c,d) heart muscle after staining by haematoxylin and eosin(H & E) while the right panels (b,d) in each case were also stained with indirect immunoperoxidase using the VH3-3H monoclonal antibody as a probe. Thus, brown dots show the presence of V. harveyi 693. Images shown are representative from three independent shrimp and examined under light microscope (x 1,000 magnification).

Those PL-30 and PL-50 shrimp from the concrete tanks and earthen pond, respectively, that had died after exposure to V. harveyi 639 were checked for evidence of internal V. harveyi infection by immunohistochemistry of the hepatopancreas-intestines and muscle tissues using the V. harveyi 639 specific VH3-3H monoclonal antibody (Figure 3), and by standard microbiological examination (V. harveyi counts were ~105 - 106 CFU g intestine, data not shown). In all cases V. harveyi infection was detected in the examined internal tissues. No shrimp mortality was detected among unchallenged CF-and PF shrimp

Immunity Indices

Only the PL-50 shrimp from the net cages in the earthen pond were collected for monitoring their immunity indices during the V. harveyi challenge test over 5 days. Before V. harveyi challenge the total haemocyte level and the antibacterial activity, from the PF shrimp were numerically greater than those of the CF shrimp (Table 3), although only the antibacterial activity was significantly different between the two groups. After V. harveyi challenge both the CF- and PF shrimp significantly increased their antibacterial activities and decreased the number of circulating haemocytes, but a significantly greater increase in the antibacterial activity upon V. harveyi challenge was noted for the PF shrimp. Also, the total circulating haemocyte numbers of the PF shrimp were lower than those of the CF shrimp.

DISCUSSION

BP11 is one of the natural bacterial flora found in the intestines of black tiger shrimp from the Andaman sea off the Thai coast and was found not to produce any detectable amount of antimicrobial substances or BDE in TSB based liquid culture. BP11 was identified as being a member of the genus Bacillus and most likely as an isolate of B. subtilis, which is not known as a human or animal pathogen, nor is it toxigenic like some other members of this genus [52]. B. subtilis has also been authorized in the list of additives for feeding stuffs published by the European Union Commission [53]. BP11 was preliminarily selected as a probiotic candidate after its inhibitory effect on Vibrio harveyi 639 growth had been detected in vitro (data not shown). BP11 level in freshly dried feed ~10^8 and 10^9 CFU g^{-1} was still found after storage for 6 months at room temperature and 4o C, respectively, showing that BP11 is fairly stable in accord with the fact that Bacillus species can resist extreme environments and can produce spores, as can be seen for the BP11 isolate (Figure 1b). This affords them a greater survival, including a longer shelf life in feed, than other groups of bacteria during storage [54]. In addition, after storage at -18o C for 2 years, BP11 was found to still be able to cause growth enhancement in P. monodon (in vivo), as shown in Table 2, supporting the efficacy of BP 11 to induce beneficial effects to hosts, to be able to retain their probiotic properties after feed processing, and to display a sufficient bacterial survival level during storage. These are all important properties for consideration as a good probiotic, as suggested previously [55,56].

More than 15 years after the first finding of Bacillus S11 (BS11) as a potential probiotic for black tiger shrimp [22], we here isolated Bacillus P11. Although both BS11 and PP11 have been identified as isolates of Bacillus subtilis, their cell size, physical appearance of their colonies on agar plates and the optimal growth conditions of BS11 and BP11 are different, suggesting they are different strains. It is thus unclear if this represents stable probiotic-shrimp relationships between different host populations or a change over time due to environmental changes.

After 80 days of being fed BP11 supplemented feed (PF shrimp) in an earthen pond, BP11 bacterial counts in the shrimp intestines were found to be ~10^5 - 10^6 CFU $g^{-1}$of shrimp intestines, and likely BP11 (cells conforming to Bacillus BP11 morphology) were found on the intestine's surface by SEM analysis (Figure 2c and d). Meanwhile, very rare detection of bacteria on the intestines of the CF shrimp (fed with regular feed) was observed by SEM analysis (Figure 2a and b), though total bacteria counts ~10^6 -10^7 CFU g^{-1} intestine of either CF- or PF shrimp were determined (data not shown). This finding supports the ability of BP11 to adhere to the mucosal surface of the shrimp's intestines more than other bacterial flora (Figure 2), which supports our previous finding of BS11 on shrimp intestines' surface using Green fluorescent protein (GFP) as a monitoring system [57]. The benefits of probiotics to the host have been reported as providing an interfering substance to inhibit pathogens, produce useful enzymes or increase the uptake of useful nutrients, provide a competitive exclusion effect on pathogens and help beneficial immune induction [23,25,27,58-60]. In our case, using the principle of balancing an indigenous microorganism in situ along the intestinal tract, BP11 may possibly function as a protective barrier against pathogens, perhaps by competitive exclusion preventing pathogens like V. harveyi from gaining occupation and becoming a major member of the intestinal flora as mentioned earlier [12,14-17, 19,20,25,61]. However, they may also produce enzymes to digest and absorb nutrients and thus lead to healthy shrimp. That BP11 adhered to the surface of the shrimp intestinal wall better than other bacterial flora may due to specific interaction (recognition) between the shrimp host and probiotic bacteria. All the above beneficial effects of probiotic bacteria on the shrimp are not mutually exclusive and could occur simultaneously. In addition, the lysis of the probiotic bacteria could release peptidoglycan, an immunogen that can induce the immune responses of the shrimp, as suggested previously [62-65].

The yield of probiotic fed shrimp (PF group) in the concrete tanks were significantly higher than those of the control shrimp (CF group). However, before BP11 can be recommended for use in commercial farms, confirmation was sought by evaluation of its effects in net caged shrimp in an earthen pond that more closely simulates an actual commercial culture environment. Our results still confirmed the same positive outcome of shrimp fed BP11, including the prophylactic response towards protection from vibriosis. In addition, B. subtilis is known to be a non- to low- virulent species and requires a high number of bacteria to cause disease in humans [66]. Therefore, in conclusion, BP11 could be a good probiotic candidate for black tiger shrimp, P. monodon, as an alternative to antibiotics or chemical agents, and may lead to a more sustainable and safe commercial shrimp culture. However, regarding the recognition as safe for human consumption, risk assessment of BP11 should be further confirmed.

ACKNOWLEDGEMENTS

The authors sincerely thank Robert Butcher for revising the manuscript and providing many valuable suggestions and comments. This research was supported by the Thai Government Stimulus Package 2 (TKK 2555) under the Project for Establishment of Comprehensive Center for Innovative food, Health products and Agriculture (PERFECTRA); the Higher Education Research Promotion and National Research University Project of Thailand, Office of the Higher Education Commission (FW643A); CU Graduate School Thesis Grant from Chulalongkorn University and the National Research Council of Thailand (NRCT).

REFERENCES

1. Pongthanapanich T, Roth E (2006) Toward environmental responsibility of Thai shrimp farming through a voluntary management scheme. IME Working paper 70/06. Department of Environment and Business Economics. University of Southern Denmark, Esbjerg and the authors. 1-44.
2. Fisheries statistics of Thailand (2001-2010). Part 5: Foreign trade of fishery community. Annual report of Information Technology Center, Department of Fisheries, Ministry of Agriculture and Cooperatives, Thailand.
3. Lebel L, Lebel P, Garden P, Giap DH, Khrutmuang S, et al. (2008) Places, chains, and plates: Governing transitions in the shrimp aquaculture production–consumption system. Globalizations 5: 211–226.
4. Lebel L, Mungkung R,Gheewala HS, Leel P (2010) Innovation cycles, niches and sustainability in the shrimp aquaculture industry in Thailand. Environ Sci Policy. 13: 291-302.
5. Amábile-Cuevas CF, Gárdenas-Garciá M, Ludgar M (1995) Antibiotic resistance. Am Sci 83: 320-329.
6. Karunasagar I, Pai R, Malathi GR, Karunasagar I (1994) Mass mortality of Penaeus monodon larvae due to antibiotic resistant Vibrio harveyi infection. Aquaculture 128: 203–209.
7. Holmström K, Gräslund S, Wahström A, Poungshompoo, S, Bengtsson BE, et al. (2003) Antibiotic use in shrimp farming and implications for environmental impacts and human health. Int J Food Sci Technol 38: 255-266.
8. Le TX, Munekage Y, Kato SI (2005) Antibiotic resistance in bacteria from shrimp farming in mangrove areas. Sci Total Environ 349: 95-105.
9. Tendencia EA, de la Peña LD (2001) Antibiotic resistance of bacteria from shrimp ponds. Aquaculture 195: 193-204.
10. Austin B, Stuckey LF, Robertson PAW, Effendi I, Griffith DRW (1995) A probiotic strain of Vibrio alginolyticus effective in reducing diseases caused by Aeromonas salmonicida, Vibrio anguillarum and Vibrio ordalii. J Fish Dis 18: 93-96.
11. Bly JE, Quiniou SM-A, Lawson LA, Clem LW (1997) Inhibition of Saprolegnia pathogenic for fish by Pseudomonas fluorescens. J Fish Dis 20: 35-40.
12. Verschuere L, Rombaut G, Sorgeloos P, Verstraete W (2000) Probiotic Bacteria as Biological Control Agents in Aquaculture. Microbiol Mol Biol Rev 64: 655-671.
13. Westerdahl A, Olsson JC, Kjelleberg S, Conway PL (1991) Isolation and characterization of turbot (Scophthalmus maximus)-associated bacteria with inhibitory effects against Vibrio anguillarum. Appl Environ Microbiol 57: 2223-2228
14. Fuller R (1989) A Review: Probiotics in Man and Animals. J Appl Bacteriol 66: 365-378.
15. Fuller R (1992) Probiotics: The Scientific Basis, 1st edn. Chapman&Hall, London.
16. Fuller R (1997) Probiotics 2: Applications and Practical Aspects 1st edn. Chapman & Hall, London.
17. Gatesoupe FJ (1999) The use of probiotics in aquaculture. Aquaculture 180: 147-165.
18. Tannock GW (1999) Introduction. In: Probiotics: A Critical Review (ed. by G. W. Tannock) pp. 1–4. Horizon Scientific Press, Norfolk, England
19. Irianto A, Austin B (2002) Use of probiotics to control furunculosis in rainbow trout, Oncorhynchus mykiss (Walbaum) J Fish Dis 25: 333–342.
20. Kesarcodi-Watson A, Kaspar H, Lategan MJ, Gibson L (2008) Probiotics in aquaculture: The need, principles and mechanisms of action and screening processes. Aquaculture 274: 1–14.
21. Gildberg A, Mikkelsen H, Sandaker E, Ringø E (1997) Probiotic effect of lactic acid bacteria in the feed on growth and survival of fry and Atlantic cod (Gadus morhua). Hydrobiologia 352: 279–285.
22. Rengpipat S, Phianphak W, Piyatiratitivorakul S, Menasveta P (1998) Effect of probiotics on Black tiger shrimp Penaeus monodon survival and growth. Aquaculture 167: 301-313.

23. Vaseeharan B, Ramasamy P (2003) Control of pathogenic Vibrio spp. By Bacillus subtilis BT23, a possible probiotic treatment for black tiger shrimp Penaeus monodon. Lett Appl Microbiol 36: 83–87.
24. Li P, Burr GS, Gatlin DM, Hume ME, Patnaik S, et al. (2007) Dietary supplementation of short-chain fructooligosaccharides influences gastrointestinal microbiota composition and immunity characteristics of Pacific white shrimp, Litopenaeus vannamei, cultured in a recirculating system. J Nutri 137: 2763– 2768.
25. Balcázer JL, Rojas-Luna T, Cunningham DP (2007) Effect of the addition of four potential probiotic strains on the survival of pacific white shrimp (Litopenaeus vannamei) following immersion challenge with Vibrio parahaemolyticus. J Invert Pathol 96: 147-150.
26. Gómez GD, Shen MA (2008) Influence of probiotics on the growth and digestive enzyme activity of white Pacific shrimp (Litopenaeus vannamei). J Ocean University of China (English Edition) 7: 215–218.
27. Tseng DY, Ho PL, Huang SY, Cheng SC, Shiu YL, et al. (2009) Enhancement of immunity and disease resistance in the white shrimp, Litopenaeus vannamei, by the probiotic, Bacillus subtilis E20. Fish Shellfish Immunol 26: 339–344.
28. Rengpipat S, Rukpratanporn S, Piyatiratitivorakul S, Menasaveta P (2000) Immunity enhancement in Black tiger shrimp (Penaeus monodon) by a probiont bacterium (Bacillus S11). Aquaculture 191: 271-288.
29. Rengpipat S, Tunyanun A, Fast AW, Piyatiratitivorakul S, Menasveta P (2003) Enhanced growth and resistance to Vibrio challenge in pond-reared black tiger shrimp (Penaeus monodon) fed a Bacillus probiotic. Dis Aquat Organ 55: 169-173.
30. Liu KF, Chiu CH, Shiu YL, Cheng W, Liu CH (2010) Effects of the probiotic, Bacillus subtilis E20, on the survival, development, stress tolerance, and immune status of white shrimp, Litopenaeus vannamei larvae. Fish Shellfish Immunol 28: 837-844.
31. Zhou X, Wang Y, Li W (2009) Effect of probiotic on larvae shrimp (Penaeus vannamei) based on water quality. Aquaculture 287: 349–353.
32. Moriarty DJW (1998) Control of luminous Vibrio species in penaeid aquaculture ponds. Aquaculture 164: 351–358.
33. Brunt J, Newaj-Fyzul A, Austin B (2007) The development of probiotics for the control of multiple bacterial diseases of rainbow trout, Oncorhynchus mykiss (Walbaum). J Fish Dis 30: 573-579.
34. Kumar R, Mukherjee SC, Ranjan R, Nayak SK (2008) Enhanced innate immune parameters in Labeo rohita (Ham.) following oral administration of Bacillus subtilis. Fish Shellfish Immunol 24: 168-172.
35. Newaj-Fyzul A, Adesiyun AA, Mutani A, Ramsubhag A, Brunt J, et al. (2007) Bacillus subtilis AB1 controls Aeromonas infection in rainbow trout (Oncorhynchus mykiss, Walbaum). J Appl Microbiol 103: 1699-1706.
36. Hill JE, Baiano JC, Barnes AC (2009) Isolation of a novel strain of Bacillus pumilus from penaeid shrimp that is inhibitory against marine pathogens. J Fish Dis 32: 1007–1016.
37. Ochoa-Solano JL, Olmos-Soto J (2006) The functional property of Bacillus for shrimp feeds. Food Microbiol 23: 519–525.
38. Naclerio G, Ricca E, Sacco M, Felice MD (1993) Antimicrobial activity of a newly identified bacteriocin of Bacillus cereus. Appl Environ Microbiol 59: 4313–4316.
39. Baumann P, Schubert RHW (1984) Vibrionaceae. In: Krieg, N.A., Holt, J.G._Eds., Bergey's Manual of Systematic Bacteriology vol. 1 William & Wilkins, Baltimore, pp. 516–549.
40. Ausubel FA, Brent RE, Moore DD, Seidman JG, Smith JA, et al. (1999) Current Protocols in Molecular Biology. 4th ed. New York, John Wiley & Sons.
41. Hauben L, Moore ER, Vauterin L, Steenackers M, Mergaert J, et al. (1998) Phylogenetic position of phytopathogens within the Enterobacteriaceae. Syst Appl Microbiol 21: 384-397.
42. Menasveta P, Fast AW, Piyatiratitivorakul S, Rungsupa S (1991) An improved, closed seawater recirculation maturation system for giant tiger prawn (Penaeus monodon Fabricius). Aquacultural Eng 10: 173-181.
43. Watson LP, Mckee AE, Merrell BR (1980) Preparation of microbiological specimens for scanning electron microscopy. Scan Electron Microsc. II: 45-56.
44. Strickland JDH, Parsons TR (1972) (2nd Edition) A practical handbook of seawater analysis. J Fish Res Board Canada. 167: 310 pp.
45. Phianphak W, Rengpipat S, Rukpratanporn S, Longyant S, Chaivisuthangkura P, et al. (2005) Production of monoclonal antibodies for detection of Vibrio harveyi. Dis Aquat Organ 63: 161-168.
46. Itami T, Takahashi Y, Tsuchihira E, Igusa H, Kondo M (1994) Enhancement of disease resistance of kuruma prawn Penaeus japonicus and increase in phagocytic activity of prawn hemocytes after oral administration of β-1,3-glucan (Schizophyllan) In:
47. Chou, L.M., Munro, A.D., Lam, T.J., Chen, T.W.,Cheong, L.K.K., et al._Eds., The 3rd Asian Fisheries Forum. Asian Fisheries Society, Manila, Philippines, pp. 375–378.
48. Van Harrevald A (1936) A physiological solution for freshwater crustaceans. Proc Soc Exp Biol Med 34: 428-432.
49. Adams A (1991) Response of penaeid shrimp to exposure to Vibrio species. Fish Shellfish Immunol 1: 59–70.
50. Statistical Analysis System (1983) SAS Introductory Guide. SAS Institute, Cary, USA, 93 pp.
51. Menasveta P, Aranyakanonda P, Rungsupa S, Moree N (1989) Maturation and larviculture of penaeid prawns in closed recirculating seawater systems. Aquacultural Engineering 8: 357-368.
52. Boyd CE (1990) Water quality in ponds for aquaculture. Birmingham Publishing Co., Birmingham, USA.482 pp.

53. U.S. Environmental Protection Agency (1997) Bacillus subtilis Final risk assessment. 2009.

54. Council Directive 70/524/EEC. (2004) List of the authorized additives in feeding stuffs published in application of Article 9t (b) of Council Directive 70/524/ EEC concerning additives in feedingstuffs. Official J Eur Union C50–C144.

55. Hong HA, Duc LH, Cutting SM (2005) The use of bacterial spore formers asprobiotics. FEMS Microbiol Rev 29: 813–835.

56. Fonseca F, Béal C, Corrieu G (2001) Operating Conditions That Affect the Resistance of Lactic Acid Bacteria to Freezing and Frozen Storage. Cryobiology 43: 189–198.

57. Wang YB, Li JR, Lin, J (2008). Probiotics in aquaculture: Challenges and outlook. Aquaculture 281: 1–4.

58. Rengpipat S, Wongtangprasert N, Palaga T (2009) The use of green fluorescent protein as a marker for monitoring a probiotic Bacillus S11 in the black tiger shrimp Penaeus monodon. Aquaculture Nutrition 15: 297-305.

59. Westerdahl A, Olsson J, Kjelleberg S, Conway P (1991) Isolation and characterization of turbot (Scophthalmus maximus) associated bacteria with inhibitory effects against Vibrio anguillarum. Appl Environ Microbiol 57: 2223–2228.

60. Gullian M, Thompson F, Rodriguez J (2004) Selection of probiotic bacteria and study of their immunostimulatory effect in Penaeus vannamei. Aquaculture 233: 1-14.

61. Ziaei-Nejad S, Rezaei MH, Takami GA, Lovett DL, Mirvaghefi AR, et al. (2006) The effect of Bacillus spp. bacteria used as probiotics on digestive enzyme activity, survival and growth in the Indian white shrimp Fenneropenaeus indicus. Aquaculture 252: 516-524.

62. Wang YB (2007) Effect of probiotics on growth performance and digestive enzyme activity of the shrimp Penaeus vannamei Aquaculture 269: 259–264.

63. SÖderhäll K, Unestam T (1979) Activation of serum prophenoloxidase in arthropod immunity. The specificity of cell wall glucan activation and activation by purified fungal glycoproteins of crayfish phenoloxidase. Can J Microbiol 25: 406-414.

64. SÖderhäll K (1981) Fungal cell wall β-1,3-glucans induce clotting and phenoloxidase attachment to foreign surfaces of crayfish hemocyte lysate. Dev Comp Immunol. 5: 565-573.

65. Itami T, Asano M, Tokushige K, Kubono K, Nakagawa A (1998) Enhancement of disease resistance of Kuruma shrimp, Penaeus japonicus, after oral administration of peptidoglycan derived from Bifidobacterium thermophilum. Aquaculture 164: 277-288.

66. Rodríguez J, Espinosa Y, Echeverría F, Cárdenas G, Román R, et al. (2007) Exposure to probiotics and β-1,3/1,6-glucans in larviculture modifies the immune response of Penaeus vannamei juveniles and both the survival to White Spot Syndrome Virus challenge and pond culture. Aquaculture 273: 405-415.

67. Edberg SC (1991) US EPA human health assessment: Bacillus subtilis. Unpublished, US. Environmental Protection Agency, Washington, DC.

CHAPTER 13

EFFECT OF PROBIOTIC ON MICROBIOLOGICAL AND HAEMATOLOGICAL RESPONSIVENESS OF CAT FISH (HETEROPNUESTES FOSSILIS) CHALLENGED WITH BACTERIA AEROMONAS HYDROPHILA AND FUNGI APHANOMYCES INVADANS

Meeran Mohideen, and Haniffa MA

Keywords: Probiotic, Microbiological, Haematological, Cat fish, Bacteria, Fungi

INTRODUCTION

Only in the last few years, aquaculture has undergone rapid advancement. The principle reasons for the increased interest and development of fish farming are due to the recent advances in the development of fish culture techniques in the world particularly in the field of husbandry and management of culture system and the development of standardized artificial breeding technologies such that supply of seeds is guaranteed and controlled by the fish farmer. A large number of reports are available on general biology in relation to food, feeding habits and breeding [1]. However, information in relation to nutrient requirements of the fish protein is limited. Development of economical feed mixture is an important factor in fish culture in which the growth of fish is influenced by the quality and quantity of the diet. Reports are available on the growth of cultivable fishes using animal and plant sources of protein diets [2]. For optimum fish growth the use of fish meal (25-65%) as higher dietary protein in fish feeds causes more expensive. Hence, the development of low cost and nutritionally balanced diet is in urgent need. The alternative sources of protein either by partial or complete replacement of the fish meal have been studied using various ingredients [3]. This study focussed on investigation attempts have been made to produce the fish feed using fish meal like anchovy, jawala and flour like soy flour, tapioca flour and wheat flour and sunflower oil, aquasavour, vitamin C.

Infectious diseases are considered as one of the main barriers to the successful development and continuation of molluscan and shrimp aquaculture as they limit production in terms of quality, quantity and regularity [4]. Although disease control is an inherent component of any intensive animal production system, controlling disease in the aquatic environment is further complicated by the intimate relationship that exists between pathogens and their host and the frequent use of open production system [5]. However, excessive antimicrobial use can lead to the emergence of bacterial resistance [6,7]. Hence the use of probiotics for disease prevention and improved nutrition in aquaculture is becoming popular due to an increasing demand for environment friendly aquaculture [6]. Probiotics act as growth promoter and reduce the substrate of pathogenic microbes. Commonly available probiotics are Lactobacillus acidophilus, Bacillus subtilis and EfinolG (mixture of microbes). Several studies have shown that probiotics improves the growth rate of fishes by improving the immune status of fishes [1,4,8-10]. The use of Probiotics to displace pathogenic bacteria by competitive process is a better remedy than administering antibiotics. Pseudomonas fluorescens (AH2) was shown to be strongly inhibitory against Vibrio anguillarum and it reduced the mortality rate

Citation: Mohideen M, Haniffa MA (2015) Effect of Probiotic on Microbiological and Haematological Responsiveness of Cat fish (Heteropnuestes fossilis) Challenged with Bacteria Aeromonas hydrophila and Fungi Aphanomyces invadans. J Aquac Res Development 6:384. doi:10.4172/2155-9546.1000384.

of rainbow trout injected by Vibrio anguillarum [5]. Improved disease resistance has also been observed in cod fry fed with dry feed containing Carnobacterium divergens [11,12] showed that the survival and growth of the black tiger shrimp (Penaeus monodom) [12]. Thus, probiotics have been shown to be effective in a wide range of species for the promotion of growth, enhanced nutrition, immunity and survival.

MATERIALS AND METHODS

Heteropneustes fossilis were collected from local market in Tirunelveli, TamilNadu, India and transported to CARE aqua farm. They were acclimatized to laboratory condition for a week. Ambient temperature 29±1°C and pH 7, 1±0.5mg/lr, was maintained, throughout the experiment and 1/3rd of water was renewed daily.

Feed Formulation

Fishes were fed regularly with an artificial balanced diet made up of wheat flour, tapioca flour, soya flour, vegetable oil, anchovy, jawala to control fishes and Bacillus subtilis was added in the diet to experimental fishes (B. Subtilis was purchased from Xi'an Lyphar Biotech co., Ltd, China). Fishes were reared in plastic troughs (capacity 40 litres). The following ingredients were used to prepare feed pellet for Heteropneustes fossilis for 100 mg fish feed.

Anchovy - 26.9 gm

Soy bean - 25 gm

Jawala Acetes - 20 gm

Tapioca meal - 10.9 gm

Wheat flour - 10 gm

Sunflower oil - 5.8ml

Mono sodium phosphate - 0.5 gm

Aquasavour - 0.3 gm

Ascorbic acid - 0.02 gm

Probiotic concentration - 2 mg

Fish meal

Anchovy and Jawala prawn were purchased from local fish market. They were powdered and sieved to required size. The protein content of the fish meal was 60%.

Flour: Wheat, soybean and tapioca flour were used. Tapioca flour acts as a binder and source of carbohydrate. Gelatinization of tapioca flour improves the stability of the feed.

Oil: 5.8 ml of sunflower oil was used in the food formulation.

Constituents

Energy - 884 cal

Total fatty acids - 100 gm

i. SFA - 12 gm

ii. MUFA - 28 gm

iii. PUFA - 60 gm

iv. Trans fatty acids - BDL

Vitamin: 0.02 gm Vitamin C was used to formulate fish feed.

i. Saturated Fatty Acid

ii. Mono Saturated Fatty Acid

iii. Poly Saturated Fatty Acid

iv. Below Detectable limit (0.05%)

Feed Preparation

All powdered ingredients were weighed separately and mixed with required amount of hot water to make semi moist dough and sterilized .Then 5.8ml sunflower oil was added under appropriate temperature and the probiotic B. subtilis was also added under aseptic condition. This mixture was pelleted and dried.

Experimental Design

The experimental fishes H. fossilis were reared in plastic trough of 40lr capacity. Twelve troughs containing six fishes in each were taken and filled with water. This was divided into two groups as Diet 1 (D1) treatment (4 troughs) and Diet 2 (D2) probiotic treatment (8 troughs). The control fishes were fed with control basal diet while the probiotic treatment fishes were fed with B. subtilis added diet for 21 days feeding trial. Before the feeding trial the gut micro flora and the haematology of the experimental fishes were analysed.One third of water was changed daily. After completion of feeding trial the fishes were divided into three treatments (D1) group fishes was kept as treatment 1 (T1) The D2 group was further divided into two treatments with four troughs of fish for each. The treatment II fishes were injected A. hydrophila in treatment III were A. invadans injected fishes. This disease challenge was done for 15 days and the microbiological and haematological changes were observed and recorded.

LD_{50}

The LD_{50} Aeromonas hydrophila (106 dilution) by one fish/ dose, six fishes each replicate, were injected intramuscularly, in 0.1 ml concentration. Aphanomyces invadans (105 dilution) by one fish/ dose, six fishes each replicate, were injected intramuscularly, in 0.1 ml concentration.

Disease Challenge

Before the treatment the control fishes were fed with control feed and the experimental fishes were fed with probiotic containing feed. After the treatment all the experimental fishes were fed with control feed including control fishes. The treatment (T1) fishes were injected with physiological saline. The treatment II (T2) and treatment III (T3) fishes were injected with 0.1 ml of A. hydrophila and A. invadans respectively.

Disease challenge was done on 31st day by intramuscular injection of A. hydrophila and A. invadans to (T2) and (T3) group fishes respectively. At the start of experiment (T=0) fishes in (T2) and (T3) of the fishes were fed with feed containing probiotic, while the remaining 1st group was fed with non-probiotic feed. The immunological parameters (RBC count, WBC count, DLC, PCV, MCV, MCHC, Haemoglobin content of blood) of the fish were recorded on (31, 30, 35 and 40 days.) in all the three treatments.

Microbiological Estimation

The samplings were made on 0, 30, 35, 40and45 days for microbial investigation of gut of experimental fish H. fossilis in order to check the changes in gut micro flora after probiotic feeding.

Enumeration of total heterotrophic bacterial count

The experimental fish H. fossilis were taken and wiped with alcohol to remove the surface bacteria. The fish were dissected and the intestine was removed and homogenized with sterile distilled water and grinded with morter and pestle. 1ml of homogenized solution was taken and added into sterile blank water representing as 10-1 dilution. Then it was serially diluted upto 10-10 dilution. For Heterotrophic bacterial count 0.1 ml was taken from 10-4 to 10-7 dilution and spread plated into sterile nutrient agar plates (10-4 to 10-7). Duplicates were maintained for each dilution. The plates were incubated at 37°C for 24 hrs. After incubation colonies were observed and recorded.

Collection of Blood Sample

Blood was obtained by puncturing the heart by using 1ml insulin syringe. Before that the syringe and blood collecting vials are coated with anti-coagulant heparin. Anticoagulated blood was used for the analysis of the blood variables except differential leucocyte count and serological parameter

Total Erythrocyte Count

Enumeration of total erythrocytes was done with Haemocytometer. Blood was diluted 200 times in the standard RBC pipette with RBC diluting fluid.

Total Leukocyte Count

Enumeration of total leucocytes was done using Haemocytometer.

Differential leucocyte count

For differential leucocyte count minimum of 100 leucocytes were classified and counted from each smear. They were indentified under 40X magnification.

Haemoglobin content

Sahlis's method had been employed for estimating haemoglobin content of blood.

Mean Corpuscular Haemoglobin (MCH)

It's used to determine the average haemoglobin content in single red cell in micro gram.

Haematocrit Value (Packed Cell Volume-PCV)

The haematocrit value (Hk) determines the ratio of the volume of the blood cells to that of blood plasma. The haematocrit value (Hk) is expressed as the percentage fraction of blood cells, in the total volume.

Mean Corpuscular Volume

The average volume of a single red cell in cubic microns. Mean Corpuscular Haemoglobin Concentration (MCHC)

To determine the haemoglobin content of 20 μm of the packed cells as percentage as opposed to the percentage of haemoglobin of whole blood.

RESULTS

The experimental fish H. fossilis readily accepted the probiotic diet (B. subtilis) and basal diet (control). No mortality occurred during the feeding trial. Length and weight gain of H. fossilis fed with probiotic diet was higher than the fishes fed with control diet. The gut micro flora of H. fossilis was found to be 6.3×10^6,5.7×10^7,5.4×10^5 and 5.1×10^5 cells in D_1 treatment fishes on 10^4,10^5,10^6 and 10^7 dilutions respectively (Table 1) on initial day (day 0). On 30th day the gut micro flora of H. fossilis fed with control diet and probiotic diet was found to be 5.7×10^7, 4.8×10^7, 5.1×10^6 and 4.3×10^5 cells on 10^4,10^5,10^6 and 10^7 dilutions respectively (Table 2). On day 35, 40 and 45 the microbial analysis was done in three treatments such as T1, T2 and T3. The gut micro flora of T2 fishes was found to be $7.4\times10_6$, $8.1\times10_5$, $7.9\times10_5$ and $4.7\times10_5$ on 35th day in 10_4,10_5,10_6 and 10_7 dilutions respectively. This showed a gradual decrease in T_3 fishes (Table 3). On 40th day its showed gradual decrease of 7.2×10^6, 7.3×10^5, 6.9×10^5 and 4.3×10^5 in 10^4,10^5,10^6 and 10^7dilutions respectively. But in T3 fishes showed fluctuation (Table 4) when compared to T3 fishes on day 35.

Table 1: Heterotrophic Bacterial Count (Day 0)

Dilution Factor	D1	D2
10^4	6.3×10^6	5.4×10^6
10^5	5.7×10^7	5.2×10^6
10^6	5.4×10^5	5.0×10^5
10^7	5.1×10^5	5.1×10^4

Table 2: Heterotrophic Bacterial Count (Day 30)

Dilution Factor	Colonies in number
10^4	5.7×10^7
10^5	4.8×10^7
10^6	5.1×10^6
10^7	4.3×10^5

Table 3: Heterotrophic Bacterial Count (Day 35)

Dilution Factor	T1	T2	
104	6.7 × 106	7.4×106	4.1 ×105
105	6.1 × 106	8.1 × 105	4.0 ×105
106	8.3 × 105	7.9 × 105	5.1 ×104
107	8.0 × 105	4.7 ×105	3.2 ×104

Table 4: Heterotrophic Bacterial Count (Day 40)

Dilution Factor	T1	T2	T3
104	6.4 × 106	7.2×106	4.3 ×105
105	6.2 × 106	7.3 × 105	4.0 ×105
106	5.8 × 106	6.9 × 105	7.9 ×104
107	7.3 × 105	4.3 ×105	5.7 ×104

The fungal and bacterial load in T2 and T3 fishes was decreased on day 45 (Table 5) when compared to days 40. The fungal colonies showed only a slight limitation in the growth recording 4.1×10^6, 5.7×10^5 , 6.2×10^5 and 4.1×10^5 on day 45 in 10^4,10^5,10^6 and 10^7 dilutions respectively. The bacterial load and fungal load were found to be least in T2 and T3 on day 45. The results of gut microflora were showed in Tables 1-5.

Table 5: Heterotrophic Bacterial Count (Day 45)

Dilution Factor	T1	T2	T3
104	6.7 × 106	4.1×106	4.0 ×105
105	6.3 × 106	5.7 × 105	7.9 ×104
106	7.1× 105	6.2 × 105	4.1 ×104
107	5.8 × 105	4.1 ×105	3.2 ×105

Intramuscular injection of Aeromonas hydrophila and Aphanomyces invadans showed slight to severe dermomuscular lesions, in the experimental fish H. fossilis. Among the three treatments external lesion initially developed as a blanched area with superficially abrading lesion and excavated lesion with swelling in both probiotic treated (T2 and T3) fishes. But in the case of probiotic fed A. hydrophila injected (T2) group the infections gradually decreased during the experimental period from day 9(+++) until day 15(+). Whereas in A. invadans injected group (T3) lesion slightly decreased on day 9(+++) and further decreased on day 15(++). In the A. hydrophila injected group external lesion was reduced in the probiotic feeding than A. invadans injected group. The control fish did not develop any lesion.

The values of various Haematological indices of T1,T2,T3 are shown in Tables 6-13 i.e, TLC, TEC, DLC, PCV, Hb content, MCH, MCV, MCHC respectively.

Table 6: Comparison of TLC (Cells/mm3) in *H. f ossilis* administered w ith *A. hydrophila* and *A. invadans*

Day	Sample	Mean	SD
30thDay	T1	3.3	1.2
	T2	4.1	1.7
	T3	3.9	1.5
35th Day	T1	3.4	1.3
	T2	3.9	0.9
	T3	2.6	0.7
40th Day	T1	3.1	0.8
	T2	3.8	1.3
	T3	3.1	0.8
45th Day	T1	3.0	1.2
	T2	4.3	1.4
	T3	4.0	0.7

TLC of T1 did not show much variation and found to be 3.3 (Cells/ mm^3) on 30th day and 3.4 (Cells/mm^3), 3.1 (Cells/ mm^3), 3.0 (Cells/ mm^3) on day 35,40,45 respectively .The TLC of A. hydrophila (T2) injected fishes showed a decrease on the day 35 and gradually increased on day 45. In A. invadans (T3) injected fishes too the leucocytes showed a decrease on 35th day 2.6 (Cells/mm^3) and a slight increase on day 45th day 4.0 (Cells/mm^3) were observed (Table 6).

Table 7: Comparison of TEC(Cells/mm3) in *H. f ossilis* administered w ith *A. hydrophila* and *A. invadans*

Day	Sample	Mean	SD
30thDay	T1	3.2	0.7
	T2	3.2	1.5
	T3	2.8	1.0
35th Day	T1	3.1	0.8
	T2	3.5	1.2
	T3	3.6	1.0
40th Day	T1	3.4	0.5
	T2	3.7	1.7
	T3	3.7	1.3
45th Day	T1	3.1	0.2
	T2	3.7	1.4
	T3	3.9	0.9

The TEC in T1 fishes were found to be normal from the 30th day 3215 (Cells/mm^3) to 45th day 3100(Cells/mm^3). In T2 fishes the red blood cells showed increased level on day 35 and it gradually decreased on day 45. In case of T3 also red blood cells increased on day 35 and slightly decreased on day 45 (Table 7).

Table 8: DLC(%) in *H. f ossilis* administered w ith *A. hydrophilia* (106 CFU/0.1ml)

Cells	Duration							
	Bacteria Injected							
	Control (T1)				Test (T2)			
	30th day	35th day	40th day	45th day	30th day	35th day	40th day	45th day
Neutrophils %	60	40	40	39	60	50	56	42
Eosinophils %	2	8	10	10	5	5	5	9
Basophils %	6	12	10	9	3	2	1	1
Lymphocytes %	30	40	39	41	32	39	38	48
Monocytes %	2	0	1	1	0	4	0	0

DLC in T1 fishes were found to be from the 30th day to 45th day showed fluctuation in T2 fishes when compared with T1 fishes (Table 8). In T3 fishes also showed a decrease level of DLC from 30th day to 45th day (Table 8.1).

Table 8.1 DLC(%) in *H. f ossilis* administered w ith *A. invadans* (105 CFU/0.1 ml)

Cells	Duration							
	Fungus Injected							
	Control (T1)				Test(T3)			
	30th day	35th day	40th day	45th day	30th day	35th day	40th day	45th day
Neutrophiles %	52	40	45	48	43	62	51	50
Eosinophiles %	3	8	10	9	5	3	4	5
Basophiles %	5	10	5	6	10	2	1	1
Lymphocytes %	33	40	40	37	42	30	41	44
Monocytes %	7	2	0	0	0	3	3	0

The PCV and total plasma level was found to be higher in T2 fishes were higher than compared to T1 and T3 fishes. Total plasma increased from 30th day to 45th day in T3 (Table 10) when compared to T1 and T2 (Table 9).

Table 9: Haemetocrit value of *H. f ossilis* administered w ith *A. hydrophilia* (106 CFU/0.1ml) and *A. invadans* (105 CFU/0.1 ml).

Duration	Bacteria Administered				Fungus Administered	
	Control (T1)		Test (T2)		Test (T3)	
	PCV%	Plasma	PCV%	Plasma	PCV%	Plasma
30th day	25	75	25	75	26	74
35th day	25	75	22	78	14.7	85.3
40th day	24	76	20	80	13.2	86.8
45th day	25	75	18.2	81.8	11.1	88.1

Haemoglobin content of T3 fishes were higher than T1 and T2 (Table 10). The MCH level of T3 was higher than that of the T1 and T2 (Table 11). The MCV level showed fluctuations (Table 12); MCHC level of T3 showed higher than that of T1 and T2 (Table 13). No mortality was observed in T1, whereas in T2 2% and T3 6% were noticed.

Table 10: Comparison of Haemoglobin content (in gram) of blood *in H. f ossilis* administered w ith *A. hydrophilia* (106 CFU/0.1ml) and *A. invadans* (105 CFU/0.1ml)

Duration	Bacteria Injected		Fungus Injected
	Control (T1)	Test (T2)	Test (T3)
30th day	19.8	19.7	20
35th day	19.7	14.5	15
40th day	19.8	10	13
45th day	19	11.1	10

Lesions induced by A. invadans were higher than A. hydrophila on H. fossilis (Table 14).Because of probiotic feeding H. fossilis gained weight in D1 was about 10.25gm, and in D2 about 8.0gm (Table 15).

Table 11: Comparision of MCH (in microgram) *in H. f ossilis* administered w ith *A. hydrophilia* (106 CFU/0.1ml) and *A. invadans* (105 CFU/0.1ml)

Duration	Bacteria Injected		Fungus Injected
	Control (T1)	Test (T2)	Test (T3)
30th day	0.0049	0.0047	0.0050
35th day	0.0051	0.0037	0.0054
40th day	0.0050	0.0034	0.0050
45th day	0.0064	0.0044	0.0045

Table 12: Comparision of MCV (in cubicmicrons) in Heteropnuestes f ossilis administered w ith Aeromonas hydrophilia (106 CFU/0.1ml) and Aphanomyces invadans (106 CFU/0.1ml)

Duration	Bacteria Injected		Fungus Injected
	Control (T1)	Test (T2)	Test (T3)
30th day	0.00622	0.00622	0.00652
35th day	0.00638	0.00556	0.00525
40th day	0.00600	0.00671	0.00510
45th day	0.00801	0.00723	0.00501

DISCUSSION

The recent intensification of aquaculture has lead to a number of fish diseases due to environmental and physiological stress in fish population. Overcrowding, handlings of fish and water quality have resulted in disease outbreaks. The present investigation reports a study of microbiological and haematological responsiveness of H. fossils fed with probiotic diet during experimental infection by A. hydrophila and A. invadans. This study reports A. hydrophila and A. invadans caused infection and severe lesions were noticed in T2 and T3.

A. hydrophila is disseminated as a cosmopolitan especially in aquatic environments which provides ample opportunity to fishes and leads to infection and then causing mortality [13]. So, the pathological symptoms made by A. hydrophila in this study, were similar to the studies made earlier [14].

TEC in A. hydrophila infected H. fossilis and A. invadans infected H. fossilis exhibited a significant decrease on prolonged exposure. Previous studies reported a hike in TEC during the unhealthy state of fish [13, 15]. Accordingly, high counts are associated with the abnormal conditions of fish. Hence, a sudden increase in the TEC is indication of a severe infection by the opportunistic bacteria and fungi. This might have been accomplished by a rapid mobilization of RBC from the haemopoietic tissue, which may transport higher amounts of Oxygen particularly to withstand stress factor caused by A. hydrophila [1,16].

TLC also showed an increasing trend in the infected the H. fossilis with A. hydrophila and A. invadans. Increase in the number of WBC's has been reported by a several investigators. The results of the present study agrees with the works of Innocent research group [17]. DLC studies showed that lymphocytes constituted maximum percentage followed by neutrophils, Ainsworth has suggested that neutrophils and macrophages are responsible for bacterial uptake [18]. Similar observation of neutrophilia an inflammatory response was record by Innocent et al., [19] in M. montanus infected by A. hydrophila [19].

Varieties of leukocyte types are involved in innate cellular defense of fish including macrophages, granulocytes and non-specific cytotoxic cells [20]. The above findings support the present study. Monocytes and macrophages are probably the single most important cell mediated immune reponse of fish [21]. They are the primary cells involved in phagocytosis and the killing of pathogens upon first recognition and subsequent infection [22]. Secombes has reported that in fish granulocytes especially neutrophils are the primary cells involved in the initial stages by inflammation. Granulocytes are highly motile, phagocytic and produce reactive oxygen species. The results of the present studies also show significant increase in neutrophils and lymphocytes. Lymphocyte numbers decreased significantly in the post challenge sample while neutrophils and monocyte remain unchanged. Hb content showed a decrease in H. fossilis experimentally infected by A. hydrophila leading to anaemia.

The administration of probitics feed in experimentally infected H. fossilis elicited alteration in haematological parameters such as TEC, TLC, DLC, MCV, MCH, MCHC, PCV. An analysis of results of present study reveals that B. sublilis plays a vital role in increasing length and weight of fish. And also, it did not cause any changes in gut micro flora.

A. hydrophila and A. invadans are potentially effective to the H. fossilis in 10-4 CFU/ml concentration. The challenging of A. hydrophila and A. invadans in H. fossilis, in terms of immunological responsiveness shows good healthy status. Post challenging of these pathogen causes disease in fishes. But B. subtilis is not having any vital role in immune mechanism of fish after post challenging

Lesions developed by A. invadans were higher than that of the A. hydrophila induced lesions. So, the probiotic was not efficient in fungal infection. During probiotic feeding weight was also gained when compared to control. So probiotic acts as a one of the growth promoter in H. fossilis.

CONCLUSION

In H. fosilis, probiotic accepted fishes gained more weight than that of the control fishes. The microbiological estimation also showed a dual increase in trial count in T2 injected fish than that of the T3 injected fish. Many factors can influence the immune response of fish. Among them are stressors and environmental factors are natural. In the present investigation behavioural symptoms to pathogenicity such as imbalance, restlessness and avoidance of food were observed. Pathological symptoms include fin necrosis and tail rot which were also observed. In some cases septicaemic ulceration was noticed. Haematological parameters elicited changes which are able to reveal some clues for diagnosis and prognosis of the disease state. T2 fishes were inflicted alterations in TEC, TLC, DLC, and Hb content which indicated decrease state of immunity, when compared with T3 fishes. Bacteria injected fishes showed good healthy status whereas fungal injected fishes showed non healthy status of fish.

REFERENCES

1. ManjuRA, Haniffa M, Singh SA, Ramakrishnan CM, Dhanaraj M, et al. (2011) Effect of dietary administration of Efinol® FG on growth and enzymatic activities of Channastriatus.Journal of Animal and Veterinary Advances 10: 796-801.
2. KhanMA, JafriAK, ChadhaNK,UsmaniN (2003) Growth and body composition of rohu (Labeorohita) fed diets containing oilseed meals: partial or total replacement of fish meal with soybean meal.Aquaculture Nutrition9: 391-396
3. GomesEF, RemaP, KaushikSJ (1995) Replacement of fish meal by plant proteins in the diet of rainbow trout (Oncorhynchusmykiss): digestibility and growth performance. Aquaculture130: 177-186.
4. NakaiT, ParkSC (2002) Bacteriophage therapy of infectious diseases in aquaculture.Research in microbiology153: 13-18.
5. GramL,LøvoldT,NielsenJ,MelchiorsenJ,SpanggaardB (2001) In vitro antagonism of the probiontPseudomonas fluorescens strain AH2 against Aeromonassalmonicida does not confer protection of salmon against furunculosis. Aquaculture199: 1-11.
6. BalcázarJL, De BlasI, Ruiz-ZarzuelaI, CunninghamD, VendrellD, et al, (2006) The role of probiotics in aquaculture.Veterinary microbiology 114: 173-186.
7. VerschuereL,RombautG,SorgeloosP,VerstraeteW (2000) Probiotic bacteria as biological control agents in aquaculture.Microbiology and Molecular Biology Reviews64: 655-671.
8. GildbergA,JohansenA,BøgwaldJ (1995) Growth and survival of Atlantic salmon (Salmosalar) fry given diets supplemented with fish protein hydrolysate and lactic acid bacteria during a challenge trial with Aeromonassalmonicida.Aquaculture138: 23-34.
9. Sundary (2007) Role of Probiotics on Water Quality Management SangramKeshari RoutAsm Nandi. Environmental Biotechnology.
10. WatanabeT, KironV (1994) Prospects in larval fish dieteticsAquaculture124: 223-251.
11. OlafsenJA (2001) Interactions between fish larvae and bacteria in marine aquaculture.Aquaculture200: 223-247.
12. RengpipatS,RukpratanpornS,PiyatiratitivorakulS,Menasavetap (2000) Immunity enhancement in black tiger shrimp (Penaeusmonodon) by a probiont bacterium (Bacillus S11). Aquaculture191: 271-288
13. TendenciaEA,Dela PeñaMR,FerminAC,Lio-PoG,ChorescaCH,InuiY (2004) Antibacterial activity

of tilapia Tilapiahornorum against Vibrio harveyi. Aquaculture232: 145-152.

14. CahillMM (1990)A review virulence factors in motile Aeromonas species.Journal of Applied Bacteriology69: 1-16.
15. FathimaKMSA,AnnalakshmiT,InnocentBX (2012) ImmunostimulantEffectofVitamin-AinChannaPunctatus Challenged with AeromonasHydrophila: Haematological Evaluation.Journal of Applied Pharmaceutical Science2: 123-126.
16. NielsenME,HoiL,SchmidtAS,QianD,ShimadaT,et al. (2001) Is Aeromonashydrophila the dominant motile Aeromonas species that causes disease outbreaks in aquaculture production in the Zhejiang Province of China? Diseases of aquatic organisms46: 23-29.
17. InnocentBX,FathimaMSA,Dhanalakshmi(2011)Studies on the immouostimulant activity of Coriandrumsativum and resistance to Aeromonashydrophila in Catlacatla. Journal of Applied Pharmaceutical Science1: 132-135.
18. AinsworthAJ (1992) Fish granulocytes: morphologydistributionand function.Annual Review of Fish Diseases2:123-148.
19. InnocentBX,MartinP(2004)Haematological studies in Mystusmontanus exposed to gram negative bacteria Aeromonashydrophila.Indian Journal of environmental protection.
20. Galindo-VillegasJ, HosokawaH (2004) Immunostimulants: towards temporary prevention of diseases in marine fishAdvances
21. ClementS,LovellR (1994) Comparison of processing yield and nutrient composition of cultured Nile tilapia (Oreochromisniloticus) and channel catfish (Ictaluruspunctatus) Aquaculture119: 299-310.
22. ShoemakerCA,EvansJJ,KlesiusPH (2000) Density and dose: factors affecting mortality of Streptococcusiniae infected tilapia (Oreochromisniloticus). Aquaculture 188: 229-235

CHAPTER 14

SURVIVAL AND IMMUNITY OF MARRON CHERAX CAINII (AUSTIN, 2002) FED BACILLUS MYCOIDES SUPPLEMENTED DIET UNDER SIMULATED TRANSPORT

Ambas I, Fotedar R and Buller N

Keywords: Simulate transport, Intestinal bacteria population, THC, Bacteraemia, Mortality

INTRODUCTION

Practices and methods used for crustacean handling and live trade may lead to serious physiological stress responses in aquatic animals [1-3] including marron [4-6]. Marron, Cherax cainii (Austin, 2002), is endemic to Western Australia and recently was introduced into South Africa, Zimbabwe, Japan, USA, China and the Caribbean as a commercial aquaculture species [6-8].

The reasons for live trade of crustaceans include for consumption, grow-out, restocking, and the aquarium trade; hence survival and quality of the animals is extremely important for welfare and economic reasons. The duration of the stressors encountered in the live trade process leads to short and long term changes in immune parameters as stress response shifts from adaptive to maladaptive [2,9]. Beyond this point the physiological stress response may reduce disease resistance and growth reduced quality and eventually death [2,10]. Therefore, improving immunity, stress tolerance and optimising health conditions of crustaceans during storage and live transport is of fundamental importance [3].

The successful culture and stocking of marron relies on better understanding the factors affecting their well-being during transport and their recovery afterwards [5]. Marron are an economically important aquaculture species in Western Australia and they show significant environmental stress tolerance post handling and live transport. Jussila observed no mortality of marron post 24 h handling and simulated transport, whereas detected no mortality of marron up to 36 h under simulated transport [5,6]. Moreover, marron may undergo live transport up to 72 h without mortality, however longer periods of transportation resulted in an average dehydration of 4.5% of body weight [4,5].

Feed additives such as probiotics and prebiotics can improve stress tolerance and immunity of aquatic animals including marron [6,11-16]. Our previous studies indicated that Bacillus mycoides, a marron origin customised probiotic, possessed a number of favourable probiotic properties [17], significantly improved gastrointestinal (GIT) health of marron [18] and improved immunity against the pathogenic bacterium Vibrio mimicus [19], a dominant bacterial pathogen of freshwater crayfish in aquaculture [20-22].

Cruz et al. suggested that aquatic animals should be treated with probiotics before exposure to transport and environmental stressors [23]. To date, improved stress tolerance and immunity by feeding probiotics have been documented in fish [11,24,25], shrimps [26-29] and molluscs [30,31] however information on probiotic-fed marron under practical transport conditions is not available. The present study evaluated the effect of simulated transport conditions on marron fed the probiotic B.

Citation: Ambas I, Fotedar R, Buller N (2015) Survival and Immunity of Marron Cherax cainii (Austin, 2002) Fed Bacillus mycoides Supplemented Diet under Simulated Transport. J Aquac Res Development 6: 390. doi:10.4172/2155- 9546.1000390.

mycoides by measuring intestinal bacterial population, total haemocyte count (THC), bacteraemia, morbidity, dehydration and mortality.

MATERIAL AND METHODS

Acclimation and Feeding Trial

Marron juveniles were supplied by Blue Ridge Western Australia Pty Ltd and then acclimated to the culture tanks, fed using basal diet for 2 weeks, and distributed into six experimental culture tanks at a density of 12 marron per tank. The experimental system consisted of three standing units of steel racks with three shelves in each unit. The experimental units were cylindrical plastic tanks (80 cm diameter and 50 cm high and 250 L in capacity) filled with freshwater running continuously at a rate of approximately 3 L/min. using a recirculating biological filtration system (Fluval 205, Askoll, Italy). Each tank was supplied with constant aeration and equipped with a submersible thermostat set to 24°C. PVC pipes of appropriate diameter were added to the tanks as shelters for the marron. Prior to the simulated transport test, a feeding trial using basal and probiotic supplemented diet was conducted for 10 weeks. Each tank was stocked with 12 marron where each treatment consisted of five replicate tanks. The test diets were given to marron every day in the late afternoon at a rate of 1% of the total biomass and adjusted weekly after determination of biomass at the end of each week.

Experimental Diet and set up

The experimental diets used in this feeding trial were (1) basal diet of a marron commercial feed supplied by Specialty Feed Pty Ltd, WA and (2) the basal diet supplemented with customised probiotic B. mycoides. Before use, the pelleted diet was homogenised with a blender to obtain a desirable pellet size before supplementation with the B. mycoides at 108 cfu/g of feed. The density of B. mycoides was based on the density used in other Bacillus species studies [32-36] and from results of our previous studies [18,19].

Supplementation of the probiotic followed established methods [37]. In brief, the basal diet was placed on tray covered with sterilised aluminium foil and sprayed with 20 mL of fish oil per kg of feed to improve attachment of probiotic bacteria. The feed was wrapped in individual sterilised aluminium foil packs containing the amount adjusted to marron biomass for each tank, and stored at 4°C until used. The diet was prepared each week and the feeding rate adjusted according to the marron weight.

Simulated Live Transport

After feeding with the test diet for 10 weeks, the animals were subjected to a simulated live transport following the "Code of Practice for the Harvest and the Post-Harvest Handling of Live Marron for Food" established by Department of Fishery Western Australia [38] and the standard packing of marron commonly used by marron growers. Feeding ceased one day before the commencement of simulated transport trial.

In brief, healthy marron of equal size (12.3 ± 0.5 g) from probiotic fed and a control basal diet were selected and placed in a polystyrene box (60 × 40 × 30 cm) for 24 h and 48 h simulated transport. Each box contained sufficient ice-gel bags covered by a moist foam layer and a temperature data logger (Onset HOBO). Marron from each treatment group were placed in a ventilated container prior to placing in the polystyrene boxes. Placing the marron in the ventilated container not only avoided the marron from mixing with different treatment groups, but also protected the marron from the moist foam layer and ice-gel bags, and was based on the method used in a previous study [6]. Subsequently, another moist foam layer and ice bag were placed over the marron ventilated containers, before the outer polystyrene box was sealed with a lid. The sealed boxes were placed on a trolley at room temperature and being moved intermittently to give simulated transport effects.

Twenty four and forty eight hours post simulated live transport, the animals were returned to the culture tanks once the parameters for intestinal bacteria population, total haemocyte count (THC), bacteraemia, morbidity, dehydration and mortality were measured and recorded.

Measurements of the Parameters

Intestinal Bacteria Population

Bacterial population of the marron was measured before and during simulated transport at 24h and 48h. In brief, marrons from each treatment group were sacrificed by placing them at -20°C for 5 minutes before aseptic removal of the GIT. The marron dorsal shell was cut-off horizontally from tail to head until the intestines were exposed. The intestine from individual animals was collected aseptically and placed in a sterilised pestle, weighed and then homogenised. The homogenates of intestines were serially (10-1, 10-2, 10-3, 10-4, 10-5 and 10-6) diluted using sterile normal saline. Fifty microliters of each serial dilution was inoculated onto a blood agar (BA) plate and incubated overnight in a CO_2 incubator at 25°C. A colony count was performed for each dilution to determine the total number of aerobic bacteria [39].

Total haemocyte count (THC)

The total haemocyte count was measured following the established methods used in western rock lobsters Panulirus cygnus George [40] and marron [6]. In brief, 0.5 mL of haemolymph withdrawn from the second last ventral segment of marron was inserted into a haemocytometer (The Neubauer Enhanced Line, Munich, Germany) counting chamber and immediately viewed under 100-fold magnification on a camera-equipped microscope and images were taken for THC counts. Cells were counted in both grids, and the mean was used as the haemocyte count. For each treatment group, the procedure was repeated using ten different animals. The total haemocyte count was calculated as THC = (cells counted x dilution factor ×1000)/volume of grid (0.1 mm^3).

Haemolymph bacteria (Bacteraemia)

Bacteraemia of marron was determined following the established method described by sang et al. [6] with a minor modification. Briefly, the haemolymph was withdrawn into a sterile syringe and placed onto a sterile glass slide to avoid bubbles before a 0.05 mL aliquot was lawn inoculated onto a BA plate. The plates were placed in a sterilised container and incubated overnight at 25°C. The total colony forming units (CFU) for each plate and CFU/mL were calculated on the basis of a total volume of 0.05 mL/ plate.

Dehydration

Dehydration of marron was measured using the established method [5]. Ten marron from each treatment group of equal size were weighed prior to the commencement of the simulated transport, then weighed at 24 h and 48 h post transport and the percentage of weight loss was recorded.

Morbidity and survival rate (%)

Morbidity (vigour index) of marron was measured following the established method proposed by Jussila et al. [5]. In this study, morbidity of marron was identified based on the response to stimuli at a time after simulated transport, and the tim 24 h and 48 h post-simulated transport up to one day they were returned to the culture tanks. Determination of survival rate following the established equation;

$$SR = (Nt/No) \times 100$$

where SR is the survival rate (%); Nt is the number of marron at time t and No is the number of marron at the commencement (0), respectively.

Data Analysis

The data were analysed using SPSS statistical package version 22.0 for Windows and Microsoft Excel. The difference between means was determined using one way analysis of variance (ANOVA) and a t-test. All significant tests were performed at $P < 0.05$ level. All data were presented as mean ± SE, unless otherwise indicated.

RESULTS

Intestinal Bacteria Population

Intestinal bacterial population of marron declined at 24 and 48 h simulated transport both for basal diet and probiotic fed marron. Reduction of the intestinal bacterial population occurred at 24 h and 48 h of transport; however a significant reduction of more than half the initial population levels were observed at 48 h, both in basal diet and probiotic fed marron (Table 1). This result suggests that the longer the stress disturbance, the greater the reduction of intestinal bacterial population. Nevertheless, at 48 h post-transport, the bacterial population (646 ± 16.4) of probiotic fed marron was comparable to the initial bacterial population (626.7 ±19.7) of basal diet fed marron.

Total haemocyte count (THC)

Prior to the simulated transport test, the THC of marron fed probiotic and basal diets for 10 weeks was measured. The THC of marron fed probiotic supplemented diet was significantly higher compared to THC of basal diet fed marron at the commencement of simulated transport test (Figure 1). This result indicates that the health status of marron fed probiotic was higher at the initiation of the simulated transport test.

After 24 h and 48 h post transport, the THC in both treatment groups declined indicating that transport affects the THC in marron.

Haemolymph bacteria (Bacteraemia)

Haemolymph bacteria were observed in basal diet and probiotic fed marron after feeding with the test diets for ten weeks prior to the simulated transport test. Total bacterial count in the haemolymph of probiotic fed marron was significantly ($P<0.05$) lower compared to basal fed marron, indicating the customised probiotic of host origin may induce protection from bacteria and other foreign particles in the haemolymph. This result was strongly related to THC of marron in each treatment group, as haemocytes play an important role in removal of bacteria and foreign particles from the haemolymph of crayfish [41,42].

Dehydration

Dehydration occurred in marron fed either the basal diet or the probiotic supplemented diet. During the first 24 h, dehydration in the basal diet fed marron was 3.8 percent, whereas in probiotic fed marron it ranged between 3.0 and 3.7%. After 48 h of transport, the dehydration still occurred in both treatment groups but in the basal diet it was 0.45% while in probiotic fed marron the extra dehydration was 0.55%. There was no significant different ($P>0.05$) in the mean dehydration between basal diet and probiotic fed marron after 24h and 48h of transport. Morbidity and survival rate (%) Marron showed a very weak response to stimuli after 24 h of transport and this condition was more noticeable after 48 h of transport. Marron started to show response to stimuli after 30 minutes during the temperature acclimation in the boxes. As most marron in the boxes were actively crawling and swimming, they were returned to the tanks for mortality observation. No mortality was observed in marron fed probiotic supplemented diet; while in basal diet fed marron mortality (6.7%) occurred at 48 h of transport. No clinical signs were observed on dead marron shells (Figure 2).

DISCUSSION

A number of immune and physiological parameters are involved in the stress response following handling and live transport of crustaceans and fish such as behaviour, morbidity and vigour, THC, blood glucose, dehydration , oxygen uptake, blood composition, pH, hormones and ion [1,3]. In marron, the common selected parameters for testing following handling and simulated transport include dehydration [4] THC, haemolymph/plasma glucose, serum protein, dehydration [5], proportion of granule cells, clotting time and bacteraemia [6]. The circulating haemocytes of crustaceans are an essential part of the immune system, and perform functions such as phagocytosis, encapsulation, and lysis of foreign cells and much research in the defensive role of haemocyte in crustacean is being conducted [26,40,43].

The results suggest total haemocyte count (THC) is a reliable indicator of stress in crustaceans [2,44] including in marron [5,6]. In the present study, THC was investigated as an indicator for stress tolerance. THC of marron fed probiotic was significantly higher compared to basal diet fed marron both at 24 h and 48 h post simulated transport, indicating that the customised probiotic B mycoides was able to improve marron immunity. Enhancement of THC in probiotic fed marron leads to increased stress tolerance and diseases resistance, which results in a significantly higher survival rate (100%) over 24 h and 48 h post live transport. Higher THC of probiotic fed marron also provides better protection to the gill from parasites and bacteria pathogens which may cover and reduce respiration efficiency. Tinh et al. [45] suggested that probiotic bacteria can also be active on the gills and skin of the host. Inefficient respiration during handling and live transport may contribute to marron mortality in basal diet fed marron.

Thus prior to transport, purging is generally essential in freshwater crayfish [3] to evacuate the GIT content and clean the gills and skin. Handling and live transport creates physiological stresses which reduce THC in many crustacean species such as American lobster Homarus americanus) [1], crab Cancer pagurus) [2], western rock lobster Panulirus cygnus [2,46] marron[5,6], and a mollusc abalone Haliotis tuberculata [47]. Therefore, increasing the THC by feeding probiotic supplemented diets may improve stress tolerance and protect the animals from pathogens. This has been demonstrated in crayfish, Pacifastacus leniusculus, where a higher THC corresponded to improved defence reactions of the animal when infected with the Oomycete fungus Aphanomyces astaci [48] and in marron where a higher THC corresponded to reduced bacteraemia and improved immunity against the pathogen Vibrio mimicus[6,19].

Table 1: Intestinal bacteria population (million CFU ± SE) of marron fed different diets at 24 h and 48 h post transport

Treatment	Intestinal bacteria population (million CFU/g of gut)		
	0h	24h	48h
Basal diet	626.7 ±19.7^{a,1}	531.1 ± 15.7^{a,1}	260.0 ±67.1^{b,1}
B. mycoides diet	1656 ± 167.7^{a,2}	1318 ± 131.5^{a,2}	646 ± 16.4^{b,2}

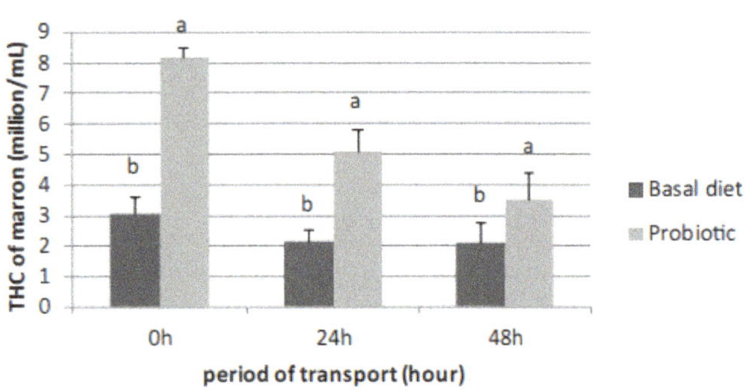

Figure 1: THC of marron fed basal and probiotic diets at 24h and 48h of transport.

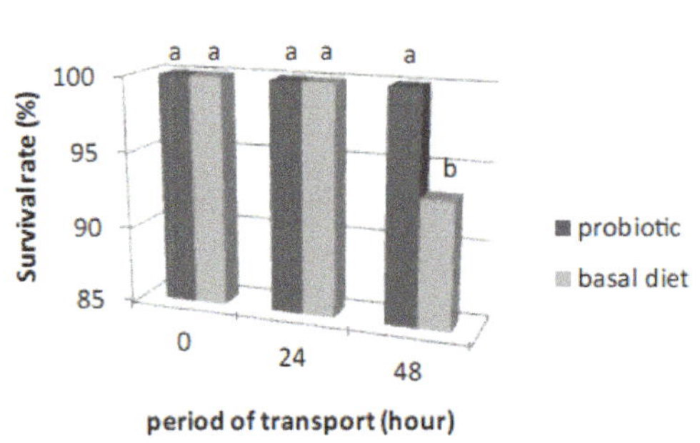

Figure 2: Survival rate (%) of marron fed basal and probiotics diets at 24 h and 48 h of transport.

Bacterial population and diversity in the GIT is an important health component for aquatic animals [49,50] significantly affected by acute stress [51]. Stress due to high stocking density could also affect the performance of probiotics [24]. In the present study, the bacterial population level decreased both in probiotic and basal diet fed marron at 24h and a significant reduction was observed at 48h of transport indicating that prolonged stress significantly reduces intestinal bacteria of the animal. However, the higher bacterial population level of probiotic fed marron prior to transport resulted in the bacterial population remaining higher at 48h of transport compared to the population level of basal diet fed marron at the initiation of the transport test. The reduction of intestinal bacterial population due to handling and transport stress was also determined in other species. In Atlantic salmon, Salmo salar and rainbow trout, Oncorhynchus mykiss (Walbaum), adherent bacterial population in the mid-gut and hindgut were significantly reduced following acute handling stress, whereas the level increased in faeces, which suggests that considerable amounts of mucus was lost following stress[51,52]. This significant reduction of the intestinal bacterial population implies decline in the health status of the animal. The microbiota within the GIT can be considered a metabolically active organ, is an essential health component, provides protection against infection and instructs mucosal immunity [50]. Beneficial bacteria such as lactobacilli and bifidobacteria decrease following a stress response [50] and this may provide opportunistic pathogens to become established [2,44,53].

Haemolymph bacteria (bacteraemia) of marron fed probiotic diet was significantly lower compared to basal diet fed marron suggesting that greater THC plays an important role in reducing bacteria and foreign particles in marron haemolymph. Once pathogens or foreign particles enter the haemocoel, the haemocytes initiate phagocytosis [54]. In crayfish, hyaline cells are chiefly involved in phagocytosis, whereas semi-granular cells are active in encapsulation [41]. Crustacean haemocytes contain antibacterial activity [55,56] which can reduce the viable count of bacteria within 4 hours, however the antibacterial potency (per unit protein) varies from species to species [42].

Morbidity and mortality post live transport often occurs as a result of stress [3]. In the present study, morbidity and weakness indicated by no response to stimuli were observed both in basal diet and probiotic fed marron at 24 h and 48 h post simulated transport. However, after returning to the culture tanks the probiotic fed marron fully recovered and swam normally in less than 30 minutes, whereas basal diet fed marron took more time to recover and had several mortalities.

Other than mortality, marron also could be losing weight through dehydration from tissue and gill chambers while out of water during handling and live transport [3-5]. The present study indicated that dehydration of marron was observed in both test diets at 24 h and 48 h post transportation, however the dehydration was not significantly different between the two treatments. The results revealed that the dehydration of marron was comparable to the previous marron handling and live transport studies indicating that dehydration of 4 to 5% of the body weight over 24 to 72 h transportation is a common phenomenon. Jussila et al. observed a minor dehydration of marron during the first 4 hours that remained at 4.0 to 4.5% up to 24 h post handling and transportation, whereas Morrissy et al. observed wet dehydration of 3.9% during the first 24 h, with a further additional loss of 0.4% over the next 24 h [4,5]. Acute stress requires high energy which could reduce the hepatosomatic indices and contribute to the dehydration of the animal [23]. In marron, hepatopancreas significantly reduced after 24 h of transport [5]. Therefore, dehydration should be considered when crayfish are going to be transported for a long distance [5].

Probiotics have improved stress tolerance and immunity of various aquatic animal groups such as fish species including sea bream, Sparus aurata [11], grouper Epinephelus coioides [57], nile tilapia Oreochromis niloticus [58]; shrimp such as western king prawns, Penaeus latisulcatus [59] shrimp, Litopenaeus vannamei [60-62], and tiger shrimp, Penaeus monodon [63].

Among the bacterial genera evaluated for probiotic use, Bacillus species have been successful in improving the stress tolerance in these aquatic animal hosts and more recently in black swordtail, Xiphophorus helleri [16]. Our previous work using B. mycoides also showed improved marron immunity when experimentally challenged with the pathogen Vibrio mimicus [19]. Overall, the present study demonstrated that supplementation with host origin customized probiotic B. mycoides significantly improved the health status of marron by increasing their tolerance to a live transport stress test, which resulted in no mortality up to 48 h of simulated live transport.

ACKNOWLEDGEMENT

The authors wish to The Directorate General of Higher Education (DIKTI) of Indonesia and The Department of Fishery, Hasanuddin University of Makassar Indonesia which financially supported this study. Thanks also are due to Mr. Peter McGinty of Aquatic Resource Management Pty Ltd Manjimup, Western Australia for providing the marron for this research. My gratitude are due to Mr. John Fielder of the Learning Centre Curtin University for the proof reading of the manuscript.

REFERENCES

1. Lorenzon S, Giulianini PG, Martinis M, Ferrero EA (2007) Stress effect of different temperatures and air exposure during transport on physiological profiles in the American lobster Homarus americanus. Comp Biochem Physiol 147: 94-102.
2. Lorenzon S, Giulianini PG, Libralato S, Martinis M, Ferrero EA (2008) Stress effect of two different transport systems on the physiological profiles of the crab Cancer pagurus. Aquaculture 278: 156-163.
3. Fotedar S, Evans L (2011) Health management during handling and live transport of crustaceans: A review. Journal of Invertebrate Pathology 106: 143-152.
4. Morrissy NM, Walker PCF, Moore W (1992) An investigation of dehydration of marron (Cherax tenuimanus) in air during live transportation to market. Bernard Bowen Fisheries Research Institute, Western Australia Marine Research Laboratories. Fisheries Department of Western Australia. Fisheries Research Report 99: 1-21.
5. Jussila J, Paganini M, Mansfield S, Evans LH (1999) On physiological responses, plasma glucose, total hemocyte count and dehydration, of marron Cherax tenuimanus (Smith) to handling and transportation under simulated conditions. Freshwater Crayfish 12: 154-167.
6. Sang HM, Le KYT, Fotedar R (2009) Dietary supplementation of mannan oligosaccharide improves the immune responses and survival of marron, Cherax tenuimanus (Smith, 1912) when challenged with different stressors. Fish shellfish immunology 27: 341-8.
7. Morrissy NM, Evans L, Huner JV (1990) Australian freshwater crayfish: Aquaculture species. Journal of the World Aquaculture Society 21: 113-122.
8. Rouse DB, Kartamulia I (1992) Influence of salinity and temperature on molting and survival of the Australian freshwater crayfish (Cherax tenuimanus). Aquaculture 105: 47-52.
9. Barton BA, Iwama GK (1991) Physiological changes in fish from stress in aquaculture with emphasis on the response and effects of corticosteroids. Annu Rev Fish Dis 1: 3-26.
10. Ige BA (2013) Probiotics use in intensive fish farming. African Journal of Microbiology Research 7: 2701-2711.
11. Rollo A, Sulpizio R, Nardy M, Silvi S, Orpianesi C, et al.(2006) Live microbial feed supplement in aquaculture for improvement of stress tolerance. Fish Physiol Biochem 32: 167-177.
12. Zhang J, Liu Y, Tian L, Yang H, Liang G, et al. (2012) Effects of dietary mannan oligosaccharide on growth performance, gut morphology and stress tolerance of juvenile Pacific white shrimp, Litopenaeus vannamei. Fish & Shellfish Immunology 33: 1027-1032.
13. Soleimani N, Hoseinifar SH, Merrifield DL, Barati M, Abadi ZH (2012) Dietary supplementation of fructooligosaccharide (FOS) improves the innate immune response, stress resistance, digestive enzyme activities and growth performance of Caspian roach (Rutilus rutilus) fry. Fish & Shellfish Immunology 32: 316-321.
14. Lund I, Skov PV, Hansen BW (2012) Dietary supplementation of essential fatty acids in larval pikeperch (Sander lucioperca); short and long term effects on stress tolerance and metabolic physiology. Comp Biochem Physiol A Mol Integr Physiol 162: 340-348.
15. Wongsasak U, Chaijamrus S, Kumkhong S, Boonanuntanasarn S (2015) Effects of dietary supplementation with β-glucan and synbiotics on immune gene expression and immune parameters under ammonia stress in the Pacific white shrimp. Aquaculture 436: 179-187.
16. Hoseinifar SH, Roosta Z, Hajimoradloo A, Vakili F(2015) The effects of Lactobacillus acidophilus as feed supplement on skin mucosal immune parameters, intestinal microbiota, stress resistance and growth performance of black swordtail (Xiphophorus helleri). Fish & Shellfish Immunology 42: 533-538.
17. Ambas I, Buller N, Fotedar R (2015a) Isolation and screening of probiotic candidates from marron, Cherax cainii (Austin, 2002) gastrointestinal tract (GIT) and

commercial probiotic products for the use in marron culture. Journal of Fish Diseases 38: 467-476.

18. Ambas I, Fotedar R, Buller N (2015b) Bacillus mycoides Improves Health of Gastrointestinal Tract in Marron (Cherax cainii, Austin 2002). J Aquac Mar Biol 2: 1-7.
19. Ambas I, Suriawan A, Fotedar R (2013) Immunological responses of customized probiotics-fed marron, Cherax tenuimanus, (Smith 1912) when challenged with Vibrio mimicus. Fish & shellfish immunology 35: 262-270.
20. Eaves LE, Ketterer PJ (1994) Mortalities in red claw crayfish Cherax quadricarinatus associated with systemic Vibrio mimicus infection Diseases of Aquatic Organism 19: 233-237.
21. Wong FYK, Fowler K, Desmarchelier PM (1995) Vibriosis due to Vibrio mimicus in Australian freshwater crayfish. Journal of Aquatic Animal Health 7: 284-291.
22. Evans L, Edgerton BF (2002) Pathogens, Parasites and Commensals, in Biology of Freshwater Crayfish, Blackwell Science Ltd, Osney OX2 0EL, UK.
23. Cruz PM, Ibanez AL, Hermosillo OAM, Saad HCR (2012) Use of Probiotics in Aquaculture. ISRN Microbiology.
24. Nayak SK (2010) Probiotics and immunity: A fish perspective. Fish and shellfish immunology 29: 2-14.
25. Tapia-Paniagua ST, Vidal S, Lobo C, Prieto-Alamo MJ, Jurado J, et al (2014) The treatment with the probiotic Shewanella putrefaciens Pdp11 of specimens of Solea senegalensis exposed to high stocking densities toenhance their resistance to disease. Fish shellfish immunology 41: 209-221.
26. Bachere E (2000) Shrimp immunity and disease control - Introduction. Aquaculture 191: 3-11.
27. Farzanfar A (2006) The use of probiotics in shrimp aquaculture. FEMS Immunology and Medical Microbiology 48: 149-158.
28. Ninawe AS, Selvin J (2009) Probiotics in shrimp aquaculture: Avenues and challenges. Critical Reviews in Microbiology 35: 43-66.
29. Lakshmi B, Viswanath B, Gopal DVRS (2013) Probiotics as Antiviral Agents in Shrimp Aquaculture. Journal of Pathogens.
30. Kesarcodi-Watson A (2009) Screening for probiotics of Green shell mussel larvae, Perna canaliculus, using a larval challenge bioassay. Aquaculture 296: 159-164.
31. Kesarcodi-Watson A, Kaspar H, Lategan MJ, Gibson L (2012) Performance of single and multi-strain probiotics during hatchery production of Greenshell™ mussel larvae, Perna canaliculus. Aquaculture 354-355: 56-63.
32. Keysami MA, Saad CR, Daud HM, Sijam K, Ar A (2007) Effect of Bacillus subtilis on growth development and survival of larvae Macrobrachium rosenbergii (de Man). Aquaculture nutrition 13: 131-136.
33. Keysami M, Mohammadpour M, Saad CR (2012) Probiotic activity of Bacillus subtilis in juvenile freshwater prawn, Macrobrachium rosenbergii (de Man) at different methods of administration to the feed. Aquaculture international 20: 499-511.
34. Li J, Tan B, Mai K (2009) Dietary probiotic Bacillus OJ and isomaltooligosaccharides influence the intestine microbial populations, immune responses and resistance to white spot syndrome virus in shrimp (Litopenaeus vannamei). Aquaculture 291: 35-40.
35. Zhang Q, Tan B, Mai K, Zhang W, Ma H, et al. (2011) Dietary administration of Bacillus (B. licheniformis and B. subtilis) and isomaltooligosaccharide influences the intestinal microflora, immunological parameters and resistance against Vibrio alginolyticus in shrimp, Penaeus japonicus (Decapoda: Penaeidae). Aquaculture research 42: 943-952.
36. Liu CH, Chiu CH, Wang SW, Cheng W (2012) Dietary administration of the probiotic, Bacillus subtilis E20, enhances the growth, innate immune responses, and disease resistance of the grouper, Epinephelus coioides. Fish shellfish immunology 33: 699-706.
37. Hai NV, Buller N, Fotedar R (2010) Effect of Customized Probiotics on the Physiological and Immunological Responses of Juvenile Western King Prawns (Penaeus latisulcatus Kishinouye, 1896) Challenged with Vibrio harveyi. Journal of Applied Aquaculture 22: 321-36.
38. Fisheries WA (2002) Code of Practice for the Harvest and the Post Harvest Handling of Live Marron for Food.
39. Buller NB (2004) Bacteria from Fish and Other Aquatic Animals: A Practical Identification Manual. CABI Publishing, Oxford shire, UK.
40. Fotedar S, Tsvetnenko E, Evans L (2001) Effect of air exposure on the immune system of the rock lobster Panulirus cygnus. Marine and Freshwater Research 52: 1351-1355.
41. Johansson MW, Keyser P, Sritunyalucksana K, Soderhall K (2000) Crustacean haemocytes and haematopoiesis. Aquaculture 191: 45-52.
42. Chisholm JRS, Smith VJ (1995) Comparison of antibacterial activity in the hemocytes of different crustacean species. Camp Biochem Physiol A Physiol 110: 39-45.
43. Soderhall K, Cerenius L (1992) Crustacean immunity. Annu Rev Fish Dis 2: 3-23.
44. Lorenz S, Francese M, Smith VJ, Ferrero EA (2001) Heavy metals affect the circulating haemocyte number in the shrimp Palaemon elegans. Fish shellfish immunology 11: 459-472.
45. Tinh NT, Dierckens K, Sorgeloos P, Bossier P (2008) A review of the functionality of probiotics in the larviculture food chain. Mar. Biotechnol 10: 1-12.
46. Jussila J, Mcbride S, Jago J, Evans LH (2001) Hemolymph clotting time as an indicator of stress in western rock lobster (Panulirus cygnus George). Aquaculture 199: 185-193.
47. Cardinaud M, Offret C, Huchette S, Moraga D, Paillard C (2014) The impacts of handling and air exposure on immune parameters, gene expression, and susceptibility to vibriosis of European abalone Haliotis tuberculata. Fish shellfish immunology 36: 1-8.
48. Persson M, Cerenius L, Soderhall K (1987) The Influence of Haemocyte Number on The Resistance of

The Freshwater Crayfish, Pasifastacus leniusculus Dana, to The Parasitic Fungus, Aphanomyces astacii. Journal of fish diseases 10: 471-477.

49. Gomez GD, Balcazar JL (2008) A review on the interactions between gut microbiota and innate immunity of fish. FEMS Immunol Med Microbiol 52:145-154.

50. Gaggìa F, Mattarelli P, Biavati B (2010) Review Probiotics and prebiotics in animal feeding for safe food production. International Journal of Food Microbiology 141: S15-28.

51. Olsen RE, Sundell K, Mayhew TM, Myklebust R, Ringø E (2005) Acute stress alters intestinal function of rainbow trout, Oncorhynchus mykiss (Walbaum). Aquacult 250: 480-495.

52. Olsen RE, Sundell K, Hansen T, Hemre GI, Myklebust R, et al. (2002) Acute stress alters the intestinal lining of Atlantic salmon, Salmo salar L.: An electron microscopical study Fish Physiology and Biochemistry 26: 211-221.

53. Ringo E, Lovmo L, Kristiansen M, Bakken Y, Salinas I, et al. (2010) Lactic acid bacteria vs. pathogens in the gastrointestinal tract of fish: a review. Aquaculture Research 41: 451-467.

54. Li CC, Yeh ST, Chen JC (2010) Innate immunity of the white shrimp Litopenaeus vannamei weakened by the combination of a Vibrio alginolyticus injection and low-salinity stress. Fish & Shellfish Immunology 28: 121-127.

55. Van de Braak CB, Botterblom MH, Liu W, Taverne N, Van der Knaap WP, et al. (2002) The role of the haematopoietic tissue in haemocyte production and maturation in the black tiger shrimp (Penaeus monodon). Fish & shellfish immunology 12: 253-72.

56. Haug T, Kjuul A, Stensvåg K, Sandsdalen E, Styrvold O (2002) Antibacterial activity in four marine crustacean decapods. Fish and Shellfish Immunology 12: 371-85.

57. Sun YZ, Yang HL, Ma RL, Lin WY (2010) Probiotic applications of two dominant gut Bacillus strains with antagonistic activity improved the growth performance and immune responses of grouper Epinephelus coioides. Fish and Shellfish Immunology 29: 803-809.

58. Pirarat N, Pinpimai K, Endo M, Katagiri T, Ponpornpisit A, et al. (2011) Modulation of intestinal morphology and immunity in nile tilapia (Oreochromis niloticus) by Lactobacillus rhamnosus GG. Research in Veterinary Science 91: e92-e97.

59. Hai NV, Buller N, Fotedar R (2009b) The use of customised probiotics in the cultivation of western king prawns (Penaeus latisulcatus Kishinouye, 1896). Fish Shellfish Immunology 27: 100-104.

60. Tseng DY, Ho PL, Huang SY, Cheng SC, Shiu YL, et al. (2009) Enhancement of immunity and disease resistance in the white shrimp, Litopenaeus vannamei, by the probiotic, Bacillus subtilis E20. Fish Shellfish Immunology 26: 339-344.

61. Li K, Zheng T, Tian Y, Xi F, Yuan J, et al. (2007) Beneficial effects of Bacillus licheniformis on the intestinal microflora and immunity of the white shrimp, Litopenaeus vannamei. Biotechnology Letters 29: 525-30.

62. Liu KF, Chiu CH, Shiu YL, Cheng W, Liu CH (2010) Effects of the probiotic, Bacillus subtilis E20, on the survival, development, stress tolerance, and immune status of white shrimp, Litopenaeus vannamei larvae. Fish Shellfish Immunology 28: 837-844.

63. Rengpipat S, Rukpratanporn, S, Piyatiratitivorakul S, Menasaveta P (2000) Immunity enhancement in black tiger shrimp (Penaeus monodon) by a probiont bacterium (Bacillus S11). Aquaculture 191: 271-288.

CHAPTER 15

USING BACILLUS AMYLOLIQUEFACIENS FOR REMEDIATION OF AQUACULTURE WATER

Fengxing Xie, Taicheng Zhu, Fengfeng Zhang, Ke Zhou, Yujie Zhao and Zhenghua Li

Keywords: *Bacillus amyloliquefaciens*, Aquaculture water remediation, Nitrogenous compound, Nitrite removal

INTRODUCTION

The rapid expansion of intensive aquaculture industries, are often companied by rotted uneaten feed, sedimentation of feces and organic residues. The water quality rapidly deteriorates as a result. In particular, nitrogenous compounds such as ammonia and nitrite quickly build up, which are both harmful to fish and shrimp even at low concentrations (Crab et al. 2007; Mohapatra et al. 2012). Water exchange can be applied to maintain good water quality. However, frequent water exchange is not only laborious and costly, but also may incur disease causing agents and pollute nearby rivers and coastal areas (Mohapatra et al. 2012). Therefore, there is an urgent demand for cost-effective and environment-friendly approaches for remediation of aquaculture water.

In recent years, the use of microorganisms to improve water quality becomes a burgeoning trend (Ninawe and Selvin 2009; Verschuere et al. 2000). *Bacillus* species are widely used for water remediation because they are stable for long period due to spore formation, easily prepared by fermentation and possess antagonistic effects on pathogens (Hong et al. 2005). Strains belonging to several *Bacillus* species, such as *Bacillus subtilis*, *Bacillus cereus*, *Bacillus licheniformis*, *Bacillus pumilus* were isolated and evaluated for their potential as biological agents for water quality enhancement. Several strains with good nitrogen removal properties were thus found. To date, screening strains with good remediation characteristics still remains a fundamental step towards developing commercial microbial agents.

Previously, we isolated a *Bacillus amyloliquefaciens* strain, named as *B. amyloliquefaciens* HN from the activated sludge of a polluted river. This strain was shown to effectively remove nitrogenous compounds and grow in broad temperature, pH, and salts concentration in preliminary studies. Moreover, no previous studied has characterized the nitrogen removal ability of *B. amyloliquefaciens*. Therefore, the aim of this study is to evaluate *B. amyloliquefaciens* HN for its remediation properties using simulated aquaculture water.

MATERIALS AND METHODS

Strains and culture media

B. amyloliquefaciens HN was isolated from a polluted pool in Xiqing District, Tianjin. The strain was identified by Agricultural Culture Collection of China, and deposited at China General Microbiological Culture Collection Center (CGMCC No. 3261).

The culture medium (also referred as basal medium) for *B. amyloliquefaciens* HN contained (per liter): peptone 8 g, beef

Citation: Fengxing XieEmail author, Taicheng Zhu, Fengfeng Zhang, Ke Zhou, Yujie Zhao and Zhenghua Li, Using Bacillus amyloliquefaciens for remediation of aquaculture water, DOI: 10.1186/2193-1801-2-119.

extract 3 g, soluble starch 5 g, NaCl 5 g (plus agar 15 g f(solid medium). The simulated aquaculture water was prepare referring to Chen (2005) and LuzE. De-Bashanet (2002 which contained (per liter): beef extract 0.5 g, sucrose 0.5 NaCl 0.25 g, KH_2PO_4 0.075 g.

Nitrite and ammonia-N tolerance tests

Sodium nitrite and ammonium sulfate were added to th culture medium to make the final concentrations of nitrite- 1, 5, 10, 20, 40, 80, and 100 mg l^{-1}, respectively, and those o ammonia-N 10, 20, 40, 80, and 160 mg l^{-1}, respectively. All the experiments were repeated thrice with the basal medium as control. Three milliliters of strain culture were inoculated into 250 ml shake flask with 100 ml culture media and cultured at 37°C.

Removal tests of nitrite-N and ammonia-N

The simulated polluted water was prepared by making the final concentrations of nitrite-N 10, 20, 40, and 80 mg l^{-1}, respectively, and the concentrations of ammonia-N 5, 10, 20, and 40 mg l^{-1}, respectively. Three milliliters of strain culture were inoculated into 250 ml shake flask with 100 ml culture media and cultured at 37°C.

Effects of different conditions on the removal of nitrite-N and ammonia-N

The model polluted water was prepared to make the initial concentration of nitrite-N was 10 mg l^{-1}, and ammonia-N was 20 mg l^{-1}. The inoculum volume was 3%, and cultured at different temperature (25°C, 30°C, and 35°C), and different initial pH (pH5.0, 6.0, 7.0, 8.0, and 9.0, adjusted by NaOH and HCl).

Analysis methods

The Nesslar method was used for ammonia determination. The 1,2-ethanediamine, N-1-naphthalenyl-,dihydrochloride spectrophotometric method was used for the nitrite measurement (APHA, 2005). The nitrate concentration was measured using the salicylic (2-hydroxybenzoic) acid method (Cataldo et al. 1975). Cell concentration was determined by plate counting.

RESULTS

Nitrite and ammonia-N tolerance tests for of *B. amyloliquefaciens* HN

For Nitrite-N tolerance test (Figure 1a), cell concentrations of *B. amyloliquefaciens* HN with all tested nitrite-N concentrations were higher that of control, illustrating that the strain could tolerate wide range of nitrite-N. The results also showed that the presence of nitrite-N could promote the growth of *B. amyloliquefaciens*, especially at low concentrations (< 10 mg/L). The highest cell concentration was achieved with the nitrite-N concentration of 10 mg l^{-1}, which is 80.1% higher than that of control.

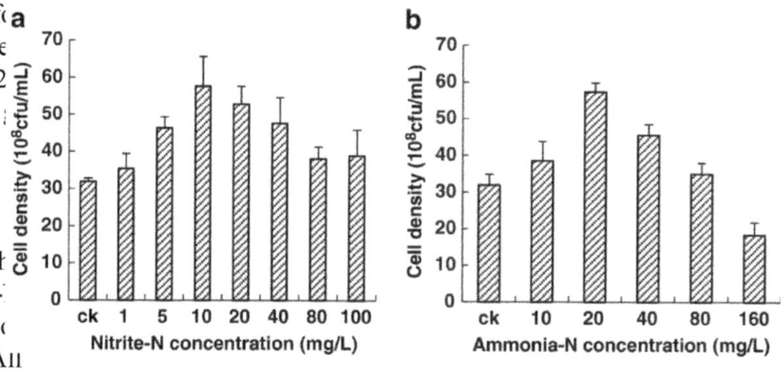

Figure 1. Nitrite and ammonia-N tolerance tests for *B. amyloliquefaciens* HN. a) Nitrite tolerance tests; **b**)

Ammonia-N tolerance tests.

For ammonia-N tolerance test (Figure 1b), all cell concentrations with tested ammonia-N concentrations were higher than that of control, except the one with 160 mg l^{-1} ammonia-N. The ammonia-N concentration of 20 mg l^{-1} resulted in the highest cell growth, 79.9% higher than that of control. The results indicated that *B. amyloliquefaciens* HN can tolerate ammonia-N ranging from 10 mg l^{-1} to 80 mg l^{-1}. An ammonia-N concentration of 160 mg l^{-1} was shown to inhibit the growth of the strain.

Nitrite-N and ammonia-N removal tests for *B. amyloliquefaciens* HN

Figure 2a showed that the nitrite-N could not be detected within 24 h with an initial nitrite-N concentration of 10 mg l^{-1}. 90% of the nitrite-N was removed in 24 h for 20 mg l^{-1}, and almost 100% was removed in 48 h. 19.2 mg l^{-1} and 20.1 mg l^{-1} of nitrite-N (representing 53.8% and 21.6% of the total nitrite-N) was eliminated for an initial nitrite-N concentration of 40 mg l^{-1} and 80 mg l^{-1}, respectively. The removal rate, however, did not significantly elevated with the time (the residue nitrite-N even increased a little for 80 mg l^{-1}), suggesting that the maximum removal concentration of nitrite-N might be around 20 mg l^{-1}.

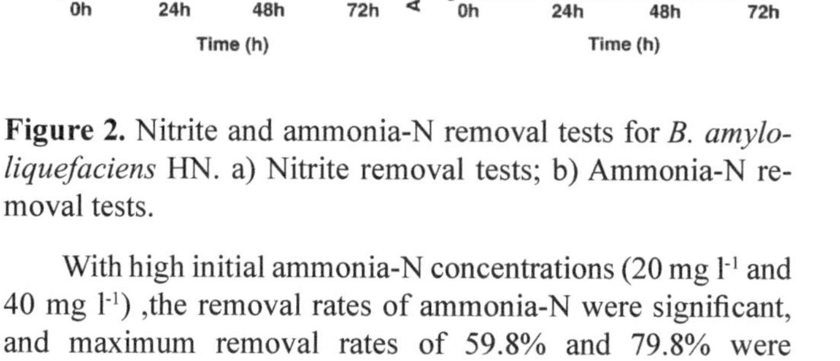

Figure 2. Nitrite and ammonia-N removal tests for *B. amyloliquefaciens* HN. a) Nitrite removal tests; b) Ammonia-N removal tests.

With high initial ammonia-N concentrations (20 mg l^{-1} and 40 mg l^{-1}) ,the removal rates of ammonia-N were significant, and maximum removal rates of 59.8% and 79.8% were reached at 48 h (Figure 2b). As for low initial ammonia-N concentrations (10 mg l^{-1} and 20 mg l^{-1}), the removal efficiencies of ammonia-N were not efficient. During the time course, the residue ammonia-N levels fluctuated around 10 mg l^{-1}, suggesting that *B. amyloliquefaciens* HN could not only utilize ammonia-N but also produce it.

Effects of temperature on the nitrite-N and ammonia-N removal abilities of B. *amyloliquefaciens*HN

The nitrite-N removal rates of *B. amyloliquefaciens* HN at 25°C, 30°C and 35°C all reached 99% (Figure 3a), suggesting temperature (over the experimental range) did not affect the removal of nitrite-N. The ammonia-N removal rates at 25°C, 30°C and 35°C were 29%, 37.5% and 39.4% respectively (Figure 3b), suggesting higher temperature might be preferable for effective removal of ammonia-N by *B. amyloliquefaciens* HN.

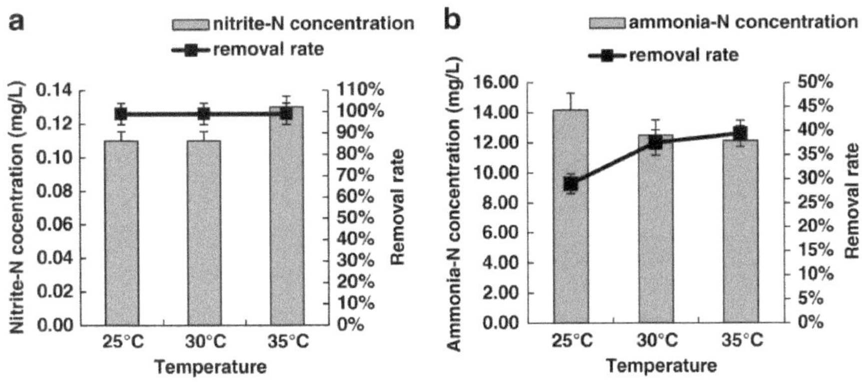

Figure 3. Effects of different temperature on the removal rates of nitrite-N and ammonia-N by *B. amyloliquefaciens* HN.a) Effects of temperature on nitrite-N removal rate; b) Effects of temperature on the ammonia-N removal rate.

Effects of pH on the nitrite-N and ammonia-N removal abilities of *B. amyloliquefaciens* HN

Figure 4a showed that pH remarkably influenced the removal of nitrite-N by *B. amyloliquefaciens* HN. Nitrite-N removal rates increased significantly with increased pH value. A maximum removal rate of 96% could be reached at pH 8.0. Further increase of pH reduced the removal rate. Ammonia-N removal rates from pH 5.0~9.0 were between 32~38%, suggesting pH had no significant effects on the removal of ammonia-N (Figure 4b).

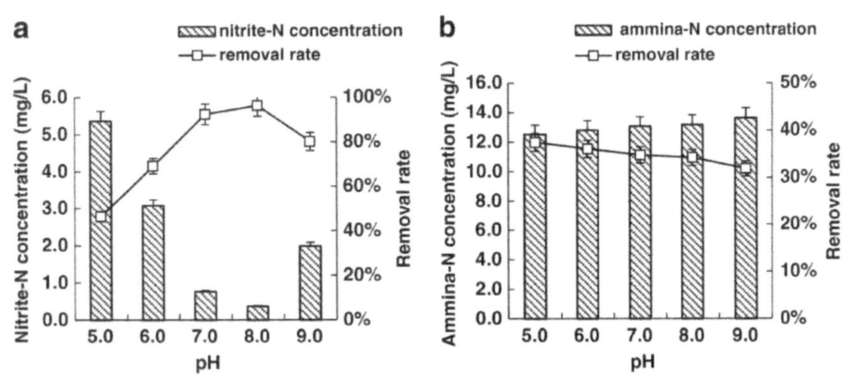

Figure 4. Effects of pH on the removal rates of nitrite-N and ammonia-N. a) Effects of pH on nitrite-N removal rate; **b**) Effects of pH on the ammonia-N removal rate.

DISCUSSION

In this work, we investigated the potential of using *B. amyloliquefaciens* in the remediation of aquaculture water. Particularly, its remediation characteristics regarding nitrite and ammonia removal were evaluated. The accumulation of nitrite and ammonia is highly toxic to aquatic fauna. 0.1~10 mg l^{-1} of nitrite can cause 50% mortality (LC 50) of a number of fish and shrimp (Philips et al. 2002). Ammonia-N can be toxic to commercially cultured fish at concentrations above 1.5 mg l^{-1} (Crab et al. 2007).

Bacillus species are important candidates for developing commercial biological agents for nitrogen removal and water quality enhancement (Hong et al. 2005). Previously, a few studies reported that some strains of *B. subtilis*(Chen and Hu 2011; Meng et al. 2009), *B. lichenformis* (Meng et al. 2009) and *B. cereus* (Lalloo et al. 2007) exhibited strong nitrite removal ability. Physiological studies on *Bacillus* spp. also showed that *Bacillus* spp. could utilize nitrate and nitrite as alternative electron acceptors and nitrogen sources (Nakano et al. 1998; Hoffmann et al. 1998).

This study showed that an *B. amyloliquefaciens* strain, isolated from the activated sludge, was also a very efficient nitrite-N cleaner, which was able to completely remove 10 mg l^{-1} of nitrite-N present in the simulated aquaculture water within 24 h. In the past, *B. amyloliquefaciens* were applied in enzyme production (Wei et al. 2011), plant disease control (Alvindia and Natsuaki 2009; Arrebola et al. 2010) and food preservation (Wang et al. 2006). This is the first study on its potential application in improving water quality. It suggested that *B. amyloliquefaciens* might be an important alternative *Bacillus* species for nitrite removal.

The ammonia-N removal ability of *B. amyloliquefaciens* was not satisfying, because when the ammonia-N concentration was below 20 mg l^{-1}, the removal rate was at best 39.4% and the ammonia-N concentration cannot be reduced below 10 mg l^{-1}. Past literature seems to suggest that *Bacillus* species are not very efficient in ammonia removal, and no ammonia removal efficiency of a single *Bacillus* strain has been reported to exceed 90%. Therefore, to create a microbal agent which can simultaneously eliminate nitrite and ammonia, *B. amyloliquefaciens*NH can be formulated with efficient ammonia reducing bacteria, such as nitrifying bacteria (Meng et al. 2009).

ACKNOWLEDGEMENT

This work is supported by Key Projects in Science and Technology Support Program of Tianjin, China (10ZCZDSY06100), Major Projects in Science and Technology Support Program of Tianjin, China (11ZCKFNC00500) and the Knowledge Innovation Program of the Chinese Academy of Sciences (KSCX2-EW-G-15-03).

REFERENCES

1. Alvindia DG, Natsuaki KT: Biocontrol activities of *Bacillus amyloliquefaciens* DGA14 isolated from banana fruit surface against banana crown rot-causing pathogens. *Crop Prot* 2009, 28(3):236-242. 10.1016/j.cropro.2008.10.011
2. Apha A WPCF: *Standard methods for the examination of water and wastewater*. Washington, DC: Public Health Association; 2005.
3. Arrebola E, Jacobs R, Korsten L: Iturin A is the principal inhibitor in the biocontrol activity of *Bacillus amyloliquefaciens*PPCB004 against postharvest fungal pathogens. *J Appl Microbiol* 2010, 108(2):386-395. 10.1111/j.1365-2672.2009.04438.x
4. Cataldo D, Haroon M, Schrader L, Youngs V: Rapid colorimetric determination of nitrate in plant-tissue by nitration of salicylic-acid. 1975, 1(6):71-80.
5. Chen GK, Chen G: Effect of synthetic wastewater on some physico-chemical properties of soil in Avicennia marina simulated wetland system. *Mar Environ Sci* 2005, 24(2):26-28.
6. Chen S, Hu Y: Use of *Bacillus subtilis* in purification of slightly-polluted water. *Acta Scienctiae Circumstantiae* 2011, 31(8):1594-1601.
7. Crab R, Avnimelech Y, Defoirdt T, Bossier P, Verstraete W: Nitrogen removal techniques in aquaculture for a sustainable production. *Aquaculture* 2007, 270(1–4):1-14.
8. De-Bashan LE, Moreno M, Hernandez JP: Removal of ammonium and phosphorus ions from synthetic wastewater by the microalgae *Chlorella vulgaris* coimmobilizedin alginate beads with the microalgae growth promoting bacterium *Azospirilum brasilense*. *Water Res* 2002, 36(12):2941-2948. 10.1016/S0043-1354(01)00522-X
9. Hoffmann T, Frankenberg N, Marino M, Jahn D: Ammonification in *Bacillus subtilis* utilizing dissimilatory nitrite reductase is dependent on resDE. *J Bacteriol* 1998, 180(1):186-189.
10. Hong HA, Duc H, Cutting SM: The use of bacterial spore formers as probiotics. *FEMS Microbiol Rev* 2005, 29(4):813-835. 10.1016/j.femsre.2004.12.001
11. Lalloo R, Ramchuran S, Ramduth D, Görgens J, Gardiner N: Isolation and selection of *Bacillus* spp. as potential biological agents for enhancement of water quality in culture of ornamental fish. *J Appl Microbiol* 2007, 103(5):1471-1479. 10.1111/j.1365-2672.2007.03360.x
12. Meng R, He LS, Xi BD, Hu X, Li YY: Experimental study on purifying aquaculture wastewater between *Bacillus* and nitrifying bacteria. *Environ Sci Technol* 2009, 32(11):28-31.
13. Mohapatra S, Chakraborty T, Kumar V, Deboeck G, Mohanta KN: Aquaculture and stress management: a review of probiotic intervention. *J Anim Physiol Anim Nutr* 2012, 1-26. 10.1111/j.1439-0396.2012.01301.x
14. Nakano MM, Hoffmann T, Zhu Y, Jahn D: Nitrogen and oxygen regulation of *Bacillus subtilis* nasDEF encoding NADH-dependent nitrite reductase by TnrA and ResDE. *J Bacteriol* 1998, 180(20):5344-5350.
15. Ninawe AS, Selvin J: Probiotics in shrimp aquaculture: avenues and challenges. *Crit Rev Microbiol* 2009, 35(1):43-66. 10.1080/10408410802667202
16. Philips S, Laanbroek HJ, Verstraete W: Origin, causes and effects of increased nitrite concentrations in aquatic environments. 2002, 1(2):115-141.
17. Verschuere L, Rombaut G, Sorgeloos P, Verstraete W: Probiotic bacteria as biological control agents in aquaculture. *Microbiol Mol Biol Rev* 2000, 64(4):655-671. 10.1128/MMBR.64.4.655-671.2000
18. Wang J, Quan C, Xu H, Fan S: Antifungal characterization of *Bacillus amyloliquefaciens* Q-12. *Food Ferment Ind* 2006, 32(6):47-50.
19. Wei X, Luo M, Xu L, Zhang Y, Lin X, Kong P, Liu H: Production of fibrinolytic enzyme from *Bacillus amyloliquefaciens* by fermentation of chickpeas, with the evaluation of the anticoagulant and antioxidant properties of chickpeas. *J Agric Food Chem* 2011, 59(8):3957-3963. 10.1021/jf1049535

CHAPTER 16

CURRENT KNOWLEDGE IN SYNBIOTIC USE FOR FISH AQUACULTURE

Rebeca Cerezuela, Jose Meseguer and Ma Angeles Esteban

Keywords: Synbiotics, Probiotics, Prebiotics, Fish, Teleosts, Aquaculture

PROBIOTICS AND PREBIOTICS: CONCEPT AND PROPERTIES

The use of probiotics and prebiotics has been regarded during recent years as an alternative viable therapy in fish culture, appearing as a promising biological control strategy and becoming as an integral part of the aquaculture practices for improving growth and disease resistance [1]. This strategy offers innumerable advantages to overcome the limitations and side effects of antibiotics and other drugs and also leads to high production [2-5].

Based on the intricate relationship that an aquatic organism has with the external environment when compared with that of terrestrial animals, the definition of a probiotic for aquatic environments needed to be modified or adapted. Probiotics are often defined as applications of entire or component(s) of a micro-organism which are beneficial to the health of the host [6]. Other probiotic definitions are more encompassing, for example, Verschuere et al. [7] suggested the definition "a live microbial adjunct which has a beneficial effect on the host by modifying the host-associated or ambient microbial community, by ensuring improved use of the feed or enhancing its nutritional value, by increase the host response towards disease, or by improving the quality of its environment". Although there is some dispute about what an aquatic probiotic actually is, All definitions differ to that of Fuller [8] in that there is no longer the requisite for the probiotic to be acting in the gastrointestinal tract. Therefore, modes of action such as competition for nutrients and production of inhibitory substances could occur in the culture water. Additional effects of probiotic action should also be considered, given the modified definition, including change of the water quality and interaction with phytoplankton [7].

In aquaculture, the range of probiotics evaluated for use is considerably wider than in terrestrial agriculture. The feasibility and efficacy of a wide range of probiotics for aquatic animals have been investigated in numerous studies, but the best documented are possibly lactic acid bacteria, Bacillus spp. and yeasts [9-11]. Moreover, several probiotics either as monospecies or multispecies supplements are commercially available for aquaculture practices [10].

The success of probiotics has laid the foundation for other concepts as prebiotics. Prebiotics offer an alternative method for manipulating endogenous microbes to improve health. Instead of introducing favourable bacteria with the diet the aim of prebiotics is to stimulate selected favourable indigenous microbial populations [11]. A prebiotic is defined as a non-digestible food ingredient that beneficially affects the host by selectively stimulating the growth and/or the activity of specific health promoting bacteria that can improve the host health

Citation: Cerezuela R, Meseguer J, Esteban MA (2011) Current Knowledge in Synbiotic Use for Fish Aquaculture. J Aquac Res Development S1:008. doi:10.4172/2155-9546.S1-008.

[12]. Prebiotics mainly consist of oligosaccharides promoting beneficial bacterial growth within the gastrointestinal tract of higher vertebrates [13,14]. In different studies from last ten years, many substances have been investigated as prebiotics. Based on the studies of Mahious and Ollevier [15], Fooks et al. [16] and Gibson et al. [17] any foodstuff that reaches the colon (e.g. non-digestible carbohydrates, some peptides and proteins, as well as certain lipids) is a candidate prebiotic [18]. However, most studies have focused on non-digestible carbohydrates, mainly oligosaccharides. According to Gibson et al. [17], only three oligosaccharides were classified as prebiotics: inulin, transgalactooligosaccharide (TOS) and lactulose. A more recent study includes fructooligosaccharides (FOS) in the list of prebiotics [19]. Mannan oligosaccharides (MOS) and fuctooligosaccharides (FOS, or oligofructose) are perhaps the most well studied oligosaccharides in fish but some information regarding applications of galactooligosaccharides, xylooligosaccharides, arabinoxylooligosaccharides and isomaltooligosaccharides in aquatic animals is also available [11,20,21]. Aditionally, inulin, a heterogeneous blend of fructose polysaccharides has also been well regarded in fish [22,23].

In recent years there has been a growing interest in understanding the mechanisms of action of probiotics and prebiotics, especially in humans and other mammals. Probiotics activity is mediated by a variety of effects that are dependent on the probiotic itself, the dosage employed, treatment duration and route and frequency of delivery. Some probiotics exert their beneficial effects by elaborating antibacterial molecules such as bacteriocins that directly inhibit other bacteria or viruses, actively participating in the fight against infections, whereas others inhibit bacterial movement across the gut wall (translocation), enhance the mucosal barrier function by increasing the production of innate immune molecules or modulate the inflammatory/immune response. Several studies have demonstrated that patter recognition receptors [PRRs, such as toll-like receptors (TLRs)], signalling pathways, immune responses and the secretion of antimicrobial peptides such as defensins and chemokines by the epithelium play important roles in these mechanisms [24,25]. On the other hand, potential mechanism of prebiotics includes selective increase/decrease in specific intestinal bacteria that modulate local cytokine and antibody production, increase in intestinal short chain fatty acids production and enhanced binding of these fatty acids to G-coupled protein receptors on leycocytes, interaction with carbohydrate receptors on intestinal epithelial and immune cells, partial absorption resulting in local and systemic contact with the immune system [26].

These alternative methods of disease prevention have been used as a means of reducing the presence of opportunistic pathogens and simultaneously stimulating the host immune responses. However, other effects not immune related have been observed, as improve growth performance, feed utilisation, digestive enzyme activity, antioxidant enzyme activity, gene expression, disease resistance, larval survival, gut morphology, alter the gut microbiota, mediate stress response, improve nutrition, reduce risk of certain cancers (colon, bladder), produce lactase, alleviate symptoms of lactose intolerance and malabsorption [1,10,11,18,21,27].

Despite numerous studies evaluating the efficacy of probiotics and prebiotics in fish, combined application of both additives (synbiotics) has received till present very little attention and available data are still scarce. Therefore, the aim of this review is to compile, for the first time, existing data on the use of synbiotics in aquaculture fish, highlighting the main effects demonstrated to date.

STUDIES OF THE ADMINISTRATION OF SYNBIOTICS IN FISH AQUACULTURE

Synbiotics refer to nutritional supplements combining probiotics and prebiotics in a form of synergism, hence synbiotics, enhancing their isolated beneficial effects. When two nutritional ingredients or supplements are given together the resulting positive effect generally follows one of three patterns: additivity, synergism or potentiation. Additive effect occurs when the effect of two ingredients used together approximates to the sum of the individual ingredient effects. In case of synergism, it is said to occur when the combined effect of the two products is significantly greater than the sum of the effects of each agent administered alone. The term potentiation is used differently, some pharmacologists use potentiation interchangeably with synergism to describe a greater than additive effect and others use it to describe the effect that is only present when two compounds are concurrently [28].

Synbiotics affects the host by improving the survival and implantation of live microbial dietary supplements in the gastrointestinal tract by selectively stimulating the growth and/or by activating the metabolism of one or a limited number of health promoting bacteria, and thus improving the host "welfare". In humans, probiotics are mainly active in the small intestine while prebiotics are only effective in the large intestine, so the combination of the two may give a synergistic effect [12].

To our knowledge, only five studies so far have investigated the effect of synbiotics in fish (Table 1). In them, probiotics evaluated correspond to two bacterial genera, Enterococcus spp. and Bacillus spp. [29,30], as well as prebiotics FOS and MOS and the studied fish species have been rainbow trout, Japanese flounder, yellow croaker and cobia. The first application of synbiotics in fish is that of RodriguezEstrada et al. [31]. In this study a commercial preparation of E. faecalis, and two kinds of ingredients, MOS and PHB (a prebiotic included in the group of biopolimers) were used as probiotics and prebiotics, respectively. Seven supplementation protocols were conducted: C) control group, without supplementation; E) supplemented with 1% E. faecalis; M) supplemented with 0.4% MOS; P) supplemented with 1% PHB; EM) supplemented with 1% E. faecalis and 0.4% MOS; EP) supplemented with 1% E. faecalis and 1% PHP; EMP) supplemented with 1% E. faecalis, 0.4% MOS and 1% PHP.

In the study of Ye et al. [32] diets were supplemented with FOS, MOS and B. clausii alone or in combination. B. clausii strain was previously isolated from grouper (Epinephelus coioides) intestine [33]. The concentrations of the administered supplements in the eight experimental diets were as follows: C) control diet (no FOS, MOS and B. clausii); F) 5 g kg^{-1} FOS; M) 5 g kg^{-1} MOS; FM) 2.5 g kg^{-1} FOS and 2.5 g kg^{-1} MOS); B) 10^7 cells g^{-1} *B. clausii*), FB) 5 g kg^{-1} FOS and 10^7 cells g^{-1} *B. clausii*; MB) 5 g kg^{-1} MOS and 10^7 cells g^{-1} *B. clausii*; FMB) 2.5 g kg^{-1} FOS, 2.5 g kg^{-1} MOS and 10^7 cells g^{-1} *B. clausii*.

Mehrabi et al. [34] assessed the dietary inclusion of a commercial symbiotic, Biomin IMBO (Biomin, Herzogenburg, Austria), in which was comprised of probiotic (E. faecium $5x10^{11}$ CFU/kg) and FOS as prebiotic, in three levels $T_1 = 0.5$, $T_2 = 1.0$ and $T_3 = 1.5$ g into each kilogram of commercial rainbow trout food.

The study of Ai et al. [35] was conducted to evaluate the effects of different concentrations of B. subtilis and FOS, combined or separated, on yellow croaker. Nine practical diets were formulated to contain three levels of B. subtilis (0.0, 0.42×10^7 cfu g^{-1} and 1.35×10^7 cfu g^{-1}), each with three FOS levels (0, 0.2% and 0.4% of dry weight).

Table 1: Summary of fish synbiotic studies

Fish species	Probiotic	Prebiotic	Parameters investigated	Reference
O. mykiss	*E. faecalis*	MOS, PHB	GP, BC, IR, HP, DR	Rodriguez-Estrada et al [31]
P. olivaceus	*B. clausii*	MOS, FOS	GP, FU, BC, DEA, HP, IR	Ye et al [32]
O. mykiss	*E. faecium*	FOS	GP, SR, BC, HP	Mehrabi et al [34]
L. crocea	*B. subtilis*	FOS	SR, GP, IR, DR	Ai et al [35]
R. canadum	*B. subtilis*	Chitosan	SR, GP, IR, DR	Geng et al [36]

Fish genera abbreviations: O. = Onchorhynchus, P. = Paralichtys, L. = Larimichthys, R. = Rachycentron.

Bacterial genera abbreviations: E. = Enterococcus, B. = Bacillus.

Prebiotic abbreviations: PHB- plyhydroxybutyrate acid, MOS- Mannan oligosaccharide, FOS- fructooligosaccharide.

Parameters investigated: GP- growth performance, BC- body composition, IR- immunological response, HP- haematological/serum biochemical parameters, DRdisease resistance, FU- feed utilisation, DEA- digestive enzyme activity, SR- survival rate.

The most recent study, conducted in cobia [36] investigated the effects of various levels of dietary B. subtilis and chitosan. Out of the control diet, six different experimental diets containing three graded levels of B. subtilis (0.0, 1.0, 2.0 g kg^{-1} diet) for each of two levels of chitosan (3.0 and 6.0 g kg^{-1} diet) were prepared for the experimental trial. All the experimental diets evaluated in the referred studies appeared summarized in Table 2.

Effects of Synbiotics on fish Survival rate, Growth Parameters, feed Utilisation and body Composition

All studies conducted to date have evaluated the effects of synbiotics on survival, growth parameters or feed utilisation. Feed cost about over 50% of the variable costs in most aquaculture operations, therefore applying the best feeding strategy can have a significant impact on optimizing profit, which is the primary goal of commercial aquaculture. Also, if more fish are able to survive until they are of marketable size, the subsequent cost of production would be reduced drastically.

Administration of E. faecalis and MOS/PHB in rainbow trout for 12 weeks not affects the survival rate of fish [31]. The experimental fish was healthy and no mortality was observed during the feeding trial. Same occurs in Japanese flounder feeding B. clausii and MOS/FOS, in which fish maintained active ingestion, exhibited proper growth and survived for all time [32].

However, in rainbow trout fingerlings [34], in all synbiotic treatments, test fish were survived significantly in higher rate toward the end of experimental period compared to those from the control. The highest average of survival rate was observed in the T_2 that was statistically different from T_1, T_3 and the control groups.

Table 2: Experimental diets and administration time evaluated in studies aimed on the effects of the administration of synbiotic in farmed fish. All the diets are explained in the text

Reference	Probiotic/Prebiotic	Diets	Prebiotic/Probiotic Concentration	Time
Rodriguez-Estrada et al. [31]	*E. faecalis*/MOS, PHB	C	0/0, 0	12 weeks
		E	1%/0, 0	
		M	0/0.4%, 0	
		P	0/0, 1%	
		EM	1%/0.4%, 0	
		EP	1%/0, 1%	
		EMP	1%/0.4%, 1%	
Ye et al. [32]	*B. clausii*/MOS, FOS	C	0/0, 0	56 days
		F	0/5 g kg^{-1}, 0	
		M	0/0, 5 g kg^{-1} g kg^{-1}	
		FM	10^7 cells g^{-1}/5 g kg^{-1}, 0	
		B	10^7 cells g^{-1}/5 g kg^{-1}, 5	
		BF	0/5 g kg^{-1},5 g kg^{-1}	
		BM	10^7 cells g^{-1}/0, 0	
		BMF	10^7 cells g^{-1}/0, 5 g kg^{-1}	
Mehrabi et al. [34]	Biomin IMBO (*E. faecium*/FOS)	T_1	0.5 g kg^{-1}	2 months
		T_2	1 g kg^{-1}	
		T_3	1.5 g kg^{-1}	
Ai et al. [35]	*B. subtilis*/FOS	C	0/0	10 weeks
		F_1	0/0.2%	
		F_2	0/0.4%	
		B_1	$0.42x10^7$/0	
		B_1F_1	$0.42x10^7$/0.2%	
		B_1F_2	$0.42x10^7$/0.4%	
		B_2	$1.35x10^7$/0	
		B_2F_1	$1.35x10^7$/0.2%	
		B_2F_2	$1.35x10^7$/0.4%	
Geng et al. [36]	*B. subtilis*/Chitosan	C	0/0	8 weeks
		Ch_1	0/3 g kg^{-1}	
		Ch_2	0/6 g kg^{-1}	
		B_1Ch_1	1 g kg^{-1}/3 g kg^{-1}	
		B_1Ch_2	1 g kg^{-1}/6 g kg^{-1}	
		B_2Ch_1	2 g kg^{-1}/3 g kg^{-1}	
		B_2Ch_2	2 g kg^{-1}/6 g kg^{-1}	

Bacterial genera abbreviations: E. = Enterococcus, B. = Bacillus.

Prebiotic abbreviations: PHB/P- plyhydroxybutyrate acid, MOS/M- Mannan oligosaccharide, FOS/F- fructooligosaccharide, Ch: chitosan.

Subindex numbers indicates different concentrations of administrated ingredient.

In yellow croaker and cobia, administration of B. subtillis/FOS or B. subtillis/chitosan respectively, not affect the survival rate, with no differences among different dietary treatments [35,36].

In terms of growth parameters, almost all references reported a positive effect of synbiotics on any parameter. Rainbow trout fed with diets M, EM and EMP recorded significantly higher weigh gain rate (WGR) and specific growth rate (SGR) than those of the rest experimental groups. Differences among experimental groups E and P were not significant with respect to the control group, whereas fish fed the EP diet showed significant better growth and SGR. In general, major effects were observed in EM and EMP, with no differences between the two groups [31]. These results suggest that dietary MOS significantly improved growth performance and nutrient utilization of fish, showing a synergist effect when administered combined with E. faecalis. Combination of E. faecalis and PHB slightly improved growth performance of rainbow trout.

In the Japanese flounder study [32], the control group exhibited the lowest final body weight (FBW) and WGR of all dietary treatments. Diets M, FM, BM and BFM increased FBW, whereas diets MB and BFM increased WGR ($P < 0.05$). However, there were no significant differences in FBW and WGR among any FOS, MOS and B. clausii treatments. In contrast, the fish that fed diets M, FM, BF, BM or BFM had significantly lower feed conversion ratio (FCR) than those fed the control diet. These results showed that a single administration of FOS, MOS and B. clausii did not improve the WGR of the Japanese flounder compared with the control diet. Feeding FOS, MOS or B. clausii alone, or in various combinations, improved WGR and feed efficiency, which was more pronounced in fish fed the synbiotics than those fed prebiotics or probiotics alone.

The study of Mehrabi et al. [34] showed that after 60 days groups fed diets containing different levels of synbiotics (0.5, 1.0 and 1.5) increased body weight gain about 50, 59 and 53%, respectively, in comparison with the control group. Even though, the highest average of final weight and weight gain was observed in group T_2, however, there was no significant difference between T_2 and T_3. The addition of synbiotic to the feed also produced the better SGR, FCR and conversion factor (CF) with values significantly higher than the control, more specifically in groups treated with 0.1% synbiotic. In this study, significant better growth performance was observed in O. mykiss fingerlings maintained on the diet supplemented with synbiotic.

The results of the study carried out by Ai et al. [35] showed that at each dietary FOS level, dietary supplementation of 1.35×10^7 cfu g^{-1} B. subtilis significantly increased the SGR and feed efficiency ratio (FER) compared with the groups without B. subtilis supplementation. However, no significant differences were observed between the groups without B. subtilis and 0.42×10^7 cfu g^{-1} B. subtilis group, and between the two B. subtilis-supplemented groups. In addition, at each B. subtilis level, dietary supplementation of FOS did not significantly affect the SGR, survival rate, FER and hepatosomatic index (HIS) of juvenile large yellow croaker. In this study, dietary supplementation of FOS did not exert beneficial effects on the growth performance, survival and feed utilization, and no significant interactions were observed between B. subtilis and FOS. The significantly enhanced SGR and FER by B. subtilis in this study clearly showed the growth improving property of B. subtilis.

Results in cobia [36] showed that SGR in fish fed with dietary treatments was significantly higher than that of the control fish except diet 6 group, fed diet with 2.0 g kg^{-1} B. subtilis and 3.0 g kg^{-1} chitosan. At 1.0 and 2.0 g kg^{-1} supplemental B. subtilis, the cobia fed diets supplemented with 6.0 g kg^{-1} chitosan had significantly higher SGR than those fed diets supplemented with 3.0 g kg^{-1} chitosan. At 6.0 g kg^{-1} supplemental chitosan, the cobia had significantly higher SGR in group fed diets supplemented with 1.0 g kg^{-1} supplemental B. subtilis. The significant and maximum SGR was observed in the fish fed with 1.0 g kg^{-1} B. subtilis and 6.0 g kg^{-1} chitosan, suggesting the supplementation level was optimal for the growth of cobia.

Body composition has been analyzed by Rodriguez-Estrada et al. [31], Ye et al. [32] and Mehrabi et al. [34]. First of these studies, reported no significant differences in moisture, crude ash, crude lipids and crude proteins contents among all the experimental groups. However, in Japanese flounder there was an increase in body protein content in fish fed a FOS-, MOS- and/or B. clausii-containing diet compared to the control. Fish fed diets BF and BFM also exhibited significantly higher body protein content than fish fed the control diet.

Body lipid content demonstrated an opposite trend to body protein content, where fish fed diets B, MB and BFM presented with significantly lower levels than fish fed the control diet. Neither body moisture nor ash content was affected by any dietary treatments [32]. In the same manner, the percentage of body protein in cobia fed with synbiotic was significantly higher than that from the control fish whereas the percentage of lipid, moisture and dry matter was not [34]. Higher body protein content in the treatment groups implies on this fact that by application of synbiotics, the ingested food was converted more effectively into the structural protein and subsequently resulted more muscle as it is a desirable aspect in fish farming

Effects of Synbiotics on fish Haematological and Biochemical Parameters

Among the haematological and biochemical parameters evaluated in fish after treatments with synbiotics we find hematocrit, triglyceride (TG), cholesterol (CHO), low-density protein cholesterol (LDL-C), high-density protein cholesterol (HDL-C), total serum protein, albumin, globulin and glucose.

Rodriguez-Estrada and colleagues [31] found that hematocrit value was higher in the EM and EMP groups than the C, E, M, P, and EP groups, and significant higher hematocrit value was recorder in the E and M groups than the C and P groups.

Serum biochemical parameters evaluated by Ye et al. [32], reported the following results: The TG level was lower or tended to be lower in fish fed the FOS, MOS and/or B. clausii-containing diets versus the control diet. Diets supplemented with FOS, MOS and/or B. clausii, except diet M, significantly decreased TG levels compared with the control diet. By comparison, the TG level in fish fed diet FM was lower than in fish fed diet M but similar to the levels in fish fed diet. There were no differences in TG levels among diets BF, BM and BFM. The LDL-C response was similar to TG levels. However, feeding diets F, BF or BFM yielded significant decreases in LDL-C levels compared with the control diet. Significant differences were not observed in CHO or HDL-C levels among all dietary treatments.

Dietary administration of commercial symbiotic in rainbow trout resulted in an increase of total serum protein content in T_2 and T_3 treatments compared to other fed treatment and the control. In terms of albumin, only T_2 showed a significant increase. Some other recorded parameters such as albumin/globulin ratio and triglycerides in fish fed with different levels of synbiotic did not show any significant difference in comparison with that from the control group at the end of the experiment. Furthermore, there was no significant difference between blood glucose in fish fed with different levels of synbiotic and control group, whereas this parameter was significantly different between T_1 and T_3. The significant increase in total serum protein and albumin confirmed the involvement of synbiotic in triggering the immune system of rainbow trout following dietary administration [34].

Effects of synbiotics on fish digestive enzyme activity only one of these five studies has regarded the effect of synbiotic administration on digestive enzyme activity. Ye et al. [32], which evaluated the activities of amylase and protease, found that protease activity was significant higher in fish fed diets BM and BFM than in fish fed the control diet, obtaining similar value in both diets. No any other supplementation could positively affect the protease activity. These results indicated that administration of FOS in combination with MOS and B. clausii not determine a major increase in protease activity than the observed with the BM diet.

Moreover, fish fed the diet BFM exhibited the highest amylase activity, which was significantly higher than in fish fed diets FM, B, and control diets. The increase in digestive enzyme activities would allow the host degrades more nutrients, improving digestion and promoting a possible increase in the WGR and/or feed efficiency

Effects of Synbiotics on Fish Immune System

Among the studies carried out about synbiotics in fish, four of them have evaluated the effect on immune system, analyzing different activities such as lysozyme, alternative complement pathway, phagocytosis, respiratory burst, superoxide dismutase and mucus production.

Lysozyme activity: Lysozyme is one of the important bactericidal enzymes of innate immunity, and constitutes an essential defence mechanism against pathogens in fish [37]. The study of Ye et al. [32] on Japanese flounder revealed that dietary treatments affected lysozyme activity differently. This activity was significantly higher in fish fed diets B, BF, BM and BFM compared with those fed the control diet, however there was no significant difference in lysozyme activity between diets F, M and FM and between diets FB, MB and FMB. Therefore, the study concludes that feeding B. clausii promoted lysozyme activity. Moreover, lysozyme activity was significantly higher in fish fed a symbiotic diet than in those fed diets supplemented with the individual prebiotics and tended to enhance activity compared to those fed the B. clausii diet, which suggests that the dietary administration of FOS or/ and MOS combined with B. clausii synergistically modulates lysozyme activity.

These results contrast with those obtained in yellow croaker by Ai et al. [35], in which no synergic effect was observed with FOS and B. subtilis administration. In this study, the serum lysozyme activities in fish fed B_1, B_1 F_1, B_1 F_2, B_2, B_2 F_1, and B_2 F_2 diets were significantly higher than fish fed control, F_1 or F_2 diets, while no significant differences were observed between the two dietary B. subtilis levels. These results showed that lysozyme activity increase could be independent of FOS, being a response to B. subtilis. Although no significant differences were observed between the two dietary B. subtilis levels, a tendency was observed towards major activity with major B. subtilis concentration.

Geng et al. [36] observed that chitosan at 6.0 g kg^{-1} significant increases lysozyme activity when administered alone or in combination with B. subtilis at any of the two levels. No significant differences were observed among groups fed diets with 3.0 g kg^{-1} chitosan with respect to control diet. Fish fed diets supplemented with high chitosan level (6.0 g kg^{-1}) had higher serum lysozyme activities than those fed the basal diet and diets supplemented with low chitosan level (3.0 g kg^{-1}) regardless of B. subtilis levels. At the same chitosan level, no significant differences were observed among B. subtilis levels, suggesting the serum lysozyme activity was not affected by dietary B. subtilis.

Alternative Complement Pathway Activity

This activity has been only evaluated in yellow croaker and cobia, following to administration of B. subtilis, combined with FOS or chitosan respectively [35, 36]. In yellow croaker, ACP activity was not significantly affected by the proand prebiotics supplementation. In cobia, this activity was significant enhanced with diets supplemented with low chitosan level (3.0 g kg^{-1}) at any B. subtilis level, whereas the high chitosan level (6.0 g kg^{-1}) only produces an increase in ACP when administered with B. subtilis at 2.0 g kg^{-1}. However, no significant differences were observed among groups fed with B. subtilis 2.0 g kg^{-1} + chitosan 3.0 g kg^{-1} and those fed with B. subtilis 2.0 g kg^{-1} + chitosan 6.0 g kg^{-1}.

Superoxide Dismutase Activity

In the yellow croaker study by Ai et al. [35] at each FOS level, serum SOD activity was significantly enhanced by the supplementation of $1.35x10^7$ cfu g^{-1} B. subtilis and there were no significant differences between the B. subtilis-deficient group and the $0.42x10^7$ cfu g^{-1} B. subtilis group, and between the two B. subtilis supplemented groups. These results seem to indicate than increase in this activity occurs in response to B. subtilis, and discard a synergic effect of this symbiotic.

Phagocytic Activity

Phagocytic activity has been evaluated in various studies. Rodriguez-Estrada et al. [31] demonstrated that phagocytic activity and index in the E, M, EM and EMP groups were significantly higher than that of the C group. Major increment was observed in EM and EMP groups, without differences between both treatments or between groups E and M. This suggests that PHB does not affect phagocytic activity, and reveals again a synergic effect of E. faecalis and MOS when administered together.

However, Ye and colleagues [32] report that phagocytic percentage and index of leucocytes in fish fed diets supplemented with FOS, MOS and/or B. clausii, were similar to those of fish fed the control diet. The study carried out by Geng et al. [36] indicates that at all experimental diets, except chitosan at 6.0 g kg^{-1}, significantly enhances phagocytic percentage with respect to the control diet. At any of the levels of B. subtilis, phagocytic percentage in fish fed diets with 3.0 g kg^{-1} chitosan were higher than that in the groups fed diets with 6.0 g kg^{-1} chitosan, and there was only statistically significant difference at 2.0 g kg^{-1} B. subtilis level. At any of the two levels of chitosan, phagocytic percentage significantly increased with the increase of B. subtilis supplementation level. Major effect was observed with chitosan at 3.0 g kg^{-1} and B. subtilis at 2.0 g kg^{-1}. These results reveal a synergic effect of chitosan and B. subtilis when administered together in adequate amounts.

Respiratory Burst Activity

The respiratory burst of head kidney leucocytes were not affected by dietary administration of FOS and B. subtilis in yellow croaker [35]. However, following to administration of chitosan and B. subtilis in cobia [36] this activity was significant enhanced by dietary treatments except the group fed high level chitosan (6.0 g kg^{-1}). When administered at low level (3.0 g kg^{-1}) chitosan enhances respiratory burst single or with any of B. subtilis level, and present a maximum with B. subtilis at low concentration. When administered at high level 6.0 g kg^{-1}, chitosan exert a possitive effect on respiratory burst if is combined with any level of B. subtilis. These results remark the synergistic activity of both prebiotic and probiotic.

Mucus production: Mucus is the first physical barrier that inhibits entry of disease organisms from the environment into the fish. It is also a chemical barrier, containing enzymes and antibodies which can kill invading disease organisms. Mucus also lubricates the fish, aiding their movement through water, and is important for osmoregulation [38]. Rodriguez-Estrada et al. [31] demonstrated that dietary E. faecalis, MOS and PHB in rainbow trout increase significantly mucus production in fish fed E, M, EM and EMP than that of the group C.Major increment was observed in EM and EMP groups, without differences between both treatments.

Effects of Synbiotics on Fish Disease Resistance

As synbiotic therapy may offer a suitable alternative for controlling pathogens, the effectiveness of synbiotics in terms of protection against infectious agents could be evaluated by a challenge test. To date, challenge test carried out in fish following to symbiotic administration, have employed Vibrio sp species as pathogens, specifically V. anguillarum and V. harveyi [32,35,36].

In 2009, Rodriguez-Estrada et al. [31] carried out for first time a challenge test in fish feed with synbiotics during 12 weeks. After 14 dpi with V. anguillarum all groups showed lower mortality than that control group, being significant in EM and EMP groups. Dietary B. subtilis supplementation (1.35×10^7 cfu g^{-1}) in yellow croaker [35] during 10 weeks elevated the resistance to V. harveyi infection, but this response was not related with presence or concentration of FOS. No significant interactions were observed between dietary B. subtilis and FOS in the cumulative mortality after challenge.

In Geng et al. [36] cobia fed diets supplemented with various levels of probiotic and chitosan for 56 days were challenged with V. harveyi. Post-challenge survival of fish was always significantly higher in high chitosan groups and increased at each chitosan level with the increase of B. subtilis supplementation level, which speculated chitosan and B. subtilis have a synergistic effect

CONCLUSIONS

In recent years, probiotics and prebiotics have become an integral parts of the aquaculture practices for improving the growth performance and disease resistance [10,21]. Many studies revealed the effects of probiotics to improve feed conversion, growth rates, weight gain, immune system and disease resistance of fish [10,11,20], whereas the effects of prebiotics on these and other parameters have been investigated to a limited extent in different species [18,21]. However, given the progress of previous studies it is clear that biotic applications are useful tools for improving the health status and production of fish, in place of more established immunological and disease control methods as vaccinations, antimicrobials and immunostimulants.

Combined application of probiotics and prebiotics, known as synbiotics, is based on the principle of providing a probiont with a competitive advantage over endogenous populations, improving the survival and implantation of the live microbial dietary supplement in the gastrointestinal tract of the host [12]. The use of synbiotics may make possible to produce greater benefits than the application of individual probionts [20].

Table 3: Summary of results obtained in fish synbiotic studies

Reference	Fish species	Probiotic/prebiotic	Results
Rodriguez-Estrada et al [31]	*O. mykiss*	*E. faecalis*/MOS, PHB	↑ body weight, SGR, hematocrit value, phagocytic activity, phago-cytic index, mucus production, ↓ mortality.
Ye et al [32]	*P. olivaceus*	*B. clausii*/MOS, FOS	↑ body weight, WGR, ↓ FCR, CF, ↑ crude protein and lipid, ↑ lisozime, ↓ TG and LDL-C, ↑ protease and amylase activities
Mehrabi et al [34]	*O. mykiss*	Biomin IMBO (E. faecium/FOS)	↑ body weight, body wieght gain, SGR, FCE, SR, ↑ crude protein, ↑ serum protein.
Ai et al [35]	*L. crocea*	*B. subtilis*/FOS	↑ SGR, FER, lisozime, SOD, ↓ mortality.
Geng et al [36]	*R. canadum*	*B. subtilis*/Chitosan	↑ SGR, lisozime, ACP, phacocytosis, respiratory burst, ↓ mortality.

Fish genera abbreviations: O. = Onchorhynchus, P. = Paralichtys, L. = Larimichthys, R. = Rachycentron.

Bacterial genera abbreviations: E. = Enterococcus, B. = Bacillus.

Prebiotic abbreviations: PHB- plyhydroxybutyrate acid, MOS- Mannan oligosaccharide, FOS- fructooligosaccharide.

SGR: specific growth rate, WGR: weight gain rate, FCR: feed conversion ratio; CF: condition factor; FER: feed efficiency ratio; TG: triglicerides, LDL-C: low-density lipoprotein cholesterol, SOD: superoxide-dismutase, ACP: alternative complement pathway.

Studies considered in this review reveal different effects of synbiotics on regarded species (Table 3). On this way, RodriguezEstrada et al. [31] concludes that dietary incorporation of MOS in a single or a combined inclusion with E. faecalis, enhances the rainbow trout growth and activates its immune system, whereas in japanese flounder the authors did not observe a clear synergistic effect when we compared the combination feeding to feeding prebiotics (MOS, FOS) and probiotics (B. clausii) alone [32]. Results of this study are accordingly with obtained by Ai et al. [35] This study demonstrates that administration of B. subtilis enhanced not only the growth performance and feed utilization of juvenile large yellow croaker, but also the non-specific immune responses and disease resistance. However, no significant interactions between dietary B. subtilis and FOS were observed, suggesting no synergic effect of these pre and probiotic under the experimental conditions. For last, Geng et al. [36] concludes that diet supplemented with the right level of chitosan and B. subtilis significantly enhanced the growth, innate immunity and protection against infection for cobia, and establish the optimal combination of both ingredients for the assayed conditions and parameters.

Dietary administration of a commercial synbiotic in Mehrabi et al. [34] has demonstrates an increase in the growth performances, survival rate, serum protein and albumin content, and improved feeding efficiency in rainbow trout. However, this study does not evaluate the capacity of pre and probiotic contained in synbiotic when administered single, so not allow consider the synergic properties of both ingredients. In conclusion, considerable variation in growth, feed utilization and health benefits with the dietary use of prebiotics and/or probiotics is likely dependent on the fish species, feeding duration and supplement dose as well as the type of prebiotics and probiotics tested. Moreover, aspects such as effects on gut microbiota, mucosal barriers, cell damage/morphology, and mechanisms which mediate host benefits remain unclear. Therefore, further studies regarding the application of synbiotics in aquatic animals, especially their appropriate inclusion levels in specific species and specific rearing conditions, are needed to obtain the expected responses in fish.

ACKNOWLEDGEMENTS

This work was partly supported by national (Ministerio de Ciencia e Innovación, AGL2008-05119-C02-01) and regional (Fundación Séneca, 04538/GERM/06) projects

REFERENCES

1. Rombout JH, Abelli L, Picchietti S, Scapigliati G, Kiron V (2010) Teleost intestinal immunology. Fish Shellfish Immunol 31: 616-626.
2. Das S, Ward LR, Burke C (2008) Prospects of using marine Actinobacteria as probiotics in aquaculture. Appl Microbiol Biotechnol 81: 419-429.
3. Mehrim AI (2009) Effect of dietary supplementation of Biogen® (Commercial probiotic) on mono-sex Nile tilapia Oreochromis niloticus under different stocking densities. J Fisher Aquat Sci 4: 261-273.
4. Mishra S, Mohanty S, Pattnaik P, Ayyappan S (2001) Probiotics: possible application in aquaculture. Fish Chimes 21: 31-37.
5. Sahu MK, Swarnakumar NS, Sivakumar K, Thangaradjou T, Kannan L (2008) Probiotics in aquaculture: importance and future perspectives. Ind J Microbiol 48: 299-308.
6. Irianto A, Austin B (2002) Probiotics in aquaculture. J Fish Dis 25: 633-642.
7. Vershuere L, Rombaut G, Sorgeloos P, Verstraete W (2000) Probiotic bacteria as biological control agents in aquaculture. Microbiol Mol Biol Rev 64: 655-671.
8. Fuller R (1989) Probiotics in man and animals. J Appl Bacteriol 66: 365-378.
9. Gatesoupe FJ (2008) Updating the importance of lactic acid bacteria in fish farming: natural occurrence and probiotic treatments. J Mol Microbiol Biotechnol 14: 107-114.
10. Nayak SK (2010) Probiotics and immunity: a fish perspective. Fish Shellfish Immunol 29: 2-14.
11. Dimitroglou A, Merrifield DL, Carnevali O, Picchietti S, Avella M, et al. (2011) Microbial manipulations to improve fish health and production--a Mediterranean perspective. Fish Shellfish Immunol 30: 1-16.
12. Gibson GR, Roberfroid MB (1995) Dietary modulation of the human colonic microbiota: introducing the concept of prebiotics. J Nutr 125: 1401-1412.
13. Yazawa K, Imai K, Tamura Z (1978) Oligosaccharides and polysaccharides specifically utilisable by bifidobacteria. Chem Pharm.Bull 26: 3306-3311.
14. Gibson GR, Rastall RA, Fuller R (2003) The health benefits of probiotics and prebiotics. In: Fuller R, Perdigón G (Eds), Gut Flora, Nutrition, Immunity and Health. Blackwell Publishing Ltd, Oxford, UK, pp. 52-76.
15. Mahious AS, Ollevier F (2005) Probiotics and Prebiotics in Aquaculture. 1st Regional Workshop on Techniques for Enrichment of Live Food for Use in Larviculture-2005, AAARC, Urmia, Iran. p. 67.
16. Fooks LJ, Fuller R, Gibson GR (1999) Prebiotics, probiotics and human gut microbiology Int Dairy J 9: 53-61.
17. Gibson GR, Probert HM, Van Loo J, Rastall RA, Roberfroid MB (2004) Dietary modulation of the human colonic microbiota: Updating the concept of prebiotics. Nutr Res Rev 17: 259-275.
18. Yousefian M, Amiri MS (2009) A review of the use of prebiotic in aquaculture for fish and shrimp. Afr J Biotechnol 8: 7313-7318.
19. Roberfroid M (2007) Prebiotics: the concept revised. J Nutr 137: 830S-837S.
20. Merrifield DL, Dimitroglou A, Foey A, Davies SJ, Baker RTM, et al. (2010) The current status and future focus of probiotic and prebiotic applications for salmonids. Aquaculture 302: 1-18.
21. Ringø E, Olsen R, Gifstad T, Dalmo R, Amlund H, et al. (2010) Prebiotics in aquaculture: a review. Aquac Nutr 16: 117-136.
22. Burr G, Hume M, Ricke S, Nisbet D, Gatlin D III (2010) In vitro and in vivo evaluation of the prebiotics GroBiotic-A, inulin, mannanoligosaccharide, and galactooligosaccharide on the digestive microbiota and performance of hybrid striped bass (Morone chrysops x Morone saxatilis). Microb Ecol 59: 187-198.
23. Ibrahem MD, Fathi M, Mesalhy S, Abd El-Aty AM (2010) Effect of dietary supplementation of inulin and vitamin C on the growth, hematology, innate immunity, and resistance of Nile tilapia (Oreochromis niloticus). Fish Shellfish Immunol 29: 241-246.
24. Ibrahem MD, Fathi M, Mesalhy S, Abd El-Aty AM (2010) Effect of dietary supplementation of inulin and vitamin C on the growth, hematology, innate immunity,

and resistance of Nile tilapia (Oreochromis niloticus). Fish Shellfish Immunol 29: 241-246.

25. Sherman PM, Ossa JC, Johnson-Henry K (2009) Unraveling Mechanisms of Action of Probiotics. Nutr Clin Pract 24: 10-14.
26. Quigley EM (2010) Prebiotics and probiotics; modifying and mining the microbiota. Pharmacol Res 61: 213-218.
27. Seifert S, Watzl B (2007) Inulin and Oligofructose: Review of Experimental Data on Immune Modulation. J Nutr 137: 2563S-2567S.
28. Magnadottir B (2010) Immunological control of fish diseases. Mar Biotechnol 12: 361-379.
29. Chou T-C, Rideout D, Chou J, Bertino JR (1991) Chemotherapeutic synergism, potentiation and antagonism. In: Dulbecco R (Ed), Encyclopedia of human biology, vol. 2. Academic Press, San Diego, California, pp. 371-379.
30. Hagi T, Hoshino T (2009) Screening and characterization of potential probiotic lactic acid bacteria from cultured common carp intestine. Biosci Biotechnol Biochem 73: 1479-1483.
31. Lauzon HL, Gudmundsdottir S, Steinarsson A, Oddgeirsson M, Petursdottir SK, et al. (2010) Effects of bacterial treatment at early stages of Atlantic cod (Gadus morhua L.) on larval survival and development. J Appl Microbiol 108: 624-632.
32. Rodriguez-Estrada U, Satoh S, Haga Y, Fushimi H, Sweetman J (2009) Effects of single and combined supplementation of Enterococcus faecalis, mannanoligosaccharide and polyhydrobutyric acid on growth performance and immune response of rainbow trout Oncorhynchus mykiss. Aquacult Sci 57: 609-617.
33. Ye JD, Wang K, Li FD, Sun YZ (2011) Single or combined effects of fructoand mannan oligosaccharide supplements and Bacillus clausii on the growth, feed utilization, body composition, digestive enzyme activity, innate immune response and lipid metabolism of the Japanese flounder Paralichthys olivaceus. Aquac Nutr 17: 902-911.
34. Sun Y, Yang H, Ling Z, Chang J, Ye J (2009) Gut microbiota of fast and slow growing grouper Epinephelus coioides. Afr J Microbiol Res 3: 713-720.
35. Mehrabi Z, Firouzbakhsh F, Jafarpour A (2011) Effects of dietary supplementation of synbiotic on growth performance, serum biochemical parameters and carcass composition in rainbow trout (Oncorhynchus mykiss) fingerlings. J Anim Physiol Anim Nutr (Berl) 24: doi: 10.1111/j.1439-0396.2011.01167.x.
36. Ai Q, Xu H, Mai K, Xu W, Wang J, et al. (2011) Effects of dietary supplementation of Bacillus subtilis and fructooligosaccharide on growth performance, survival, non-specific immune response and disease resistance of juvenile large yellow croaker, Larimichthys crocea. Aquaculture 317: 155-161.
37. Geng X, Dong XH, Tan BP, Yang QH, Chi SY, et al. (2011) Effects of dietary chitosan and Bacillus subtilis on the growth performance, non-specific immunity and disease resistance of cobia, Rachycentron canadum. Fish Shellfish Immunol 31: 400-406.
38. Lindsay GJH (1986) The significance of chitinolytic enzymes and lysozyme in rainbow trout (Salmo gairdneri) defence. Aquaculture 51: 169-173.
39. Shephard KL (1994) Functions for fish mucus. Review in Fish Biol and Fisheries. 4: 401-429.

CHAPTER 17

ANTIMICROBIAL ACTIVITY, ANTIBIOTIC SUSCEPTIBILITY AND VIRULENCE FACTORS OF LACTIC ACID BACTERIA OF AQUATIC ORIGIN INTENDED FOR USE AS PROBIOTICS IN AQUACULTURE

Estefanía Muñoz-Atienza, Beatriz Gómez-Sala, Carlos Araújo, Cristina Campanero, Rosa del Campo, Pablo E Hernández, Carmen Herranz and Luis M Cintas

Keywords: Lactic Acid Bacteria, Aquatic animals, Aquaculture probiotics, Anti-fish pathogens activity, Antibiotic resistance and virulence factors, Qualified Presumption of Safety

BACKGROUND

Aquaculture has the potential to make a significant contribution to the increasing demand for aquatic food in most world regions; however, in order to achieve this goal, the sector will have to face significant challenges, including the production intensification, the disease control and the prevention of the environmental deterioration [1]. In fish farming, the widespread use of antibiotics as prophylactic and therapeutic agents to control bacterial diseases has been associated with the emergence of antibiotic resistance in bacterial pathogens and with the alteration of the microbiota of the aquaculture environment [2, 3]. This resulted in the ban of antibiotic usage as animal growth promoters in Europe and stringent worldwide regulations on therapeutical antibiotic applications. This scenario has led to an evergrowing interest in the search and development of alternative strategies for disease control, within the frame of good husbandry practices, including adequate hygiene conditions, vaccination programmes and the use of probiotics, prebiotics and immunostimulants [4, 5, 6]. Recently, novel strategies to control bacterial infections in aquaculture have emerged, such as specific killing of pathogenic bacteria by bacteriophages, growth inhibition of pathogen by short-chain fatty acids and polyhydroxyalkanoates, and interference with the regulation of virulence genes (quorum sensing disruption), which have been reviewed by Defoirdt *et al.*[7]. With regard to probiotics, they are defined as live microbial adjuncts which have a beneficial effect on the host by: (i) modifying the host-associated or ambient microbial community; (ii) improving feed use or enhancing its nutritional value; (iii) enhancing the host response towards disease; and/or (iv) improving its environment [8]. To date, most probiotics proposed as biocontrollers and bioremediation agents for aquaculture belong to the LAB group (mainly to the genera *Lactobacillus*, *Lactococcus*, *Leuconostoc*, *Enterococcus* and *Carnobacterium*), to the genera *Vibrio*, *Bacillus*, and *Pseudomonas* or to the species *Saccharomyces cerevisiae*[8, 9]. Recently, a probiotic culture (Bactocell®, *Pediococcus acidilactici* CNCM MA18/5 M) has been authorized for the first time for use in aquaculture in the European Union.

According to the FAO/WHO [10], the development of commercial probiotics requires their unequivocal taxonomic identification, as well as their *in vitro* and *in vivo* functional characterization and safety assessment. In Europe, the European Food Safety Agency (EFSA) proposed a system for a pre-market safety assessment of selected groups of microorganisms used in food/feed and the production of food/feed additives leading to a Qualified Presumption of Safety (QPS) status [11, 12, 13]. The QPS approach propose that the

Citation: Estefanía Muñoz-Atienza, Beatriz Gómez-Sala, Carlos Araújo, Cristina Campanero, Rosa del Campo, Pablo E Hernández, Carmen Herranz and Luis M Cintas, "Antimicrobial activity, antibiotic susceptibility and virulence factors of Lactic Acid Bacteria of aquatic origin intended for use as probiotics in aquaculture," DOI: 10.1186/1471-2180-13-15.

safety assessment of a defined taxonomic group could be made based on establishing taxonomic identity, body of knowledge, possible pathogenicity and commercial end use. According to the EFSA approach [13], most LAB species are included in the QPS list and, therefore, demonstration of their safety only requires confirmation of the absence of determinants of resistance to antibiotics of human and veterinary clinical significance. However, in the case of enterococci, a more thorough, strain-specific evaluation is required to assess the risk associated to their intentional use in the food chain. In this work, we present the antimicrobial activity against fish pathogens and the *in vitro* safety assessment beyond the QPS approach of a collection of 99 LAB belonging to the genera *Enterococcus*, *Lactobacillus*, *Lactococcus*, *Leuconostoc*, *Pediococcus* and *Weissella*, previously isolated from aquatic animals regarded as human food [14] and intended for use as probiotics in aquaculture.

RESULTS

Direct Antimicrobial Activity of the 99 LAB of Aquatic Origin

The 99 LAB strains isolated from fish, seafood and fish products displayed direct antimicrobial activity against, at least, four of the eight tested indicator microorganisms (Table 1). The most sensitive indicators were *Listonella anguillarum* CECT4344, *Ls. anguillarum* CECT7199 and *Aeromonas hydrophila* CECT5734, followed by *Lactococcus garvieae* JIP29-99, *Streptococcus iniae* LMG14521 and *Streptococcus agalactiae* CF01173. On the contrary, *Photobacterium damselae* CECT626 and *Vibrio alginolyticus* CECT521 were the less sensitive indicator microorganisms.

Table 1. Origin and direct antimicrobial activity against fish pathogens of LAB isolated from aquatic animals

Origin	Strain		Indicator microorganisms[a]							
			Lactococcus garvieae- JIP29-99	*Streptococcus agalacti-ae*CF01173	*Streptococ-cus iniae* LMG14521	*Aeromo-nas hydrophila* CECT5734	*Listonella anguillarum-* CECT4344	*Ls. anguil-larum-* CECT7199	*Photo-bacterium damselae-* CECT626	*Vibrio algi-nolyticus-* CECT521
Albacore (*Thunnus alalunga*)	*Enterococcus faecium*	BNM58	+	+	+	++	++	+++	+	-
	Weissella cibaria	BNM69	+	+	+	+++	+++	+++	-	-
Atlantic salmon (*Salmo salar*)	*Enterococcus faecalis*	SMF10	+	+	+	++	+++	++	-	+
		SMF28	+	+	++	++	+++	+	-	+
		SMF37	+	+	+	+	++	+++	-	+
		SMF69	+	+	++	++	+++	+++	+	+
		SMM67	+	+	++	++	+++	+++	-	-
		SMM70	+	+	+	+	+++	+++	-	-
	E. faecium	SMA1	+	+	+	++	++	+++	+	-
		SMA7	+	+	+	+	++	+++	+	+
		SMA8	+	+	+	++	++	+++	+	+
		SMA101	+	+	+	++	+++	++	+	+
		SMA102	+	+	+	++	+++	+	+	+
		SMA310	++	+	+	++	+++	++	+	+
		SMA320	++	+	+	++	++	+++	+	+
		SMA361	+	+	+	++	++	+++	+	+
		SMA362	+	+	+	++	++	+++	+	-
		SMA384	+	+	+	++	++	+++	+	-
		SMA389	+	+	+	++	++	+++	-	+
		SMF8	+	+	++	++	++	++	+	-
		SMF39	+	+	++	++	++	+++	+	+
	Lactobacillus sakei subsp. *carnosus* (*Lb. carnosus*)	SMA17	+	-	+	++	+++	+++	-	-
	Lactococcus lactis subsp. *cremoris* (*L. cremoris*)	SMF110	+	+	+	+	+++	+++	+	+
		SMF161	+	+	+	++	+++	+++	+	++
		SMF166	+	+	+	++	++	+++	+	++
	Leuconostoc mesenteroides- subsp. *cremoris* (*Lc. cremoris*)	SMM69	+	+	+	++	+++	+++	-	-
	Pediococcus pentosaceus	SMF120	++	++	++	++	+++	+++	-	+
		SMF130	++	+	++	++	+++	+++	-	+
		SMM73	++	+	+	+++	+++	+++	+	++
	W. cibaria	SMA14	++	+	+	++	+++	+++	+	++

		SMA25	+	+	+	+++	+++	+++	-	-
Cod (*Gadus morhua*)	*E. faecalis*	BCS27	++	++	++	++	+++	+++	-	-
		BCS32	+	+	+	+	++	+++	-	+
		BCS53	+	++	+	+	+++	+++	+	-
		BCS67	+	+	-	++	+++	++	-	+
		BCS72	+	+	+	++	+++	+++	+	-
		BCS92	+	+	+	++	+++	++	+	+
	E. faecium	BCS59	++	+	++	++	+++	+++	-	+
		BCS971	+	+	+	+	+++	+++	-	+
		BCS972	+	+	+	+	+++	+++	-	+
	*Lactobacillus curvatus*subsp. *curvatus* (*Lb. curvatus*)	BCS35	-	-	+	++	+++	+++	-	-
	Lc. cremoris	BCS251	+	+	++	+	+++	+++	-	+
		BCS252	+	+	++	+	+++	+++	-	+
	P. pentosaceus	BCS46	++	+	++	+++	+++	+++	-	+
	W. cibaria	BCS50	++	+	++	++	+++	+++	-	+
Common cockle (*Cerastoderma edule*)	*E. faecium*	B13	+	+	++	++	+++	+++	-	-
		B27	+	+	+	++	+++	++	+	+
	Lb. carnosus	B43	+	+	+	++	+++	+++	-	-
	P. pentosaceus	B5	++	+	++	++	+++	+++	-	-
		B11	++	+	++	+++	+++	+++	+	-
		B41	++	++	++	+++	+++	+++	+	++
		B260	++	+	++	++	+++	+++	-	++
	W. cibaria	B4620	++	+	++	++	+++	+++	-	++
Common ling (*Molva molva*)	*E. faecium*	MV5	+	+	+	++	++	+++	+	+
Common octopus (*Octopus vulgaris*)	*E. faecalis*	P77	++	+	++	++	+++	+++	-	+
	E. faecium	P68	++	+	+++	++	+++	+++	-	+
		P623	+	+	+	+	+++	++	-	+
	P. pentosaceus	P63	++	+	++	+++	+++	+++	-	+
		P621	++	+	++	+	+++	+++	-	+
	W. cibaria	P38	++	++	++	++	+++	+++	-	+
		P50	++	+	+	++	+++	+++	-	+
		P61	++	+	+	++	+++	+++	-	-
		P64	++	+	+	+++	+++	+++	+	++
		P69	++	+	+	++	+++	+++	+	++
		P71	+	+	++	++	+++	+++	+	+
		P73	++	++	++	++	+++	+++	-	+
		P622	++	++	++	+	+++	+++	+	+
European seabass (*Dicentrarchus labrax*)	*E. faecium*	LPP29	+	+	+	+	++	+++	+	-
	P. pentosaceus	LPM78	++	+	++	++	+++	+++	-	-
		LPM83	++	+	++	++	+++	+++	-	-
		LPP32	++	++	++	++	+++	+++	-	+
		LPV46	++	+	++	++	+++	+++	-	+
		LPV57	++	+	++	+++	+++	+++	-	-
European squid (*Loligo vulgaris*)	*E. faecium*	CV1	+	+	+	+	+++	+++	-	+
		CV2	++	+	+	+	+++	++	+	+
Megrim (*Lepidorhombus boscii*)	*E. faecalis*	GM22	-	-	+	++	++	+++	+	++
		GM26	-	-	+	+	++	++	+	-
		GM33	-	-	++	+	++	+++	+	-
	E. faecium	GM23	+	+	+	++	++	+++	+	+
		GM29	++	++	+	++	++	+++	+	+
		GM351	-	-	+	+	++	++	+	-
		GM352	++	+	+	++	++	+++	+	+
Norway lobster (*Nephrops norvegicus*)	*E. faecalis*	CGM16	++	+	++	++	+++	+++	-	+
		CGM156	+	+	++	++	+++	+++	-	-

		CGM1514	+	+	+	++	+++	++	+	+
		CGV67	++	+	+	+	+++	+++	+	+
	E. faecium	CGM171	+	+	+	+	+++	+++	+	+
		CGM172	+	+	+	+	+++	+++	+	+
Rainbow trout (*Oncorhynchus mykiss*)	*E. faecium*	TPM76	+	+	+	+	++	+++	+	+
		TPP2	+	+	+	+	++	+++	+	+
	P. pentosaceus	TPP3	++	+	+	++	+++	+++	-	++
Sardine (*Sardina pilchardus*)	*E. faecalis*	SDP10	+	+	+	+	+++	+++	-	+
	W. cibaria	SDM381	++	+	++	++	+++	+++	-	-
		SDM389	+	+	++	++	+++	+++	-	-
Swimcrab (*Necora puber*)	*E. faecium*	NV50	+	+	+	++	++	++	+	-
		NV51	++	+	+	+	++	++	+	++
		NV52	++	+	+	+	++	+++	+	+
		NV54	++	+	+	+	++	+++	+	+
		NV56	++	+	+	++	++	++	+	-

[a]Direct antimicrobial activity was determined by a SOAT and the scores reflect different degrees of growth inhibition (diameter in mm); -, no inhibition; +, 3–5 mm inhibition zone; ++, 6–9 mm inhibition zone; +++, ≥10 mm inhibition zone.

Preliminary Safety Evaluation of Enterococci: Presence of Virulence Factors, Production of Gelatinase and Hemolysin and Antibiotic Susceptibility

Concerning *E. faecalis*, most of the strains (20 strains, 95%) harboured, at least, one relevant virulence factor: *efaAfs* (95%), *gelE* (71%), or *agg* (67%) genes (Table 2). A positive gelatinase reaction was found in 15 *E. faecalis*strains (71%) which harboured *gelE*, from which 12 also harboured *agg* gene. Only one *E. faecalis* strain (*E. faecalis* SDP10) (5%), harbouring *cylL* $_L$ -*cylL* $_S$ -*cylM*, exerted hemolytic activity, while none of the strains amplified *hyl* or *esp* genes. With regard to *E. faecium*, 20 strains (53%) harboured, at least, one relevant virulence factor: *efaAfs* (45%), *gelE* (24%) or *agg* (8%), but only 4 strains (11%) exerted gelatinase activity. None of the *E. faecium* strains exerted hemolytic activity nor amplified *hyl* or *esp* genes. The results of the antibiotic susceptibility tests revealed that 39 enterococccal strains (66%) displayed acquired antibiotic resistance to antibiotics other than penicillin G, chloramphenicol and high-level gentamicin. In this respect, 13 *E. faecalis*strains (62%) showed acquired resistance to (i) second generation quinolones (ciprofloxacin and/or norfloxacin) (12 strains, 57%), (ii) rifampicin (5 strains, 24%), (iii) nitrofurantoin (5 strains, 24%), (iv) glycopeptides (vancomycin and teicoplanin) (4 strains, 19%), and/or (v) erythromycin (1 strain, 5%). However, 26 *E. faecium* strains (68%), including 17 strains that encode virulence factors and nine strains without these traits, displayed acquired resistance to (i) erythromycin (14 strains, 37%), (ii) nitrofurantoin (11 strains, 29%), (iii) second generation quinolones (ciprofloxacin and/or norfloxacin) (10 strains, 26%), (iv) rifampicin (4 strains, 11%), (v) tetracycline (2 strains, 5%), and/or (vi) glycopeptides (vancomycin and teicoplanin) (1 strain, 3%). Moreover, multiple antibiotic resistance (two to six antibiotics) was found in *E. faecalis* (10 strains, 48%) and, to a lesser extent, in *E. faecium* (12 strains, 32%) (Table 2). According to the results above, 21 *E. faecalis* strains were discarded for further studies based on the presence of virulence factors (8 strains, 38%), acquired antibiotic resistance (1 strain, 5%) or both (12 strains, 57%). Regarding *E. faecium* strains, 29 (76%) were dropped from further screening based on acquired antibiotic resistance (9 strains, 24%), the presence of virulence factors (3 strains, 8%) or both (17 strains, 45%).

Table 2. Preliminary safety evaluation of enterococci

***Enterococcus*spp.**	**Strain**	**Virulence Factors**		**Antibiotic resistance phenotype[c]**
		Genotype[a]	**Phenotype[b]**	
E. faecalis	SMF10	*efaAfs*$^+$, *gelE*$^+$, *agg*$^+$	GelE$^+$, Hly$^-$	CIP, NOR
	SMF28	*efaAfs*$^+$, *gelE*$^+$	GelE$^+$, Hly$^-$	CIP, NOR
	SMF37	*efaAfs*$^+$, *gelE*$^+$, *agg*$^+$	GelE$^+$, Hly$^-$	-
	SMF69	*efaAfs*$^+$, *gelE*$^+$, *agg*$^+$	GelE$^+$, Hly$^-$	CIP, RIF
	SMM67	n.d.	GelE$^-$, Hly$^-$	CIP, NIT, NOR, TEC, VAN
	SMM70	*efaAfs*$^+$, *gelE*$^+$	GelE$^+$, Hly$^-$	ERY, NIT
	BCS27	*efaAfs*$^+$, *gelE*$^+$, *agg*$^+$	GelE$^+$, Hly$^-$	CIP, NIT, NOR, RIF, TEC, VAN
	BCS32	*efaAfs*$^+$, *gelE*$^+$, *agg*$^+$	GelE$^+$, Hly$^-$	NOR
	BCS53	*efaAfs*$^+$, *gelE*$^+$, *agg*$^+$	GelE$^+$, Hly$^-$	-

	BCS67	*efaAfs* $^{+}$	GelE^{-}, Hly^{-}	CIP
	BCS72	*efaAfs*$^{+}$, *agg*$^{+}$	GelE^{-}, Hly^{-}	-
	BCS92	*efaAfs* $^{+}$	GelE^{-}, Hly^{-}	-
	P77	*efaAfs*$^{+}$, *gelE*$^{+}$	GelE^{+}, Hly^{-}	NIT, NOR, RIF, TEC, VAN
	GM22	*efaAfs*$^{+}$, *gelE*$^{+}$, *agg*$^{+}$	GelE^{+}, Hly^{-}	CIP, NOR
	GM26	*efaAfs*$^{+}$, *gelE*$^{+}$, *agg*$^{+}$	GelE^{+}, Hly^{-}	-
	GM33	*efaAfs*$^{+}$, *gelE*$^{+}$, *agg*$^{+}$	GelE^{+}, Hly^{-}	-
	CGM156	*efaAfs* $^{+}$	GelE^{-}, Hly^{-}	CIP, NIT, NOR, RIF, TEC, VAN
	CGM1514	*efaAfs*$^{+}$, *agg*$^{+}$	GelE^{-}, Hly^{-}	-
	CGM16	*efaAfs*$^{+}$, *gelE*$^{+}$, *agg*$^{+}$	GelE^{+}, Hly^{-}	CIP, NOR, RIF
	CGV16	*efaAfs*$^{+}$, *gelE*$^{+}$, *agg*$^{+}$	GelE^{+}, Hly^{-}	NOR
	SDP10	*efaAfs*$^{+}$, *gelE*$^{+}$, *agg* $^{+}$, $cylL_L L_S^{+}$, $cylL_L L_S M^{+}$	GelE^{+}, Hly^{+}	-
E. faecium	BNM58	n.d.	GelE^{-}, Hly^{-}	-
	SMA1	n.d.	GelE^{-}, Hly^{-}	CIP
	SMA7	n.d.	GelE^{-}, Hly^{-}	-
	SMA8	n.d.	GelE^{-}, Hly^{-}	-
	SMA101	n.d.	GelE^{-}, Hly^{-}	ERY, NIT
	SMA102	*efaAfs* $^{+}$	GelE^{-}, Hly^{-}	ERY, NIT
	SMA310	n.d.	GelE^{-}, Hly^{-}	ERY, NIT
	SMA320	*efaAfs* $^{+}$	GelE^{-}, Hly^{-}	ERY, NIT
	SMA361	*efaAfs* $^{+}$	GelE^{-}, Hly^{-}	ERY
	SMA362	n.d.	GelE^{-}, Hly^{-}	ERY, NIT
	SMA384	*gelE* $^{+}$	GelE^{-}, Hly^{-}	NIT
	SMA389	*gelE* $^{+}$	GelE^{-}, Hly^{-}	CIP, NIT, NOR
	SMF8	n.d.	GelE^{-}, Hly^{-}	-
	SMF39	*efaAfs*$^{+}$, *gelE*$^{+}$	GelE^{-}, Hly^{-}	-
	BCS59	n.d.	GelE^{-}, Hly^{-}	NIT
	BCS971	n.d.	GelE^{-}, Hly^{-}	ERY
	BCS972	n.d.	GelE^{-}, Hly^{-}	ERY
	B13	*gelE* $^{+}$	GelE^{+}, Hly^{-}	CIP
	B27	*efaAfs*$^{+}$, *gelE*$^{+}$	GelE^{+}, Hly^{-}	CIP
	MV5	*efaAfs*$^{+}$, *gelE*$^{+}$, *agg*$^{+}$	GelE^{-}, Hly^{-}	CIP, NIT
	P68	*efaAfs*$^{+}$, *gelE*$^{+}$, $cylL_L L_S^{+}$	GelE^{+}, Hly^{-}	CIP, NIT, NOR, RIF, TEC, VAN
	P623	*efaAfs* $^{+}$	GelE^{-}, Hly^{-}	ERY
	LPP29	n.d.	GelE^{-}, Hly^{-}	-
	CV1	n.d.	GelE^{-}, Hly^{-}	-
	CV2	n.d.	GelE^{-}, Hly^{-}	-
	GM23	*efaAfs* $^{+}$	GelE^{-}, Hly^{-}	CIP, NOR, RIF, TET
	GM29	*efaAfs*$^{+}$, *gelE*$^{+}$, $cylL_L L_S^{+}$	GelE^{-}, Hly^{-}	CIP, NOR, RIF
	GM351	*efaAfs*$^{+}$, *gelE*$^{+}$, *agg*$^{+}$	GelE^{+}, Hly^{-}	CIP, NOR
	GM352	*efaAfs* $^{+}$	GelE^{-}, Hly^{-}	CIP, NIT, NOR, RIF, TET
	CGM171	n.d.	GelE^{-}, Hly^{-}	ERY

	CGM172	*efaAfs* $^{+}$	GelE^{-}, Hly^{-}	ERY
	TPM76	n.d.	GelE^{-}, Hly^{-}	-
	TPP2	n.d.	GelE^{-}, Hly^{-}	-
	NV50	*efaAfs*$^{+}$, *agg*$^{+}$	GelE^{-}, Hly^{-}	-
	NV51	*efaAfs* $^{+}$	GelE^{-}, Hly^{-}	ERY
	NV52	n.d.	GelE^{-}, Hly^{-}	ERY
	NV54	*efaAfs* $^{+}$	GelE^{-}, Hly^{-}	ERY
	NV56	*efaAfs* $^{+}$	GelE^{-}, Hly^{-}	-

[a]n.d., not detected.

[b]GelE and Hly refer to gelatinase and cytolysin/hemolysin activity, respectively.

[c]Abbreviation of antibiotics: CIP, ciprofloxacin; ERY, erythromycin; NIT, nitrofurantoin; NOR, norfloxacin; RIF, rifampicin; TEC, teicoplanin; TET, tetracycline; VAN, vancomycin.

Extracellular Antimicrobial Activity of the 49 pre-Selected LAB

The antimicrobial activity of supernatants from the 49 pre-selected LAB (9 *E. faecium* selected based on their preliminary safety assessment and 40 non-enterococcal strains) with direct antimicrobial activity against fish pathogens was assayed against three indicator microorganisms by an ADT (Table 3). In this regard, 24 (49%) and 10 (20%) strains displayed extracellular antimicrobial activity in their supernatants and/or 20-fold concentrated supernatants against *Pediococcus damnosus* CECT4797 and *L. garvieae* JIP 29–99, respectively, but none of the strains inhibited the Gram-negative strain *A. hydrophila* CECT5734. Interestingly, the antimicrobial activity of the respective supernatants was sensitive to proteinase K treatment, but was not affected by the heat treatment, revealing the proteinaceous nature and heat stability of the secreted antimicrobial compounds (i.e., heat-stable bacteriocins). The 24 LAB strains secreting bacteriocins into the liquid growth medium belong to the species *P. pentosaceus* (15 strains), *E. faecium* (8 strains), and *Lb. curvatus* (1 strain).

Table 3. Extracellular antimicrobial activity of the 49 pre-selected LAB [a]

LAB species[b]	**Strain**	**Indicator microorganisms**					
		***P. damnosus*CECT4797**		***L. garvieae*-JIP29-99**		***A. hydrophila*CECT5734**	
		S	**CS**	**S**	**CS**	**S**	**CS**
Enterococci							
E. faecium	BNM58	22.4	26.8	14.0	15.0	-	-
	SMA7	-	-	-	-	-	-
	SMA8	19.0	19.6	9.4	10.2	-	-
	SMF8	19.0	21.8	10.3	10.8	-	-
	LPP29	20.5	24.4	12.6	13.1	-	-
	CV1	15.0	19.2	-	-	-	-
	CV2	19.8	23.7	12.7	11.4	-	-
	TPM76	17.0	21.2	-	8.7	-	-
	TPP2	19.7	23.5	12.8	12.4	-	-
Non-enterococci							
Lb. curvatus	BCS35	18.2	24.7	-	-	-	-
P. pentosaceus	SMF120	-	-	-	-	-	-
	SMF130	7.4	9.7	-	-	-	-
	SMM73	-	9.5	-	-	-	-
	BCS46	-	9.4	-	-	-	-
	B5	8.1	9.0	-	-	-	-
	B11	-	9.0	-	-	-	-
	B41	7.3	11.7	-	-	-	-

	B260	7.3	10.6	-	-	-	-
	P63	-	9.8	-	-	-	-
	P621	-	10.5	-	-	-	-
	LPM78	-	8.3	-	-	-	-
	LPM83	7.9	11.0	-	-	-	-
	LPP32	8.5	11.3	-	8.9	-	-
	LPV46	8.2	11.3	-	8.2	-	-
	LPV57	7.6	10.5	-	-	-	-
	TPP3	9.0	11.7	7.5	9.2	-	-

[a]Antimicrobial activity (mm) of supernatants (S) and 20-fold concentrated supernatants (CS) as determined by an ADT.

[b]*Lb. carnosus*, *L. cremoris*, *Lc. cremoris* and *W. cibaria* strains did not show extracellular antimicrobial activity against any of the tested indicator microorganisms.

*In vitro*safety assessment of the 49 pre-selected LAB

The 49 pre-selected LAB were further submitted to a comprehensive safety assessment by different *in vitro*tests.

Hemolysin Production, bile salts Deconjugation and Mucin Degradation

None of the non-enterococcal strains showed hemolytic activity, similarly as found for the 9 enterococci. Moreover, bile salts deconjugation and mucin degradation abilities were not found in any of the tested strains.

Enzymatic Activities

The results of the analysis of enzymatic activity profiles of the tested LAB are shown in Table 4. None of the strains showed lipolytic activity, except *E. faecium* LPP29, TPM76, SMA7, and SMF8 which produced esterase (C4) and esterase lipase (C8). Moreover, none of the LAB strains showed protease activity (trypsin and α-chymotrypsin). Nevertheless, peptidase activity (leucine, valine or cystine arylamidase) was found in all the species. All strains showed acid phosphatase (except *E. faecium* TPM76 and *Lc. cremoris*) and naphthol-AS-BI-phosphohydrolase activities, but none displayed alkaline phosphatase activity. β-Galactosidase was found in most species (but not in all strains) except *Lb. curvatus* and *L. cremoris*. However, α-glucosidase was only found in the three *Lc. cremoris* strains. β-Glucosidase and N-acetyl-β-glucosaminidase activities were observed in most *E. faecium*, *Lactobacillus* spp., *L. cremoris*, and *P. pentosaceus* strains, but only in two *W. cibaria*strains, while the three *Lc. cremoris* strains showed β-glucosidase but lacked N-acetyl-β-glucosaminidase activity. On the other hand, α-galactosidase, β-glucuronidase, α-mannosidase, and α-fucosidase activities were not detected in any of the tested LAB strains.

Table 4. Enzymatic activity profiles of the 49 pre-selected LAB [a]

Species	Strain	Esterase (C4)	Esterase lipase (C8)	Leucine arylamidase	Valine arylamidase	Cystine arylamidase	Acid phosphatase	Naphthol-AS-BI-phosphohydrolase	β-Galactosidase	α-Glucosidase	β-Glucosidase	N-acetyl-β-glucosaminidase
Enterococci												
E. faecium	BNM58	0	0	≥40	10	10	20	10	0	0	0	0
	SMA7	20	20	≥40	30	20	30	10	0	0	0	0
	SMA8	0	0	≥40	≥40	5	5	5	5	0	20	≥40
	SMF8	5	5	10	5	5	20	10	0	0	30	0
	LPP29	10	10	30	5	20	10	10	0	0	0	0
	CV1	0	0	≥40	≥40	5	10	20	20	0	30	≥40
	CV2	0	0	≥40	≥40	10	10	20	0	0	10	≥40
	TPM76	30	10	20	0	0	0	10	10	0	0	0
	TPP2	0	0	≥40	20	10	10	10	5	0	30	0
Non-enterococci												
Lb. carnosus	SMA17	0	0	≥40	≥40	0	30	20	30	0	30	30
	B43	0	0	≥40	≥40	0	5	5	10	0	0	0
Lb. curvatus	BCS35	0	0	≥40	10	5	10	20	0	0	5	10
L. cremoris	SMF110	0	0	≥40	≥40	0	20	20	0	0	30	30
	SMF161	0	0	20	0	5	≥40	20	0	0	0	0
	SMF166	0	0	≥40	≥40	0	20	20	0	0	10	10
Lc. cremoris	SMM69	0	0	10	0	0	0	10	≥40	30	≥40	0
	BCS251	0	0	5	0	0	0	5	20	20	10	0
	BCS252	0	0	10	0	0	0	10	30	20	10	0
P. pentosaceus	SMF120	0	0	≥40	≥40	20	≥40	≥40	0	0	20	20
	SMF130	0	0	≥40	≥40	20	30	≥40	20	0	≥40	≥40
	SMM73	0	0	≥40	30	10	20	30	20	0	30	≥40
	BCS46	0	0	≥40	≥40	5	20	30	30	0	≥40	≥40
	B5	0	0	30	≥40	10	10	20	10	0	30	≥40
	B11	0	0	≥40	30	0	5	20	0	0	30	≥40
	B41	0	0	30	≥40	0	5	20	5	0	20	≥40
	B260	0	0	≥40	≥40	10	20	30	0	0	20	30
	P63	0	0	≥40	≥40	5	20	20	30	0	30	≥40
	P621	0	0	≥40	≥40	0	5	30	0	0	30	≥40
	LPM78	0	0	30	30	5	10	20	20	0	30	≥40

	LPM83	0	0	30	30	5	10	20	30	0	10	≥40
	LPP32	0	0	≥40	≥40	5	5	20	0	0	30	≥40
	LPV46	0	0	≥40	≥40	5	20	30	5	0	30	30
	LPV57	0	0	≥40	≥40	5	20	30	30	0	≥40	≥40
	TPP3	0	0	≥40	≥40	5	5	5	10	0	0	0
W. cibaria	BNM69	0	0	0	0	0	30	10	30	0	0	0
	SMA14	0	0	0	0	0	20	5	10	0	0	0
	SMA25	0	0	≥40	≥40	0	30	20	≥40	0	30	30
	BCS50	0	0	0	0	0	30	20	30	0	0	0
	B4620	0	0	20	20	0	30	20	30	0	5	5
	P38	0	0	0	0	0	≥40	20	≥40	0	0	0
	P50	0	0	0	0	0	≥40	20	0	0	0	0
	P61	0	0	0	0	0	20	10	0	0	0	0
	P64	0	0	0	0	0	30	10	0	0	0	0
	P69	0	0	0	0	0	≥40	20	≥40	0	0	0
	P71	0	0	0	0	0	≥40	10	0	0	0	0
	P73	0	0	0	0	0	30	20	30	0	0	0
	P622	0	0	0	0	0	≥40	10	0	0	0	0
	SDM381	0	0	10	5	0	20	10	30	0	0	0
	SDM389	0	0	0	0	0	≥40	20	≥40	0	0	0

[a]Enzymatic activities determined by an APIZYM test. Relative activity between 0 and≥40 nmol.

Antibiotic Susceptibility Determined by the Broth Microdilution Test

The distribution of MICs of the tested antibiotics is summarized in Tables 5 and 6. Microbiological breakpoints for ampicillin, vancomycin, gentamicin, kanamycin, streptomycin, erythromycin, clindamycin, tetracycline, and chloramphenicol reported by the FEEDAP document on the assessment of bacterial products used as feed additives in relation to antibiotic resistance [15] were used to categorise the 49 LAB as susceptible or resistant strains. In this document, the genus *Weissella*, which is considered a group of heterofermentative *Leuconostoc*-like LAB [16], is not included. For this reason, the respective MICs were interpreted by using the breakpoints given for the genus *Leuconostoc*. Besides, due to the lack of microbiological breakpoints for penicillin and linezolid on the FEEDAP document, we interpreted our results on these antibiotics according to the cut-off levels proposed by Klare *et al.*[17] for pediococci, namely 1 and 2 mg/L for penicillin and linezolid, respectively. According to our results, the percentages of strains showing antibiotic resistance in the genera *Weissella*, *Pediococcus*, *Lactobacillus* and *Enterococcus* were 60, 44, 33 and 11%, respectively, while none of the leuconostocs and lactococci showed this phenotype. In summary, 97.5% of the 40 non-enterococal strains resulted susceptible to ampicillin, 100% to gentamicin, 72.5% to kanamycin, 100% to streptomycin, 95% to erythromycin, 87.5% to clindamycin, 95% to tetracycline, and 100% to chloramphenicol. For vancomycin, it is known that facultative and obligate heterofermentative *Lactobacillus, Pediococcus* spp. and *Leuconostoc* spp. are intrinsically resistant. In contrast, the three lactococci were clearly susceptible to these antibiotics, showing a MIC of 0.5 mg/L. On the other hand, according to the cut-off values proposed by Klare *et al.*[17], 93% of *P. pentosaceus* strains were susceptible to penicillin and linezolid. With regard to *E. faecium*, all the tested strains were susceptible to ampicillin, vancomycin, gentamicin, kanamycin, streptomycin, tetracycline, chloramphenicol, and erythromycin except *E. faecium* BNM58 against the latter antibiotic (MIC=8 mg/L). Moreover, multiple antibiotic resistance (three antibiotics) was only detected in *P. pentosaceus* LPM78 (6.2%) and *W. cibaria* SMA25 (6.7%).

Table 5. MICs distribution of 10 antibiotics for the 9 enterococcal strains

Antibiotics	Number of strains with the indicated MIC (mg/L)[a]																EFSA breakpoints (mg/L)[b]
	0.06	0.12	0.25	0.5	1	2	4	8	16	32	64	128	256	512	1024	2048	
Ampicillin			5	3	1												**2**
Vancomycin				9													**4**
Gentamicin					4		5										**32**
Kanamycin								1		2	4	2					**1024**
Streptomycin							1		3	5							**128**
Erythromycin			5				3	**1**									**4**
Tetracycline					9												**4**
Chloramphenicol							8	1									**16**
Linezolid						9											**n.a.**
Narasin			1	8													**n.a.**

[a]MICs determined by a VetMIC test. The antibiotic dilution ranges were: 0.25-32 mg/L (ampicillin), 1-128 mg/L (vancomycin), 2-256 mg/L (gentamicin), 16-2048 mg/L (kanamycin), 8-1024 mg/L (streptomycin), 0.5-64 mg/L (erythromycin, tetracycline and chloramphenicol), 0.25-16 mg/L (linezolid) and 0.12-16 (narasin). MICs which exceeded the upper or lower limit of the

tested range are listed in the next dilution series. MICs higher than the EFSA breakpoints are indicated in bold.
[b]LAB with MICs higher than the EFSA breakpoints are considered as resistant strains [15]. n.a., not available.

Table 6. MICs distribution of 15 antibiotics for the 40 non-enterococcal strains

Antibiotics	Species (no. of tested isolates)	Number of strains with the indicated MIC (mg/L)[a]																		EFSA breakpoints (mg/L)[b]
		0.016	0.03	0.06	0.12	0.25	0.5	1	2	4	8	16	32	64	128	256	512	1024	2048	
Ampicillin	*Lb. carnosus* (2)									1	**1**									**4**
	Lb. curvatus (1)						1													**4**
	L. cremoris (3)				1	2														**2**
	Lc. cremoris (3)				1	2														**2**
	P. pentosaceus (16)								15	1										**4**
	W. cibaria (15)						15													**n.a.**
Vancomycin	*Lb. carnosus* (2)										2									**n.r.**
	Lb. curvatus (1)											1								**n.r.**
	L. cremoris (3)						3													**4**
	Lc. cremoris (3)															3				**n.r.**
	P. pentosaceus (16)															16				**n.r.**
	W. cibaria (15)															15				**n.a.**
Gentamicin	*Lb. carnosus* (2)						1		1											**16**
	Lb. curvatus (1)									1										**16**
	L. cremoris (3)					3														**32**
	Lc. cremoris (3)					3														**16**
	P. pentosaceus (16)					1		1	9	3	2									**16**
	W. cibaria (15)					6		7	1		1									**n.a.**
Kanamycin	*Lb. carnosus* (2)								1		1									**64**
	Lb. curvatus (1)											1								**64**
	L. cremoris (3)								2	1										**64**
	Lc. cremoris (3)										1	2								**16**
	P. pentosaceus (16)										1			13	**2**					**64**
	W. cibaria (15)									1	1	4	4	4	1					**n.a.**
Streptomycin	*Lb. carnosus* (2)										1		1							**64**
	Lb. curvatus (1)												1							**64**
	L. cremoris (3)										2	1								**32**
	Lc. cremoris (3)										1	2								**64**
	P. pentosaceus (16)											1	5	10						**64**
	W. cibaria (15)									2		7	5	1						**n.a.**
Erythromycin	*Lb. carnosus* (2)				2															**1**
	Lb. curvatus (1)				1															**1**
	L. cremoris (3)			2	1															**1**
	Lc. cremoris (3)			1	2															**1**
	P. pentosaceus (16)			1	4	7		3				**1**								**1**
	W. cibaria (15)					9	5				1									**n.a.**
Clindamycin	*Lb. carnosus* (2)		1		1															**1**
	Lb. curvatus (1)	1																		**1**
	L. cremoris (3)	2			1															**1**

	Lc. cremoris (3)	2		1										**1**
	P. pentosaceus (16)	3	2	7					**1**	**3**				**1**
	W. cibaria (15)	2			6	5	1		1					**n.a.**
Tetracycline	*Lb. carnosus* (2)							1	1					**8**
	Lb. curvatus (1)							1						**8**
	L. cremoris (3)					1	1	1						**4**
	Lc. cremoris (3)							1	2					**8**
	P. pentosaceus (16)								1		13	**2**		**8**
	W. cibaria (15)									15				**n.a.**
Chloramphenicol	*Lb. carnosus* (2)							1	1					**4**
	Lb. curvatus (1)								1					**4**
	L. cremoris (3)								1	2				**8**
	Lc. cremoris (3)								3					**4**
	P. pentosaceus (16)							1	5	10				**4**
	W. cibaria (15)									15				**n.a.**
Neomycin	*Lb. carnosus* (2)							1		1				**n.a.**
	Lb. curvatus (1)									1				**n.a.**
	L. cremoris (3)					2		1						**n.a.**
	Lc. cremoris (3)					3								**n.a.**
	P. pentosaceus (16)					1			9	4	2			**n.a.**
	W. cibaria (15)					4		6	4		1			**n.a.**
Penicillin	*Lb. carnosus* (2)								1	1				**n.a.**
	Lb. curvatus (1)					1								**n.a.**
	L. cremoris (3)					3								**n.a.**
	Lc. cremoris (3)				1	2								**n.a.**
	P. pentosaceus (16)						7	8	1					**n.a.**
	W. cibaria (15)							7	7		1			**n.a.**
Linezolid	*Lb. carnosus* (2)							2						**n.a.**
	Lb. curvatus (1)								1					**n.a.**
	L. cremoris (3)							1	2					**n.a.**
	Lc. cremoris (3)						1	2						**n.a.**
	P. pentosaceus (16)								15	1				**n.a.**
	W. cibaria (15)								15					**n.a.**
Ciprofloxacin	*Lb. carnosus* (2)								2					**n.a.**
	Lb. curvatus (1)										1			**n.a.**
	L. cremoris (3)								2	1				**n.a.**
	Lc. cremoris (3)								1	2				**n.a.**
	P. pentosaceus (16)												16	**n.a.**
	W. cibaria (15)									5	10			**n.a.**
Rifampicin	*Lb. carnosus* (2)					1	1							**n.a.**

	Lb. curvatus (1)	1										**n.a.**
	L. cremoris (3)							1	2			**n.a.**
	Lc. cremoris (3)		1	2								**n.a.**
	P. pentosaceus (16)			2	13	1						**n.a.**
	W. cibaria (15)						12	3				**n.a.**
Trimethoprim	*Lb. carnosus* (2)								1		1	**n.a.**
	Lb. curvatus (1)						1					**n.a.**
	L. cremoris (3)										3	**n.a.**
	Lc. cremoris (3)				1	2						**n.a.**
	P. pentosaceus (16)						8	8				**n.a.**
	W. cibaria (15)										15	**n.a.**

[a]MICs determined by a VetMIC test. The antibiotic dilution ranges were: 0.03-16 mg/L (ampicillin, clindamycin, penicillin and linezolid), 0.25-128 mg/L (vancomycin and ciprofloxacin), 0.5-256 mg/L (gentamicin, streptomycin and neomycin), 2-1024 mg/L (kanamycin), 0.016-8 mg/L (erythromycin), 0.12-64 (tetracycline, chloramphenicol, rifampicin and trimethoprim). MICs which exceeded the upper or lower limit of the tested range are listed in the next dilution series. MICs higher than the EFSA breakpoints are indicated in bold.

[b]LAB with MICs higher than the EFSA breakpoints are considered as resistant strains [15]. n.r., not required; n.a., not available.

Detection of Antibiotic Resistance Genes

The non-enterococcal strains showing antibiotic resistances in the VetMIC assays (17 strains) were further submitted to PCR in order to identify the presence of the respective antibiotic resistance genes. The tested strains were the following: *Lb. carnosus* B43 (ampicillin resistant), *P. pentosaceus* TPP3 and SMF120 (tetracycline resistant), *P. pentosaceus* LPP32, LPM83 and B5 (clindamycin resistant), *P. pentosaceus* LPV57 and *W. cibaria* P50, P61, P64, P73, SDM381, SDM389, SMA14 and BCS50 (kanamycin resistant), and *P. pentosaceus* LPM78 and *W. cibaria* SMA25 (kanamycin, erythromycin and clindamycin resistant). Acquired antibiotic resistances likely due to added genes were only found in strains within the genera *Pediococcus*(12.5%) and *Weissella* (6.7%). The genes involved in the horizontal transfer of resistance to tetracycline [*tet*(K), *tet*(L) and *tet*(M)], kanamycin [*aac(6′)-Ie-aph(2′ ′)-Ia*] and erythromycin [*erm*(A), *erm*(B) and *erm*(C)] were not detected. However, *P. pentosaceus* LPM78 and *W. cibaria* SMA25 harboured the erythromycin resistance gene *mef*(A/E). The obtained amplicons were sequenced and found to have 99% homology with the macrolide-efflux protein (*mef*E) gene described for *Streptococcus pneumoniae* and other *Streptococcus* spp. Moreover, *P. pentosaceus* LPM78 and LPM83 harboured the *lnu*(A) gene encoding the lincosamide *O*-nucleotidyltransferase that inactivates lincomycin and clindamycin. Sequencing of both amplicons showed 97% and 93% homology with lincosamide nucleotidyltransferase [*lnu*(A)] gene described for *Staphylococcus haemolyticus* and *S. aureus*, respectively. Nevertheless, *lnu*(B) was not detected in any of the tested strains. With regard to *E. faecium* BNM58, which was phenotypically resistant to erythromycin, none of the respective genes [*erm*(A), *erm*(B), *erm*(C) and *mef*(A/E)] were detected.

DISCUSSION

In this work, the antimicrobial activity against fish pathogens and the *in vitro* safety of 99 LAB previously isolated from fish, seafood and fish products [14] have been assayed by using microbiological, biochemical and genetic assays in order to identify and select the most suitable candidates to be further evaluated as probiotics for a sustainable aquaculture. LAB are widely known for their ability to inhibit bacterial pathogens by the production of antimicrobial compounds such as organic acids, oxygen peroxide and ribosomally-synthesized peptides referred to as bacteriocins, which constitutes a desirable property for probiotics and a sustainable alternative to antibiotics [9, 18]. In this respect, most of the LAB of aquatic origin tested in this work displayed a broad antimicrobial spectrum against the main Gram-positive and Gram-negative fish pathogens, being remarkable that a high number of strains (24 out of 49 strains, 49%) were identified as potential bacteriocin producers. Recently, bacteriocin production ability has been proposed as a key property for selection of probiotic LAB to be used in aquaculture as an alternative to antibiotics to fight against fish pathogen infections [19], similarly as proposed for human and farm animal probiotics [20, 21, 22]. In aquaculture farming, lactococcosis produced by the zoonotic agent *L. garvieae*, causing hemorrhagic septicaemia and meningoencephalitis, is one of the most serious diseases affecting several marine and fresh water fish species [23]. With regard to this, our work shows that putative bacteriocinogenic LAB active against this relevant fish pathogen are common amongst the microbiota isolated from aquatic animals (10 strains, 20%).

The application of probiotics in aquaculture may modify the microbial ecology of the aquatic hosts and their surrounding environment, and thus the assessment of their safety to the target aquatic species, the environment and humans constitutes an essential issue [24]. To date, several studies describing the screening and evaluation of LAB as probiotic candidates for aquaculture have been reported [25, 26, 27, 28]; however, the safety assessment of the strains is generally limited to *in vivo* challenge tests and rearing trials in order to confirm their lack of toxicity to the aquatic hosts [24, 25, 28, 29, 30, 31]. Strikingly, *in vitro* safety assessment studies have not been generally addressed, despite they have lower economic and ethic costs and result very effective to evaluate the safety of a high number of candidate probiotic strains not only for the host species, but also for humans and the environment. According to EFSA [13], most of the LAB species tested in this work (*P. pentosaceus*, *Lb. curvatus*, *L. lactis*, *Lc. mesenteroides*) are included in the QPS list and, therefore, demonstration of their safety only requires confirmation of the absence of determinants of resistance to antibiotics of human and veterinary clinical significance. However, in the case of enterococci, a more thorough, strain-specific evaluation is required to assess the risk associated to their intentional use in the food chain, while no guidelines are given for the safety assessment of the species *W. cibaria*[13].

Our results show that enterococcal virulence factors were more frequently found in *E. faecalis* than in *E. faecium*, which is in concordance with previous reports [32, 33, 34]. In this respect, most of the *E. faecalis*(95%) and a large percentage of the *E. faecium* (53%) strains evaluated in this work showed, at least, one virulence factor, being *efaAfs*, *gelE* and *agg* the most frequently detected genes. With regard to *gelE*, which encodes for an extracellular zinc endopeptidase that hydrolyzes gelatin, collagen, hemoglobin, and other bioactive compounds, this gene was detected at high frequency in *E. faecalis*, with all the *gelE* $^{+}$ strains showing gelatinase activity. However, five out of nine *E. faecium* strains harbouring *gelE* were unable to degrade gelatin, suggesting the carriage of a non-functional gene, as previously reported [32, 33]. Likewise, in the case of *E. faecium* P68 and *E. faecium* GM29 harbouring $cylL_L$ $cylL_S$, the lack of hemolytic activity may be explained by the absence of *cylM*, whose product is involved in the post-translational modification of cytolysin. On the other hand, *esp* and *hyl*, which encode a cell wall-associated protein involved in immune evasion and an hyaluronidase enzyme, respectively, were not found in any of the tested LAB. Previous studies have reported that *esp* and *hyl* are more common in ampicillin-resistant/vancomycin-resistant *E. faecium* (VREF) than in ampicillin-susceptible/VREF strains [35]. In this context, the increase in the incidence of VREF at hospital settings has been attributed mainly to the spread of ampicillin-resistant VREF exhibiting *esp* and/or *hyl*[36, 37]. Therefore, the fact that the *E. faecium* strains evaluated in this work lack these genes might be related with their non-clinical origin and absence of ampicillin resistance.

The use and frequent overuse of antibiotics, including those used in human medicine, in fish farming has resulted in the emergence and spread of antibiotic-resistant bacteria in the aquaculture environment. This possesses a threat to human and animal health due to the increase of acquired antibiotic resistance in fish pathogens, the transfer of their genetic determinants to bacteria of terrestrial animals and to human pathogens, and the alterations of the bacterial microbiota of the aquatic environment [11, 29]. In our study, the percentage of enterococcal strains showing acquired antibiotic resistance was 68%. Interestingly, the results found in *E. faecium* (71%) and *E. faecalis* (62%) were similar, however, higher percentages of resistance to ciprofloxacin and/or norfloxacin, rifampicin, and glycopeptides were observed in *E. faecalis*. Nevertheless, the occurrence of erythromycin and tetracycline resistance was frequently detected amongst *E. faecium* (45%) but only in one *E. faecalis* strain (5%). In spite of the high prevalence of acquired antibiotic resistance found in enterococci of aquatic origin, they showed low incidence or absence of resistance to the clinically relevant antibiotics vancomycin (8.5%) and ampicillin, penicillin and gentamicin, respectively, which is in agreement with previous studies [33, 38]. Moreover, the percentages of strains showing antibiotic resistance in the genera *Weissella*, *Pediococcus* and *Lactobacillus* were 60, 44 and 33%, respectively, while none of the leuconostocs and lactococci showed this phenotype. In this regard, our results indicate that the LAB susceptibility patterns of MIC values to clinically relevant antibiotics are species-dependent, similarly as previously described by other authors [39, 40]. Moreover, multiple antibiotic resistance was commonly found in strains within the genus *Enterococcus* (37%), mainly in *E. faecalis*, while being very infrequent in the non-enterococcal strains (5%).

According to EFSA [29], the determination of MICs above the established breakpoint levels, for one or more antibiotic, requires further investigation to make the distinction between added genes (genes acquired by the bacteria via gain of exogenous DNA) or to the mutation of indigenous genes. According to our results, acquired antibiotic resistance likely due to added genes is not a common feature amongst the non-enterococcal LAB of aquatic origin (7.5%). In this respect, this genotype was only found in the genera *Pediococcus* (12.5%) and *Weissella* (6.7%). Although *P. pentosaceus* LPV57 and LPM78 showed resistance to kanamycin (MIC of 128 mg/L), the respective resistance gene *aac(6′)-Ie-aph(2′ ′)-Ia* was not found in these strains. Similarly, *P. pentosaceus* TPP3 and SMF120 were phenotypically resistant to tetracycline (MIC of 16 mg/L), but did not contain *tet(K)*, *tet(L)* or *tet(M)*. In this respect, Ammor *et al.*[41] reported that pediococci are intrinsically resistant to the latter two antibiotics, as well as to glycopeptides (vancomycin and teicoplanin), streptomycin, ciprofloxacin and trimethoprim-sulphamethoxazole. Other authors proposed a MIC for tetracycline in pediococci ranging between 8 and 16 mg/L [42], or of 32 mg/L for oxytetracycline in *P. pentosaceus*[17]. The tetracycline breakpoints suggested for pediococci by EFSA are lower than the MICs observed in our work and others [17, 42]. On the other hand, the only antibiotic resistance detected in *Leuconostoc* strains was for vancomycin, which is an intrinsic property of this genus. It has been previously reported that *Leuconostoc* strains display poor, if any, resistance to antibiotics of clinical interest [38]. With regard to lactococci, the three *L. cremoris* strains evaluated were susceptible to all the antibiotics; however, relatively high MICs for rifampicin (16–32 mg/L) and trimethoprim (≥ 64 mg/L) were detected. In fact, most lactococcal species are resistant to trimethoprim [41]. As expected, all strains of heterofermentative *Lactobacillus* spp. were resistant to vancomycin but susceptible to the rest of the assayed antibiotics, except *Lb. carnosus* B43, which showed the highest MIC for ampicillin and penicillin (MICs of 8 and 4 mg/L, respectively). In this context, the presence of modifications in the low affinity penicillin-binding protein (PBP) that confers resistance to penicillin and β-lactams in *E. faecium* and *Streptococcus pneumoniae*, has been reported [43, 44]. Moreover, nine PBPs have been described in *Lb. casei* ATCC 393 [45], which leads us to suggest that a similar mechanism may be also responsible for the ampicillin and penicillin resistance found in *Lb. carnosus* B43. The resistance to vancomycin detected in *Pediococcus*, *Leuconostoc* and *Lactobacillus*species in this study might be due to the presence of D-Ala-D-Lactate in their peptidoglycan rather than D-Ala-D-Ala dipeptide [46].

In this context, all tested *W. cibaria* strains showed MICs$\geq$128 mg/L for vancomycin, suggesting that vancomycin resistance is an intrinsic property of this species. In relation to *Weissella* spp., studies on antibiotic resistance profiles are very limited [47] and breakpoints have not been defined by EFSA [15]. In our study, most *W. cibaria* strains showed low MIC values; however *W. cibaria* BCS50 showed relatively high MICs for penicillin (8 mg/L) and kanamycin (64 mg/L), and *W. cibaria* SMA25 showed MICs of 128 mg/L for kanamycin, 8 mg/L for gentamicin, erythromycin and neomycin, and 2 mg/L for clindamycin. Therefore, these two strains were discarded of this study, while *W. cibaria* P50, P61, P64, P73, SMA14, SDM381 and SDM389 were not included in the final selection due to their MICs for kanamycin (32–64 mg/L). According to these results, as a rule of thumb, we propose for *W. cibaria* the breakpoints assigned to *Leuconostoc* spp. by EFSA [15], until further studies establish the wild-type MIC ranges within this species. In spite of that, different MICs for rifampicin and trimethoprim for *W. cibaria* and *Lc. cremoris* were found in this study. The reduced susceptibility of *W. cibaria* towards trimethoprim could indicate an intrinsic resistance to this antibiotic [48]. In our work, the only antibiotic resistance genes found were *mef*(A/E), which encodes a drug efflux pump conferring a low to moderate level of resistance to 14 (erythromycin and clarithromycin)- and 15 (azithromycin)-membered macrolides but not to lincosamide or streptogramin B antibiotics [49], and *lnu*(A), encoding the lincosamide *O*-nucleotidyltransferase that inactivates lincomycin and clindamycin [50]. In this respect, *P. pentosaceus* LPM78 and *W. cibaria* SMA25, displaying erythromycin resistance (MIC=8 and$\geq$8 mg/L, respectively), carried the gene *mef*(A/E), which can be found in a variety of Gram-positive bacteria, including corynebacteria, enterococci, micrococci, and several streptococcal species [51, 52]. On the other hand, two pediococci (*P. pentosaceus* LPM78 and LPM83) that showed resistance to clindamycin (MIC=4 and 2 mg/L, respectively) carried the gene *lnu*(A), which had been only previously found in staphylococci, streptococci, enterococci and lactobacilli of animal origin and in staphylococci isolated from humans [50, 53]. Strikingly, the clindamycin resistant strains *P. pentosaceus* LPP32 and B5 and *W. cibaria* SMA25 (MIC=4 and 2 mg/L, respectively) did not harbour this gene nor *lnu*(B). To our knowledge, this is the first description of *mef*(A/E) in the genera *Pediococcus* and *Weissella*, and *lnu*(A) in the genus *Pediococcus*. The detection of resistance genes for macrolide and lincosamide in non-enterococcal strains suggests a wider distribution of this group of genes than previously anticipated.

The *in vitro* subtractive screening proposed in this work also include the assessment of bile salts deconjugation, mucin degradation, biogenic amine production and other potentially detrimental enzymatic activities such as the β-glucuronidase activity, which should be absent in probiotic candidates [54, 55, 56]. Excessive deconjugation of bile salts may be unfavourable in animal production since unconjugated bile acids are less efficient than their conjugated counterparts in the emulsification of dietary lipids. In addition, the formation of micelles, lipid digestion and absorption of fatty acids and monoglycerides could be impaired by deconjugated bile salts [57]. Similarly, excessive degradation of mucin may be harmful as it may facilitate the translocation of bacteria to extraintestinal tissues [55]. In this respect, it is worthy to note that none of the 49 tested LAB deconjugated bile salts nor exhibited mucinolytic activity, the latter indicating their low invasive and toxigenic potential at the mucosal barrier. These results are in accordance with previous findings showing that LAB do not degrade mucin *in vitro*[58, 59]. Moreover, β-glucuronidase activity has been associated with the generation of potential carcinogenic metabolites [56]; however, none of the LAB tested in our study displayed this harmful enzymatic activity. In a previous work [60], we demonstrated that none of the 40 non-enterococcal strains evaluated herein produced histamine, tyramine or putrescine. With regard to enterococci, the nine *E. faecium* strains only produced tyramine, being *E. faecium* CV1 a low producer of this biogenic amine. Although the lack of biogenic amine production by probiotic strains is a desirable trait, it should be borne in mind that tyramine production by enterococci is a very frequent trait [60, 61]. Finally, several studies have suggested that probiotic microorganisms might exert a beneficial effect in the digestion process of fish due to the production of extracellular enzymes [62, 63, 64, 65]. In our work, the LAB strains of aquatic origin within the genera *Pediococcus*, *Enterococcus* and *Lactobacillus* showed a higher number of enzymatic activities than *Lactococcus*, *Leuconostoc* and *Weissella*, being the enzymatic profiles similar amongst strains within the same genus. In this respect, nearly all the strains produced phosphatases, which might be involved in nutrient absorption [64], and peptidases and glucosidases that breakdown peptides and carbohydrates, respectively. However, the tested LAB showed weak lipolytic activity and no proteolytic activity.

CONCLUSIONS

This work shows that antimicrobial/bacteriocin activity against fish pathogens is a widespread probiotic property amongst LAB isolated from aquatic animals regarded as human food. However, particular safety concerns based on antibiotic resistances and virulence factors were dominant within *E. faecalis* (100%) and *E. faecium* (79%), and acquired antibiotic resistance genes were not commonly found (7.5%; erythromycin and clindamycin) amongst the non-enterococcal isolates of aquatic origin. To our knowledge, this is the first large-scale study describing the antimicrobial activity against fish pathogens and the safety assessment beyond the QPS approach of LAB isolated from aquatic animals. The *in vitro* subtractive screening presented herein, which allowed the selection of 33 strains (8 *E. faecium*, 11 *P. pentosaceus*, 1 *Lb. carnosus*, 1 *Lb. curvatus*, 3 *L. cremoris*, 3 *Lc. cremoris* and 6 *W. cibaria*) out of 99 LAB isolates of aquatic origin, constitutes a valuable strategy for the large-scale preliminary selection of putatively safe LAB intended for use as probiotics in aquaculture and to avoid the spreading of bacterial cultures with harmful traits into the aquatic environment. Nevertheless, a comprehensive *in vivo* assessment of their lack of toxicity and undesirable effects must be also carried out using cell lines, live food and, ultimately, aquatic animals before their unequivocal consideration as safe probiotics for a sustainable aquaculture.

METHODS

Bacterial Strains and Growth Conditions

A total of 99 LAB (59 enterococci and 40 non-enterococci) of aquatic origin with antimicrobial activity against spoilage and food-borne pathogenic bacteria of concern for the fish industry, previously isolated and identified by our group from fish, seafood and fish products [14], were used in this study (Table 1). The LAB strains were isolated on non-supplemented MRS (Oxoid, Ltd., Basingstoke, United Kingdom) or KAA (Oxoid) agar (1,5%, w/v) at 25°C, and taxonomically identified [14] by sequencing of the genes encoding 16S rRNA (*16S rDNA*) [66] and/or superoxide dismutase (*sodA*) [67]. Unless otherwise stated, LAB were grown aerobically in MRS broth at 32°C.

Direct Antimicrobial Activity Assay

The antimicrobial activity of the 99 LAB against the main Gram-positive and Gram-negative fish pathogens was assayed by a qualitative stab-on-agar test (SOAT) as previously described by Cintas *et al.* [68]. Briefly, pure cultures were stabbed onto MRS or Tryptone Soya Agar (TSA) (Oxoid) plates supplemented with glucose (2%, w/v) and incubated at 32°C for 5 h, and then 40 ml of the corresponding soft agar (0.8%, w/v) medium containing about 1×10^5 CFU/ml of the indicator strain was poured over the plates. After incubation at 28-37°C for 16–24 h depending on the indicator strain, the plates were checked for inhibition zones (absence of visible microbial growth around the stabbed cultures), and only inhibition halos with diameters >3 mm were considered positive. *L. garvieae* JIP29-99 was grown aerobically in Tryptone Soya Broth (TSB) (Oxoid) at 37°C. *S. agalactiae* CF01173 and *S. iniae* LMG14521 were grown aerobically in Brain Heart Infusion (BHI) broth (Oxoid) at 37°C. *A. hydrophila* CECT5734, *Ls. anguillarum* CECT4344, *Ls. anguillarum* CECT7199, and *Ph. damselae*CECT626 strains were grown aerobically in TSB at 28°C. *V. alginolyticus* CECT521 was grown aerobically in TSB supplemented with NaCl (1%, w/v; Panreac Química S.A.U, Barcelona, Spain) at 28°C.

Extracellular Antimicrobial Activity Assay

The antimicrobial activity of supernatants from LAB cultures grown in MRS broth at 32°C for 16 h was determined by an agar well-diffusion test (ADT) as previously described by Cintas *et al.*[68]. Supernatants were obtained by centrifugation of cultures at 10,000 × *g* at 4°C for 10 min, adjusted to pH 6.2 with 1 M NaOH, filter-sterilized through 0.22 µm-pore-size filters (Millipore Corp., Bedford, Massachussets, USA) and stored at −20°C until use. Fifty-µl aliquots of cell-free culture supernatants were placed into wells (6-mm diameter) cut in cooled MRS or TSB agar (0.8%, wt/vol) plates previously seeded (1×10^5 CFU/ml) with the indicator microorganisms *Pediococcus damnosus* CECT4797, *L. garvieae* JIP29-99 or *A. hydrophila* CECT5734. After 2 h at 4°C, the plates were incubated under the same conditions mentioned above to allow for the growth of the target microorganisms and then analyzed for the presence of inhibition zones around the wells. To determine the proteinaceous nature of the antimicrobial compounds, supernatants showing antimicrobial activity were subjected to proteinase K treatment (10 mg/ml) (AppliChem GmbH, Germany) at 37°C for 2 h. After proteinase K inactivation by heat treatment (100°C, 10 min), samples were assayed for residual antimicrobial activity by an ADT as described above using *P. damnosus* CECT4797 as indicator microorganism. Supernatants with no added enzyme were treated as indicated above and used as controls. For further characterization of the antimicrobial compounds, 7 ml of supernatants from an overnight culture of LAB were subjected to peptide concentration by ammonium sulphate precipitation. Ammonium sulphate was gradually added to the supernatants to achieve 50% saturation. Samples were kept at 4°C with stirring for 3 h, and then centrifuged at 10,000 × *g* at 4°C for 30 min. Pellets and floating solid material were combined and solubilized in 350 µl of 20 mM sodium phosphate (pH 6.0), and antimicrobial activity of the resulting 20-fold concentrated supernatants was determined by an ADT as described above.

PCR detection of Potential Virulence Factors in Enterococci

Detection of genes encoding potential virulence factors in the 59 enterococci was performed by PCR. The following primer pairs were used: TE3/TE4 for detection of *agg* (aggregation substance), TE9/TE10 for *gelE*(gelatinase), TE34/TE36 for *esp* (enterococcal surface protein), TE5/TE6 for *efaAfs* (*Enterococus faecalis*endocarditis antigen) [32], HYLn1/HYLn2 for *hyl* (hyaluronidase) [35], $CYLL_L$–R1/$CYLL_S$–R2 for $cylL_L$–$cylL_S$(cytolysin precursor) [69], and RHCT1/RHCT2 for $cylL_L$–$cylL_S$-*cylM* (cytolysin precursor and posttranslational modifier) [70]. Oligonucleotide primers were obtained from Sigma-Genosys Ltd. (Cambridge, United Kingdom). The positive control strains for detection of potential virulence factors were the following: *E. faecalis* P4 for $cylL_L$–$cylL_S$, $cylL_L$–$cylL_S$–*cylM*, *agg*, *gelE* and *efaAfs*, *E. faecalis* P36 for *esp*[32], and *E. faecium* C68 for *hyl*[35]. PCR-amplifications were performed from total bacterial DNA obtained using the Wizard DNA Purification Kit (Promega, Madrid, Spain) in 25 µl reaction mixtures with 1 µl of purified DNA, 0.7 µM of each primer, 0.2 mM of each dNTP, buffer 1×, 1.5 mM $MgCl_2$ and 0.75 U of Platinum Taq DNA polymerase (Invitrogen, Madrid, Spain). Samples were subjected to an initial cycle of denaturation (97°C for 2 min), followed by 35 cycles of denaturation (94°C for 45 s), annealing (48 to 64°C for 30 s) and elongation (72°C for 30 to 180 s), ending with a final extension step at 72°C for 7 min in an Eppendorf Mastercycler thermal cycler (Eppendorf, Hamburg, Germany). PCR products were analyzed by electrophoresis on 1-2% (w/v) agarose (Pronadisa, Madrid, Spain) gels stained with Gel red (Biotium, California, USA), and visualized with the Gel Doc 1000 documentation system (Bio-Rad, Madrid, Spain). The molecular size markers used were HyperLadder II (Bioline GmbH, Germany) and 1Kb Plus DNA ladder (Invitrogen).

Production of Gelatinase by Enterococci

Gelatinase production was determined using the method previously described by Eaton and Gasson [32]. Briefly, enterococci were grown in MRS broth overnight at 32°C, and streaked onto Todd-Hewitt (Oxoid) agar plates (1.5%, w/v) containing 30 g of gelatine per litre. After incubation overnight incubation at 37°C, the plates were placed at 4°C for 5 h before examination for zones of turbidity (protein hydrolysis) around the colonies. *E. faecalis* P4 was used as positive control.

Production of Hemolysin

To investigate hemolysin production by the 99 LAB, the strains grown in MRS broth were streaked onto layered fresh horse blood agar plates (BioMérieux, Marcy l'Étoile, France) and grown at 37°C for 1–2 days [32]. β-hemolysis was revealed by the formation of clear zones surrounding the colonies on blood agar plates. *E. faecalis* P4 was used as positive control.

Determination of Antibiotic Susceptibility

Antibiotic susceptibility of the 59 enterococci was determined by overlaying antibiotic-containing disks (Oxoid) on Diagnostic Sensitivity Test Agar (Oxoid) previously seeded with approximately 1×10^5 CFU/ml of each enterococcal isolate. The antibiotics tested were ampicillin (10 μg), chloramphenicol (30 μg), ciprofloxacin (5 μg), erythromycin (15 μg), gentamicin (120 μg), nitrofurantoin (300 μg), norfloxacin (10 μg), penicillin G (10 IU), rifampicin (5 μg), teicoplanin (30 μg), tetracycline (30 μg), and vancomycin (30 μg). Inhibition zone diameters were measured after overnight incubation of the plates at 37°C. Resistance phenotypes were recorded as recommended by the Clinical and Laboratory Standards Institute [71]. *E. faecalis* CECT795 and *Staphylococcus aureus* CECT435 were used for quality control. The minimum inhibitory concentration for the 49 pre-selected LAB was determined by a broth microdilution test using e-cocci (for enterococci), and Lact-1 and Lact-2 (for non-enterococcal strains) VetMIC microplates (National Veterinary Institute, Uppsala, Sweden). The antibiotics evaluated for enterococci were ampicillin, vancomycin, gentamicin, kanamycin, streptomycin, erythromycin, tetracycline, chloramphenicol, narasin, and linezolid, while for the non-enterococcal strains, the tested antibiotics were ampicillin, vancomycin, gentamicin, kanamycin, streptomycin, erythromycin, clindamycin, tetracycline, chloramphenicol, neomycin, penicillin, linezolid, ciprofloxacin, rifampicin, and trimethoprim. Individual colonies were suspended in a sterile glass tube containing 5 ml saline solution (0.85% NaCl) to a turbidity of 1 in the McFarland scale (approx. 3×10^8 CFU/ml) and further diluted 1000-fold. Iso-sensitest (IST) broth (Oxoid) was used for enterococci, while LSM medium (IST:MRS, 9:1) was used for all the non-enterococcal strains except *Lactobacillus curvatus* subsp. *curvatus* BCS35, that required LSM broth supplemented with 0.03% (w/v) L-cysteine (Merck KGaA) [72]. Fifty or 100 μl of the diluted enterococcal and non-enterococcal suspensions, respectively, was added to each microplate well which was then sealed with a transparent covering tape and incubated at 37°C for 18 h (in the case of *Lb. curvatus* BCS35, the plates were incubated anaerobically at 32°C for 18 h). After incubation, MICs were established as the lowest antibiotic concentration that inhibited bacterial growth, and interpreted according to the breakpoints identified by the FEEDAP Panel and adopted by EFSA to distinguish between susceptible and resistant strains [15]. Accordingly, strains showing MICs higher than the respective breakpoint were considered as resistant. *E. faecalis* CECT795 and *S. aureus* CECT794 were used for quality control of e-cocci, and Lact-1 and Lact-2 VetMIC microplates, respectively.

Deconjugation of Bile Salts

The ability of the 49 pre-selected LAB to deconjugate primary and secondary bile salts was determined according to Noriega *et al.*[73]. Bile salt plates were prepared by adding 0.5% (w/v) sodium salts of taurocholate (TC) and taurodeoxycholate (TDC) (Sigma-Aldrich Corporation, St. Louis, Missouri, USA) to MRS agar (1.5%, w/v) supplemented with 0.05% (w/v) L-cysteine (Merck KGaA, Darmstadt, Germany). Overnight liquid cultures of strains (10 μl) were spotted onto agar plates and incubated under anaerobic conditions (Anaerogen, Oxoid) at 37°C for 72 h. The presence of precipitated bile acid around the colonies (opaque halo) was considered as a positive result. A fresh fecal slurry of a healthy adult horse was used as positive control for bile salts deconjugating activities.

Degradation of Mucin

The capacity of the 49 pre-selected LAB to degrade gastric mucin was determined as described by Zhou *et al.*[58]. Mucin from porcine stomach type III (Sigma-Aldrich Corp.) and agar were added to medium B without glucose at concentrations of 0.5% (w/v) and 1.5% (w/v), respectively. A volume of 10 μl of 24 h viable bacterial cultures was inoculated onto the surface of medium B. The plates were incubated anaerobically at 37°C for 72 h, subsequently stained with 0.1% (w/v) amido black (Merck KGaA) in 3.5 M acetic acid for 30 min, and then washed with 1.2 M acetic acid (Merck KGaA). A discoloured zone around the colony was considered as a positive result. A fresh fecal slurry of a healthy adult horse was used as positive control for mucin degradation ability.

Determination of Enzymatic Activities

The APIZYM test (BioMérieux, Montallieu Vercieu, France) was used for determination of enzymatic activities of the 49 pre-selected LAB. Cells from cultures grown at 32°C overnight were harvested by centrifugation at 12,000 g for 2 min, resuspended in 2 ml of API Suspension Medium (BioMérieux) and adjusted to a turbidity of 5–6 in the McFarland scale (approx. $1.5\text{-}1.9 \times 10^9$ CFU/ml). Aliquots of 65 μl of the suspensions were added to each of the 20 reaction cupules in the APIZYM strip. The strips were incubated at 37°C for 4.5 h and the reactions were developed by addition of one drop each of the APIZYM reagents A and B. Enzymatic activities were graded from 0 to 5, and converted to nanomoles as indicated by the manufacturer´ s instructions.

PCR Detection of Antibiotic Resistance Genes

The presence of genetic determinants conferring resistance to aminoglycosides except streptomycin *aac(6´)-Ie-aph(2´´)-Ia*, to erythromycin *erm*(A), *erm*(B), *erm*(C) and *mef*(A/E)], to tetracycline *tet*(K), *tet*(L) and *tet*(M)], and to lincosamides *lnu*(A) and *lnu*(B)] was determined by PCR in the LAB strains showing antibiotic resistance by the VetMIC assay. PCR-amplifications and PCR-product visualization and analysis were performed as described above using the following primer-pairs: aacF/aacR for detection of *aac(6´)-Ie-aph(2´´)-Ia*[74], ermAI/ermAII for *erm*(A) [75, 76], ermBI/ermBII for *erm*(B) [17], ermCI/ermCII for *erm*(C) [17, 77], mef(A/E)I/mef(A/E)II for *mef*(A/E) [75, 76], tetKI/ tetKII for *tet*(K) [17], tetLI/tetLII for *tet*(L) [17, 78], tetMI/tetMII for *tet*(M) [17, 78], lnuA1/lnuA2 for *lnu*(A) [79], lnuB1/lnuB2 for *lnu*(B) [50]. *E. faecalis* C1570 was used as positive control for amplification of *erm*(C), *lnu*(A) and *tet*(K) and *E. faecalis* C1231 for *erm*(A). *E. faecium* 3Er1 (clonal complex of hospital-associated strain CC9) and *E. faecium* RC714 were used as positive controls for amplification of *aac(6')-Ie-aph(2'')-Ia*, *tet*(M) and *tet*(L), and for *erm*(B) and *mef*(A/E), respectively. The amplicons obtained with *mef*(A/E) and *lnu*(A) specific primers were purified by using the NucleoSpin Extract II Kit (Macherey-Nagel GmbH & Co. KG, Düren, Germany) and both DNA strands were sequenced at the Unidad de Genómica (Parque Científico de Madrid, Facultad de Ciencias Biológicas, Universidad Complutense de Madrid, Spain). Analysis of DNA sequences was performed with the BLAST program available at the National Center for Biotechnology Information (NCBI).

DECLARATIONS

Acknowledgements

This work was partially supported by projects AGL2009−08348-ALI from Ministerio de Ciencia y Tecnología (MCYT), Spain; GR35/10-A from Banco Santander-Central Hispano-Universidad Complutense de Madrid (UCM), Spain; S−2009/AGR−1489 from Dirección General de Universidades e Investigación, Consejería de Educación, Comunidad de Madrid, Spain, and Spanish-Portuguese Integrated Action HP2008-0070 from Ministerio de Ciencia e Innovación (MICINN), Spain. E. Muñoz-Atienza is recipient of a predoctoral fellowship from UCM, Spain. C. Araújo is financially supported by a predoctoral fellowship from Fundação da Ciência e Tecnologia, Portugal. C. Campanero holds a predoctoral fellowship from UCM, Spain. The authors express their gratitude to Dr. C. Michel, INRA, Jouy-en-Josas, France, for providing a number of fish pathogens strains used as indicators and to Dr. C. Torres, Universidad de la Rioja, Spain; Dr. T.J. Eaton, Institute of Food Research, Norwich, United Kingdom, and Dr. V. Vankerckhoven, University of Antwerp, Belgium, for supplying strains used as PCR controls.

Authors' contributions

EMA carried out the phenotypic and genetic analyses, prepared the manuscript draft and participated in the design of the experiments. BGS carried out the isolation of the LAB strains and collaborated in the genetic studies. CA contributed to the phenotypic analyses and to prepare the manuscript draft. CC participated in the phenotypic analyses. RC collaborated in the antibiotic susceptibility tests. LMC conceived the study and, together with CH and PEH, designed the experiments, analyzed the results and revised the manuscript. All authors read and approved the final version of the manuscript.

REFERENCES

1. FAO: FAO Fisheries Department. State of world aquaculture 2006. FAO Fish Tech Pap. 2006, 500: 1-134.
2. FAO: Responsible use of antibiotics in aquaculture. FAO Fish Tech Pap. 2005, 469: 1-97.
3. Cabello FC: Heavy use of prophylactic antibiotics in aquaculture: a growing problem for human and animal health and for the environment. Environ Microbiol. 2006, 8: 1137-1144. 10.1111/j.1462-2920.2006.01054.x.
4. Austin B: The bacterial microflora of fish, revised. ScientificWorldJournal. 2006, 6: 931-945.
5. Robertson PAW, O'Dowd C, Burrells C, Williams P, Austin B: Use of Carnobacterium sp. as a probiotic for Atlantic salmon (Salmo salar L.) and rainbow trout (Oncorhynchus mykiss, Walbaum). Aquaculture. 2000, 185: 235-243. 10.1016/S0044-8486(99)00349-X.
6. Wang Y-B, Li J-R, Lin J: Probiotics in aquaculture: challenges and outlook. Aquaculture. 2008, 281: 1-4. 10.1016/j.aquaculture.2008.06.002.
7. Defoirdt T, Sorgeloos P, Bossier P: Alternatives to antibiotics for the control of bacterial disease in aquaculture. Curr Opin Microbiol. 2011, 14: 251-258. 10.1016/j.mib.2011.03.004.
8. Verschuere L, Rombaut G, Sorgeloos P, Verstraete W: Probiotic bacteria as biological control agents in aquaculture. Microbiol Mol Biol Rev. 2000, 64: 655-671. 10.1128/MMBR.64.4.655-671.2000.
9. Gatesoupe FJ: Updating the importance of lactic acid bacteria in fish farming: natural occurrence and probiotic treatments. J Mol Microbiol Biotechnol. 2008, 14: 107-114. 10.1159/000106089.
10. FAO/WHO: Probiotics in food. Health and nutritional properties and guidelines for evaluation. FAO Food Nutr Pap. 2006, 85: 1-50.
11. EC: On a generic approach to the safety assessment of microorganisms used in feed/food and feed/food production - A working paper open for comment. 2003, http://ec.europa.eu/food/fs/sc/scf/out178_en.pdf,
12. EFSA: Introduction of a Qualified Presumption of Safety (QPS) approach for assessment of selected microorganisms referred to EFSA. The EFSA Journal. 2007, 587: 1-16.
13. EFSA: Maintenance of the list of QPS biological agents intentionally added to food and feed (2011 update). The EFSA Journal. 2011, 9: 1-82.
14. Gómez-Sala B, Basanta A, Sánchez J, Martín M, Criado R, Gutiérrez J, Citti R, Herranz C, Hernández PE, Cintas LM: Antimicrobial activity of lactic acid bacteria isolated from aquatic animals and fish products. 13éme Colloque du Club des Bactéries Lactiques, p 45 Abstracts. 2004, Nantes, France: ENITIAA and French National Institute for Agricultural Research (INRA)
15. EFSA: Guidance on the assessment of bacterial susceptibility to antimicrobials of human and veterinary importance. EFSA Journal. 2012, 10: 2740-2749.
16. Collins MD, Samelis J, Metaxopoulos J, Wallbanks S: Taxonomic studies on some leuconostoc-like organisms from fermented sausages: description of a new genus Weissella for the Leuconostoc paramesenteroides group of species. J Appl Bacteriol. 1993, 75: 595-603. 10.1111/j.1365-2672.1993.tb01600.x.
17. Klare I, Konstabel C, Werner G, Huys G, Vankerckhoven V, Kahlmeter G, Hildebrandt B, Müller-Bertling S, Witte W, Goossens H: Antimicrobial susceptibilities of Lactobacillus, Pediococcus and Lactococcus human isolates and cultures intended for probiotic or nutritional

use. J Antimicrob Chemother. 2007, 59: 900-912. 10.1093/jac/dkm035.

18. Ringø E, Gatesoupe FJ: Lactic acid bacteria in fish: a review. Aquaculture. 1998, 160: 177-203. 10.1016/S0044-8486(97)00299-8.
19. Desriac F, Defer D, Bourgougnon N, Brillet B, Le Chevalier P, Fleury Y: Bacteriocin as weapons in the marine animal-associated bacteria warfare: inventory and potential applications as an aquaculture probiotic. Mar Drugs. 2010, 8: 1153-1177. 10.3390/md8041153.
20. O'Shea EF, Cotter PD, Stanton C, Ross RP, Hill C: Production of bioactive substances by intestinal bacteria as a basis for explaining probiotic mechanisms: Bacteriocins and conjugated linoleic acid. Int J Food Microbiol. 2012, 152: 189-205. 10.1016/j.ijfoodmicro.2011.05.025.
21. Gillor O, Etzion A, Riley MA: The dual role of bacteriocins as anti- and probiotics. Appl Microbiol Biotechnol. 2008, 81: 591-606. 10.1007/s00253-008-1726-5.
22. Corr SC, Li Y, Riedel CU, O'Toole PW, Hill C, Gahan CG: Bacteriocin production as a mechanism for the antiinfective activity of Lactobacillus salivarius UCC118. Proc Natl Acad Sci USA. 2007, 104: 7617-7621. 10.1073/pnas.0700440104.
23. Vendrell D, Balcazar JL, Ruiz-Zarzuela I, de Blas I, Girones O, Muzquiz JL: Lactococcus garvieae in fish: a review. Comp Immunol Microbiol Infect Dis. 2006, 29: 177-198. 10.1016/j.cimid.2006.06.003.
24. Decamp O, Moriarty D: Aquaculture species profit from probiotics. Feed Mix. 2007, 15: 20-23.
25. Dimitroglou A, Merrifield DL, Carnevali O, Picchietti S, Avella M, Daniels C, Güroy D, Davies SJ: Microbial manipulations to improve fish health and production - A Mediterranean perspective. Fish Shellfish Immunol. 2011, 30: 1-16. 10.1016/j.fsi.2010.08.009.
26. Nikoskelainen S, Salminen S, Bylund G, Ouwehand AC: Characterization of the properties of human- and dairy-derived probiotics for prevention of infectious diseases in fish. Appl Environ Microbiol. 2001, 67: 2430-2435. 10.1128/AEM.67.6.2430-2435.2001.
27. Balcázar JL, Vendrell D, de Blas I, Ruiz-Zarzuela I, Muzquiz JL, Girones O: Characterization of probiotic properties of lactic acid bacteria isolated from intestinal microbiota of fish. Aquaculture. 2008, 278: 188-191. 10.1016/j.aquaculture.2008.03.014.
28. Merrifield DL, Dimitroglou A, Foey A, Davies SJ, Baker RTM, Bøgwald J, Castex M, Ringø E: The current status and future focus of probiotic and prebiotic applications for salmonids. Aquaculture. 2010, 302: 1-18. 10.1016/j.aquaculture.2010.02.007.
29. Das S, Ward LR, Burke C: Screening of marine Streptomyces spp. for potential use as probiotics in aquaculture. Aquaculture. 2010, 305: 32-41. 10.1016/j.aquaculture.2010.04.001.
30. Wang Y-B, Tian Z-Q, Yao J-T, Li W: Effect of probiotics, Enteroccus faecium, on tilapia (Oreochromis niloticus) growth performance and immune response. Aquaculture. 2008, 277: 203-207. 10.1016/j.aquaculture.2008.03.007.
31. Olmos J, Ochoa L, Paniagua-Michel J, Contreras R: Functional feed assessment on Litopenaeus vannamei using 100% fish meal replacement by soybean meal, high levels of complex carbohydrates and Bacillus probiotic strains. Mar Drugs. 2011, 9: 1119-1132. 10.3390/md9061119.
32. Eaton TJ, Gasson MJ: Molecular screening of Enterococcus virulence determinants and potential for genetic exchange between food and medical isolates. Appl Environ Microbiol. 2001, 67: 1628-1635. 10.1128/AEM.67.4.1628-1635.2001.
33. Gomes BC, Esteves CT, Palazzo IC, Darini AL, Felis GE, Sechi LA, Franco BD, De Martinis EC: Prevalence and characterization of Enterococcus spp. isolated from Brazilian foods. Food Microbiol. 2008, 25: 668-675. 10.1016/j.fm.2008.03.008.
34. López M, Sáenz Y, Rojo-Bezares B, Martínez S, del Campo R, Ruiz-Larrea F, Zarazaga M, Torres C: Detection of vanA and vanB2-containing enterococci from food samples in Spain, including Enterococcus faecium strains of CC17 and the new singleton ST425. Int J Food Microbiol. 2009, 133: 172-178. 10.1016/j.ijfoodmicro.2009.05.020.
35. Vankerckhoven V, Van Autgaerden T, Vael C, Lammens C, Chapelle S, Rossi R, Jabes D, Goossens H: Development of a multiplex PCR for the detection of asa1, gelE, cylA, esp, and hyl genes in enterococci and survey for virulence determinants among European hospital isolates of Enterococcus faecium. J Clin Microbiol. 2004, 42: 4473-4479. 10.1128/JCM.42.10.4473-4479.2004.
36. Klare I, Konstabel C, Mueller-Bertling S, Werner G, Strommenger B, Kettlitz C, Borgmann S, Schulte B, Jonas D, Serr A, et al: Spread of ampicillin/vancomycin-resistant Enterococcus faecium of the epidemic-virulent clonal complex-17 carrying the genes esp and hyl in German hospitals. Eur J Clin Microbiol Infect Dis. 2005, 24: 815-825. 10.1007/s10096-005-0056-0.
37. Werner G, Coque TM, Hammerum AM, Hope R, Hryniewicz W, Johnson A, Klare I, Kristinsson KG, Leclercq R, Lester CH, et al: Emergence and spread of vancomycin resistance among enterococci in Europe. Euro Surveill. 2008, 13: 1-11.
38. Ogier JC, Serror P: Safety assessment of dairy microorganisms: the Enterococcus genus. Int J Food Microbiol. 2008, 126: 291-301. 10.1016/j.ijfoodmicro.2007.08.017.
39. Danielsen M, Wind A: Susceptibility of Lactobacillus spp. to antimicrobial agents. Int J Food Microbiol. 2003, 82: 1-11. 10.1016/S0168-1605(02)00254-4.
40. Vay C, Cittadini R, Barberis C, Hernán Rodríguez C, Perez Martínez H, Genero F, Famiglietti A: Antimicrobial susceptibility of non-enterococcal intrinsic glycopeptide-resistant Gram-positive organisms. Diagn Microbiol Infect Dis. 2007, 57: 183-188. 10.1016/j.diagmicrobio.2006.08.014.
41. Ammor MS, Flórez AB, Mayo B: Antibiotic resistance in non-enterococcal lactic acid bacteria and bifidobacteria. Food Microbiol. 2007, 24: 559-570. 10.1016/j.

fm.2006.11.001.

42. Danielsen M, Simpson PJ, O'Connor EB, Ross RP, Stanton C: Susceptibility of Pediococcus spp. to antimicrobial agents. J Appl Microbiol. 2007, 102: 384-389.

43. Klare I, Konstabel C, Badstübner D, Werner G, Witte W: Occurrence and spread of antibiotic resistances in Enterococcus faecium. Int J Food Microbiol. 2003, 88: 269-290. 10.1016/S0168-1605(03)00190-9.

44. Albarracín Orio AG, Piñas GE, Cortes PR, Cian MB, Echenique J: Compensatory evolution of pbp mutations restores the fitness cost imposed by beta-lactam resistance in Streptococcus pneumoniae. PLoS Pathog. 2011, 7: e1002000-10.1371/journal.ppat.1002000.

45. Piuri M, Sanchez-Rivas C, Ruzal SM: Cell wall modifications during osmotic stress in Lactobacillus casei. J Appl Microbiol. 2005, 98: 84-95. 10.1111/j.1365-2672.2004.02428.x.

46. Klein G, Hallmann C, Casas IA, Abad J, Louwers J, Reuter G: Exclusion of vanA, vanB and vanC type glycopeptide resistance in strains of Lactobacillus reuteri and Lactobacillus rhamnosus used as probiotics by polymerase chain reaction and hybridization methods. J Appl Microbiol. 2000, 89: 815-824. 10.1046/j.1365-2672.2000.01187.x.

47. Ayeni FA, Sánchez B, Adeniyi BA, de Los Reyes-Gavilán CG, Margolles A, Ruas-Madiedo P: Evaluation of the functional potential of Weissella and Lactobacillus isolates obtained from Nigerian traditional fermented foods and cow's intestine. Int J Food Microbiol. 2011, 147: 97-104. 10.1016/j.ijfoodmicro.2011.03.014.

48. Ayeni FA, Adeniyi BA, Ogunbanwo ST, Tabasco R, Paarup T, Peláez C, Requena T: Inhibition of uropathogens by lactic acid bacteria isolated from dairy foods and cow's intestine in western Nigeria. Arch Microbiol. 2009, 191: 639-648. 10.1007/s00203-009-0492-9.

49. Del Grosso M, Iannelli F, Messina C, Santagati M, Petrosillo N, Stefani S, Pozzi G, Pantosti A: Macrolide efflux genes mef(A) and mef(E) are carried by different genetic elements in Streptococcus pneumoniae. J Clin Microbiol. 2002, 40: 774-778. 10.1128/JCM.40.3.774-778.2002.

50. Bozdogan B, Berrezouga L, Kuo MS, Yurek DA, Farley KA, Stockman BJ, Leclercq R: A new resistance gene, linB, conferring resistance to lincosamides by nucleotidylation in Enterococcus faecium HM1025. Antimicrob Agents Chemother. 1999, 43: 925-929.

51. Roberts MC, Sutcliffe J, Courvalin P, Jensen LB, Rood J, Seppala H: Nomenclature for macrolide and macrolide-lincosamide-streptogramin B resistance determinants. Antimicrob Agents Chemother. 1999, 43: 2823-2830.

52. Leclercq R: Mechanisms of resistance to macrolides and lincosamides: nature of the resistance elements and their clinical implications. Clin Infect Dis. 2002, 34: 482-492. 10.1086/324626.

53. Achard A, Villers C, Pichereau V, Leclercq R: New lnu(C) gene conferring resistance to lincomycin by nucleotidylation in Streptococcus agalactiae UCN36. Antimicrob Agents Chemother. 2005, 49: 2716-2719. 10.1128/AAC.49.7.2716-2719.2005.

54. Marteau P, Gerhardt MF, Myara A, Bouvier E, Trivin F, Rambaud JC: Metabolism of bile salts by alimentary bacteria during transit in the human small intestine. Microb Ecol Health D. 1995, 8: 151-157. 10.3109/08910609509140093.

55. Ruseler-van Embden JG, van Lieshout LM, Gosselink MJ, Marteau P: Inability of Lactobacillus casei strain GG, L. acidophilus, and Bifidobacterium bifidum to degrade intestinal mucus glycoproteins. Scand J Gastroenterol. 1995, 30: 675-680. 10.3109/00365529509096312.

56. Heavey PM, Rowland IR: Microbial-gut interactions in health and disease, Gastrointestinal cancer. Best Pract Res Clin Gastroenterol. 2004, 18: 323-336. 10.1016/j.bpg.2003.10.003.

57. Begley M, Gahan CG, Hill C: The interaction between bacteria and bile. FEMS Microbiol Rev. 2005, 29: 625-651. 10.1016/j.femsre.2004.09.003.

58. Zhou JS, Gopal PK, Gill HS: Potential probiotic lactic acid bacteria Lactobacillus rhamnosus (HN001), Lactobacillus acidophilus (HN017) and Bifidobacterium lactis (HN019) do not degrade gastric mucin in vitro. Int J Food Microbiol. 2001, 63: 81-90. 10.1016/S0168-1605(00)00398-6.

59. Delgado S, O'Sullivan E, Fitzgerald G, Mayo B: Subtractive screening for probiotic properties of Lactobacillus species from the human gastrointestinal tract in the search for new probiotics. J Food Sci. 2007, 72: M310-M315. 10.1111/j.1750-3841.2007.00479.x.

60. Muñoz-Atienza E, Landeta G, De Las Rivas B, Gómez-Sala B, Muñoz R, Hernández PE, Cintas LM, Herranz C: Phenotypic and genetic evaluations of biogenic amine production by lactic acid bacteria isolated from fish and fish products. Int J Food Microbiol. 2011, 146: 212-216. 10.1016/j.ijfoodmicro.2011.02.024.

61. Ladero V, Fernández M, Calles-Enríquez M, Sánchez-Llana E, Canedo E, Martín MC, Alvarez MA: Is the production of the biogenic amines tyramine and putrescine a species-level trait in enterococci?. Food Microbiol. 2012, 30: 132-138. 10.1016/j.fm.2011.12.016.

62. Ringø E, Strom E, Tabachek JA: Intestinal microflora of salmonids: a review. Aquac Res. 1995, 26: 773-789. 10.1111/j.1365-2109.1995.tb00870.x.

63. Bairagi A, Sarkar Ghosh K, Sen SK, Ray AK: Enzyme producing bacterial flora isolated from fish digestive tracts. Aquacult Int. 2002, 10: 109-121. 10.1023/A:1021355406412.

64. Ramirez RF, Dixon BA: Enzyme production by obligate intestinal anaerobic bacteria isolated from oscars (Astronotus ocellatus), angelfish (Pterophyllum scalare) and southern flounder (Paralichthys lethostigma). Aquaculture. 2003, 227: 417-426. 10.1016/S0044-8486(03)00520-9.

65. Fjellheim AJ, Klinkenberg G, Skjermo J, Aasen IM, Vadstein O: Selection of candidate probionts by two different screening strategies from Atlantic cod (Gadus morhua L.) larvae. Vet Microbiol. 2010, 144: 153-159. 10.1016/j.vetmic.2009.12.032.

66. Kullen MJ, Sanozky-Dawes RB, Crowell DC, Klaenhammer TR: Use of the DNA sequence of variable regions of the 16S rRNA gene for rapid and accurate identification of bacteria in the Lactobacillus acidophilus complex. J Appl Microbiol. 2000, 89: 511-516. 10.1046/j.1365-2672.2000.01146.x.

67. Poyart C, Quesnes G, Trieu-Cuot P: Sequencing the gene encoding manganese-dependent superoxide dismutase for rapid species identification of enterococci. J Clin Microbiol. 2000, 38: 415-418.

68. Cintas LM, Rodríguez JM, Fernández MF, Sletten K, Nes IF, Hernández PE, Holo H: Isolation and characterization of pediocin L50, a new bacteriocin from Pediococcus acidilactici with a broad inhibitory spectrum. Appl Environ Microbiol. 1995, 61: 2643-2648.

69. Gilmore MS, Segarra RA, Booth MC, Bogie CP, Hall LR, Clewell DB: Genetic structure of the Enterococcus faecalis plasmid pAD1-encoded cytolytic toxin system and its relationship to lantibiotic determinants. J Bacteriol. 1994, 176: 7335-7344.

70. Hickey RM, Twomey DP, Ross RP, Hill C: Production of enterolysin A by a raw milk enterococcal isolate exhibiting multiple virulence factors. Microbiology. 2003, 149: 655-664. 10.1099/mic.0.25949-0.

71. CLSI: Performance Standards for Antimicrobial Susceptibility Testing: Twenty–first Informational Supplement M100–S21. 2011, Wayne, PA, USA: CLSI

72. Klare I, Konstabel C, Müller-Bertling S, Reissbrodt R, Huys G, Vancanneyt M, Swings J, Goossens H, Witte W: Evaluation of new broth media for microdilution antibiotic susceptibility testing of Lactobacilli, Pediococci, Lactococci, and Bifidobacteria. Appl Environ Microbiol. 2005, 71: 8982-8986. 10.1128/AEM.71.12.8982-8986.2005.

73. Noriega L, Cuevas I, Margolles A, Reyes-Gavilán CG Dl: Deconjugation and bile salts hydrolase activity by Bifidobacterium strains with acquired resistance to bile. Int Dairy J. 2006, 16: 850-855. 10.1016/j.idairyj.2005.09.008.

74. Donabedian SM, Thal LA, Hershberger E, Perri MB, Chow JW, Bartlett P, Jones R, Joyce K, Rossiter S, Gay K, et al: Molecular characterization of gentamicin-resistant Enterococci in the United States: evidence of spread from animals to humans through food. J Clin Microbiol. 2003, 41: 1109-1113. 10.1128/JCM.41.3.1109-1113.2003.

75. Sutcliffe J, Grebe T, Tait-Kamradt A, Wondrack L: Detection of erythromycin-resistant determinants by PCR. Antimicrob Agents Chemother. 1996, 40: 2562-2566.

76. Sutcliffe J, Tait-Kamradt A, Wondrack L: Streptococcus pneumoniae and Streptococcus pyogenes resistant to macrolides but sensitive to clindamycin: a common resistance pattern mediated by an efflux system. Antimicrob Agents Chemother. 1996, 40: 1817-1824.

77. Strommenger B, Kettlitz C, Werner G, Witte W: Multiplex PCR assay for simultaneous detection of nine clinically relevant antibiotic resistance genes in Staphylococcus aureus. J Clin Microbiol. 2003, 41: 4089-4094. 10.1128/JCM.41.9.4089-4094.2003.

78. Werner G, AWillems RJ, Hildebrandt B, Klare I, Witte W: Influence of transferable genetic determinants on the outcome of typing methods commonly used for Enterococcus faecium. J Clin Microbiol. 2003, 41: 1499-1506. 10.1128/JCM.41.4.1499-1506.2003.

79. Lina G, Quaglia A, Reverdy ME, Leclercq R, Vandenesch F, Etienne J: Distribution of genes encoding resistance to macrolides, lincosamides, and streptogramins among staphylococci. Antimicrob Agents Chemother. 1999, 43: 1062-1066.

CHAPTER 18

USE OF IMMUNOSTIMULANTS AND NUCLEOTIDES IN AQUACULTURE

Einar Ringø, Rolf Erik Olsen, Jose L. Gonzalez Vecino, Simon Wadsworth and Seong Kyu Song

Keywords: Fish, Immunostimulants, Nucleotides

IMMUNOSTIMULANTS

On a worldwide scale, fisheries landings remain constant at about 90 million tons of fish whereas commercial aquaculture supplies per se is approximately 60 million tons and is increasing at a rate of about 8% per annum. Use of immunostimulants (natural immune stimulants) is a unique approach for fish culturists to control disease losses in their facilities and numerous polysaccharides from a variety of sources have the ability to stimulate the immune system, and thus behaving as immunostimulants. Immunostimulants have an architecture consisting of repeating units of single molecular forms such as glucose in β-glucans and (deoxy) riboses in DNA/RNA, fatty acid chains in bacterial lipopolysaccharides (LPS) and certain lipoproteins. Such patterns are abundant in microbial communities of prokaryotes, and can be termed pathogen-associated molecular patterns (PAMPs) if they initiate inflammatory responses. It was turned out that commential microbiota also contain those molecular patterns in them and now scientists tend to use a broader term, microbe-associated molecular pattern (MAMP).

Sakai indicated that immunostimulants could be grouped depending on their sources; bacterial, algae-derived, animal-derived, nutritional factors and hormones/cytokines [1]. This sub-grouping changed the concept that immunostimulants instead are divided according to their modes of actions. Previously, an immunostimulant was defined based upon its activity on mononuclear phagocytic system only [2]. However, due to recent discoveries of the pattern recognition receptors (PPRS) of the phagocytic cells, the former definition of immunostimulant needs to be redefined. Different leucocytes possess different PPRs and the immune cells bring about different immune responses depending on the types of ligands PPRs interact with. To define an immunostimulant, all elements of the immune system must be considered, and according to Bricknell and Dalmo [3] the definition could be; "An immunostimulant is a naturally occurring compound that modulates the immune system by increasing the host's resistance against diseases that in most circumstances are caused by pathogens". Immunostimulants have been used as feed additives for several years in aquaculture, and yeast β-glucan may be the one with the longest track record. In nature, β-glucans are widespread and have been characterized in microorganisms, algae, fungi and plants [4]. The chemical structure of β-glucan varies with respect to molecular weight and degree of branching. For example, β-glucan from yeast contains a particular carbohydrate consisting of glucose and mannose residues and is a major constituent in the cell membrane.

Nucleotide-supplemented diets are not strictly immunostimulants by definition but provide a dietary supplement that allows improved resistance to a pathogen insult. Although such diets have been reported not to induce measurable immunostimulatory effects [5-7], they nevertheless seems to be able to up-regulate several immune genes in turbot (Scopthalamus maximus L.), which contradicts the earlier claims that these diets are not immunostimulatory [7].

The biological effects of immunostimulants are highly dependent on the receptors on the target cells recognising them as potential highrisk molecules thus triggering various defence pathways. Thus, it is important to increase the knowledge of receptor specificity and the inflammatory processes the different receptors, upon antigen binding, are induced. However, many mammalian receptors reported to bind immunostimulants such as NLR (NOD-like receptors) have yet to be reported in

Citation: Ringø E, Olsen RE, Vecino JLG, Wadsworth S, Song SK (2012) Use of immunostimulants and nucleotides in aquaculture. J Marine Sci Res Development 1:104. doi:10.4172/2155-9910.1000104.

fish. Nevertheless, assuming that fish and mammalian cells share many similar receptors, one may predict the biological outcome of immunostimulants in fish.

In order to evaluate whether immunostimulants may act in synergy with pro– and prebiotics we recommend this option merits further investigations. Probiotics appear to modulate immunity of the host by improving the barrier properties of mucosa and modulating production of cytokines (protein mediators produced by immune cells and contribute to cell growth, differentiation and defence mechanisms of the host) [8]. Viable live probionts are better than the non-viable heat-killed probionts in inducing higher immune responses in rainbow trout (Oncorhynchus mykiss Walbaum), especially enhancing head kidney leucocyte phagocytosis, serum complement activity etc., [9]. In recent years, several in vivo and In vitro studies have investigated the interaction between dietary probiotics and immunocompetence in humans as well as in fish and aquatic animals [8,10-17]. By increasing the host's adaptive and innate immune mechanisms, lactic acid bacteria (LAB) can protect the host against infection by enteric pathogens and tumour development. Immunological and other mechanisms behind the probiotic action may include;

* Stimulation of antibody secreting cell response [18]

* Enhancement of phagocytosis of pathogens [9,19]

* Modification/enhancement of cytokine production/ natural complement activity [20,21]

* Improvement of the host innate or acquired immune responses, direct effect on other microorganism in the digestive tract, adhesion sites, microbial action or response stemming from microbial products, host products or food components [22]

Consequently, probiotic bacteria may influence both adaptive and innate immune responses. Probiotics may reverse the increased intestinal permeability induced by antigens, but no information is available about long-term effects.

Nutritional factors and the immune response

It is well known that nutrition may affect fish health and immune responses. Nutritional factors are components of the diet essential for normal growth and development of the fish. Many of these undoubtedly have roles in the immune response. Vitamins C, B6, E and A, and the minerals iron and fluoride have been identified as micronutrients that can affect disease resistance. With respect to the strict definition of immunostimulants; vitamins and minerals are not immunostimulants as they enhance the immune system by providing substrates and serving cofactors necessary for the immune system to work properly, but controversy exists [1,23].

Readers with special interest on the effect of nutritional factors on immune response of fish are referred to the comprehensive reviews published [1,24-37]. To our knowledge no information is available about nutritional balance and the immune response and as a natural consequence this topic merits further investigations.

β-1,3/1,6-glucans

Glucans are high molecular-weight substances composed of glucose as building blocks, usually isolated from cell walls of bacteria, mushrooms, algae, cereal grains, yeast and fungi [38]. Glucans in general comprise a great variety of substances common in nature (such as cellulose, glycogen, and starch), most of which do not interact with the immune system. Pharmacologically, they are classified as biological response modifiers (BRM). The common feature of immunomodulatory glucans is a chain of glucose residues linked by β-1.3-linkages, also called beta-glucans. Of the different β-glucans, the products known as β-1.3/1.6-glucans (read as "beta-one-three-one-six-glucans"), derived from baker's yeast, are suggested to be the most potent immune-system enhancers. β-1.3/1.6-glucans is characterized by side-chains attached to the backbone that radiate outward like branches on a tree.

The primary structure of the β-1,3/1,6-glucan is determinant for its immune-enhancing ability. The frequency and nature of side-chains strongly affect the ability of the glucan to mediate binding to surface receptors on the target cells influencing the effectiveness of the glucan as an immunostimulant.

The effectiveness of MacroGard® in stimulating the immune system combined with improved growth and feed conversion has been proven in some experiments [39,40]. MacroGard® has been suggested to be used in feed for all aquatic organisms and also to enrich live feed such as Artemia. Some studies involving shrimp have also suggested that the product contributes to a better growth in crustaceans [40,41].

MacroGard® is an environmentally sound alternative to feed antibiotics and chemotherapeutics for livestock, pets and cultured aquatic organisms. MacroGard® is based on a well-characterized β-1,3/1,6-glucan that has been in use world-wide for almost 25 years as an immune modulating agent in animal husbandry and aquaculture (see Table 1).

When included in feed or administered on mucosal surfaces, MacroGard® modulates immune responses and affects biological functions in a favourable way. The products are extracted and purified from food-grade baker's yeast by patented processes. The use of MacroGard® may be regarded as a natural input that compensates for a possible sub-optimal function of the immune system of farm animals, companion animals and aquatic species in intensive culture. The product has also been suggested to reduce the use of antibiotics in animal husbandry and consequently the concern that developing antibiotics resistance that may undermine the health security effective antibiotics represents. In aquaculture, glucans have been successfully been used to enhance the resistance of fish and crustaceans against bacterial and viral infections and (Table 1) present an overview of papers using glucans as immunostimulants in aquaculture.

In Norway, β-glucan, as prebiotics and nucleotides, has been used in the pancreas disease (PD) diets; to fed fish suffering from PD (Gonzalez Vecino, personal communication). In this case, the β-glucans are a very small part of the diet composition, and the benefits of the diets come mainly from a very specific formulation concentrated to recover pancreas and reduce inflammation in muscle and heart.

Baker's yeast, Saccharomyces cerevisiae, is the 2nd major by-product from brewing industry and contains various immunostimulating compounds such as β-glucans (the cell walls are constructed almost entirely by β-1,3-D-glucan, β-1,6-D-glucan, mannoproteins and chitin bound together by covalent linkages), nucleic acids and oligosaccharides [42] and it has the capacity to enhance growth and increase both humoral (myeloperoxidase and antibody titer) and cellular (phagocytosis, respiratory burst and cytotoxicity) immune responses, and to increase or confer resistance against pathogenic bacteria in various fish species (Table 1).

Table 1: Glucans as immunostimulants in fish shrimp and scallop

Immunostimulant	Species	Administration and dose	Length of administ-ration	Mechanism of action/results	References
β-1,3 glucans	Carp	i.p; 2-10 mg kg^{-1} BW	12 days	↑ phagocytic activity of kidney leucocytes	[152]
	Channel catfish	Injected; 50 and 70 µg $fish^{-1}$ (100 g BW) suspended in 0.2 ml PBS	2 weeks	↑ phagocytic, bacterial activity and serum antibody concentration	[153]
	Marron	0.08, 0.1, 0.2, 0.4 and 0.8 % supplementations	12 weeks	↑ haemocyte count → on physiological parameters	[154]
β-glucan	Atlantic halibut (larvae)	Immersion (25 mg l^{-1})	5 days	Absorbed laminaran	[155]
	Carp	i.p (10 mg kg^{-1})	15 days	↑ superoxide dismutase and catalase activities	[156]
	Coho	i.p (5 and 15 mg kg^{-1})	36 days	→ immune response	[157]
	Trout	Oral	4 weeks	↑ stress resistance	[158]
	Turbot	Oral (2 g kg^{-1})	5 weeks	↑ leukocyte number	[159]
	Trout	i.p	18 days	↑ resistance to IHNV	[39]
	Trout	Oral (0, 0.2 and 0.4 %)	37 days	↓ expression of genes involved In acute inflammation reaction	[160]
	Sea bass	Oral (2% wet body weight)	2 weeks	↑ complement activation	[161]
	Sea bass	Oral (0.1%)	60 days	↑ serum complement activity, serum lysozyme, gill and liver HSP	[162]
	Sea bass	Oral (250, 500, 1,000 ppm)	25 days	↑ respiratory burst activity of head kidney macrophages	[163]
	Snapper	Oral (10g kg^{-1})	84 days	↑ macrophage superoxide Anion production and growth → complement activity	[164]
	Tilapia and Japanese eels	i.p (10 mg kg^{-1})	2 days	↑ lysozyme activity, phagocytic activities in anterior kidney and peripheral blood phagocytes and classical complement path-way	[165]
	Nile tilapia	Oral	6 weeks	↑ stress resistance	[166]
	Red drum	Oral (2% of diet)	6 weeks	→ on stress resistance	[124]
	Flounder	Oral (3 g kg^{-1})	n.i	↑ respiratory burst activity	[167]
	Channel catfish	Oral (1g/kg)	4 and 6 weeks	→ growth performance hematology or immune function Some improvement in stress resistance	[168]
	Climbing perch	B-glucan suspension (0, 5, 10 15 mg l^{-1})	7 days	↑ lysozyme and bactericidal activities and survival of spawns immersed in 15 mg l^{-1} when challenged with *A. hydrophila*	[169]
	American White shrimp	Immersed	120 hours	↑ total haemocyte counts and soluble haemocyte protein after 48-120 hours	[170]
	Shrimp	Oral (0.2% w/w)	7 days	↑ prophenoloxidase and reactive oxygen intern-mediate activity	[171]
	White shrimp	Oral (2000 mg kg^{-1})	6 weeks	↑ total haemocyte counts, phenoloxidase, superoxide anion and superoxide dismu-tase until day 27	[172]
	Carpet shell clam	i.v (0.05, 0.1, 0.5 and 1.0 mg ml^{-1})	Different sampling times	↑ nitric oxide production Hemolymph treated with β-glucan inhibited growth of *Vibrio algynolyticus, Vibrio splendidus* and *Escherichia coli*	[173]
	Scallop	i.p	7 days	↑six enzymes in haemocytes. β-glucan-induced activation was stronger at 15°C than 25°C .	[174]
β-glucan + mannose	Snapper	Oral (0.1-1.0% w/w)	n.i *in vitro* study	↑ macrophage activation	[175]

β-glucan + LPS	Salmon	10 μg ml⁻¹ᵛ	n.i *in vitro* study	↑ lysozyme activity	[176]
β-1,3 and β-1,6 glucans	Sea urchin eggs	0.01, 0.05, 0.1, 0.25, 0.5 and 1.0 mg ml^{-1}	Incubation	↑ survival of embryos	[177]
β-1,3 and β-1,6 inked yeast glucan (M-glucan)	Salmon	Injected i.p (1 ml of 0.5% (w/v) M-glucan in 0.9% saline)	22 days	↑ macrophage and neutrophil numbers, head kidney macrophages and ability to kill *A. salmonicida*	[178]
	Salmon	0, 500 and 1000 mg kg^{-1}	70 days	→ no detrimental effect ↓lice- infected fish when the fish were fed 14 % soybean meal + 14 % sunflower and yeast extract	[179]
	Trout	Injected i.p 1 ml of 1% M-glucan suspension	3 weeks	↑macrophages ability to kill *A. salmonicida* and serum lysozyme activity	[180]
β-1,3 glucan from *Schizophyllum commune*	Black tiger shrimp	Oral (0 and 2 g kg^{-1})	40 days	↑ survival, haemocyte activity, cell adhesion and superoxide anion production	[181]
Glucan (barley extract (Sigma)	Trout	Injected i.p. 100 μg glucan dissolved in PBS. Immersed the concentration of 100 μg ml^{-1} glucan for 30 min	10 days	↑ phagocytic activity and numbers of circulatory glass-adherent cells	[182]
Yeast glucan	Salmon	i.p	20 days	↑ resistance to *V. anguillarum*, *V. salmonicida* and *Y. rückeri*	[183]
	Salmon	i.p	4 weeks	↑ complement and lysozyme activity	[184]
	Salmon	i.p (0.5 mg/fish)	43 weeks	↑ survival against *A. salmonicida* infection	[185]
	Salmon	i.p	7 weeks	↑ antibody → resistance to *A. salmonicida*	[186]
	Salmon	Oral and anal (150 mg kg^{-1})	2 days	↑ acid phosphatase	[187]
	Trout	i.p	n.i	↑ lysozyme activity	[188]
	Trout	Diet (0, 0.125 and 0.25 g k^{-1})	4 + 4 weeks	4 weeks ↑ survival and growth 4 – 8 weeks ↑ survival (0.25 g k^{-1}) → feed conversion	[189]
	Indian major carp		15 + 30 days	↑ superoxide anion, *in vitro* phagocytic activity and nitrite production of leucytes (10 days) → heamatocrit	[190]
	Indian major carp and rohu	Diet (0 and 5 g kg^{-1})	15 + 20 days	↑ phagocytosis and prolifer-ation of lymphocytes and oxidative radical and nitrate production	[191]
	Pacu	Diet (2.5 and 5 %)	86 days	↑ feed efficiency → plasma glucose and cortisol	[192]
	Fathead minnows	Diet (10 g kg^{-1})	14 days	↑ degranulation of primary granules in fish neutrophils	[193]
	Shrimp	immersion	43 days	↑ growth at 0.5; 1 and 2 mg ml^{-1}, but not at 0.25 mg ml^{-1} ↑ phenoloxidase activity and resistance to *V. vulnificus*	[194]
	Black tiger shrimp	Oral (0.2 %)	3 days	↑ phenoloxidase, no. of haemocytes and bacterial killing activity against *Vibrio harveyi*	[195]
	Pacific white shrimp	Oral (0, 1 and 2 g kg^{-1})	4 weeks	↑ total haemocyte – and granular haemocyte counts → growth	[196]

Yeast glucan + Vitamin C	Trout	Diets containing yeast glucan and vitamin C at 150, 1,000 and 4,000 p.p.m	4 weeks	↑ alternative pathway of complement – and macrophage activity and specific Ab response following vaccination with *Y. ruckeri*	[197]
Saccharomyces cerevisiae	Sea bass	Diet (0, 10, 20, 30 and 40 %)	12 weeks	↑ feed conversion (10, 20 and 30 %), N retention (% N intake) → growth and energy retention (% E intake)	[198]
	Sea bream	0 and 10g kg⁻¹	6 weeks	↑ serum peroxidases and complement activity	[199]
	Sea bream	*In vitro* experiment, head kidney leucocytes	0 – 30 min	↑ degranulation	[200]
	Sea bream	0, 1, 5 and 10g kg⁻¹	6 weeks	↑ total serum IgM level	[201]
	Sea bream	*In vitro* experiment, blood leucocytes	30 min	↑ inhibition of the phagocytic ability	[202]
	Sea bream	0, 11.5 and 23 % (Y2)	10 weeks	↑ growth, feed intake, lipid content (Y2) and arginase activity → body composition	[203]
	Gilthead seabream	Lyophilised *S. cerevisiae* (0, 1, 5 or 10g kg⁻¹)	4 weeks	↑ cellular parameters	[204]
	Gilthead seabream	*S. cerevisiae* (0, 10, and 20 %) (BDY protein) instead of fish meal	12 weeks	→ growth, alkaline phospha-tase, blood urea Nitrogen, serum protein, cholesterol, triglyceride, albumin and amylase ↓ plasma glucose (10 % BDY)	[205]
	Nile tilapia	Diet (0 and 0.1 %)	9 weeks	↑growth and feed efficiency	[206]
	Nile tilapia	*S. cerevisiae* (0, 25, 50, 75 and 100 %) (BDY protein) instead of soy bean meal (SBM) protein	6 weeks + 5 days	→ growth and feed utilization when 50 % BDY protein instead of SBM protein	[207]
	Nile tilapia	*S. cerevisiae* (0, 25, 50, 75 and 100 %) (BDY protein) instead of soy bean meal (SBM) protein	122 days	→ growth and feed utilization when 25 % BDY protein instead of SBM protein	[208]
	Hybrid striped bass	Diet (0, 1, 2 and 4 %)	8 weeks	↑growth and feed efficiency	[209]
	Hybrid striped bass	Diet (0, 1 and 2 %)	7 weeks (trial 1) and 4 weeks (trial 2)	→ growth (trial 1) Trial 2 ↑ resistance against *Streptococcus iniae* and extra-cellular superoxide anion → growth and feed efficiency	[210]
	Hybrid striped bass	Diet (0, 1 and 2 %)	16 and 21 weeks	↑growth, feed efficiency, serum peroxidase and extracellular superoxide anion of head kidney macrophages (16 weeks) → resistance against mycobacterial infection	[211]
	Pacu	*S. cerevisiae* (0, 30, 35, 50, 70 and 100 %) (BDY protein) instead of fish meal	54 days	↑growth and feed utilization (until 50 % replacement). Protein retention was higher in fish fed 35 and 50 5 replace-ment → protein digestibility	[212]
	Galilee tilapia	0 and 10g kg⁻¹	6 weeks	↑growth, feed utilization and resistance against waterborne Cu toxicity	[213]
	Beluga	Diet (0, 1 and 2 %)	6 weeks	2 % inclusion ↑final weight, weight gain, SGR and FCR ↑ autochthonous LAB levels →survival rate, haematological and serum biochemical parameters resist	[97]

Symbols represent an increase (↑) in the specified response; no change (→); decrease (↓); n.i – no information given.

Salmon – Atlantic salmon; Trout – rainbow trout; Coho – Coho salmon (*Oncorhynchus kisutch*), Turbot – *Scophthalmus maximus*; Sea bass - *Dicentrachus labrax*, Snapper– pink snapper (*Pagrus auratus*); Flounder – Japanese flounder (*Paralychthis olivaceus*), Nile tilapia - *Oreochromis niloticus*; Red drum - *Sciaenops ocellatus*; Channel catfish - *Ictalurus punctatus*

ROS – reactive oxygen species; IHNV – infectious hematopoietic necrosis virus

Bioactive alginate (high-M alginate) and Ergosan

The adaptive immune system is poorly developed in early developmental stages of fish [43,44], which is why immunostimulants and probiotics have been used in an attempt to increase survival of larvae against microbial pathogens [45-47]. In this respect, alginate has been proposed as a potential candidate. Alginate is a polysaccharide composed of β-1,4-D-mannuronic acid (M) and C5-epimer α-Lglucuronic acid (G) [48]. The monomers are usually arranged in M-blocks, G-blocks and alternating MG-blocks [49]. Alginate also binds various cations found in the seawater such as Mg2+, Sr2+, Ba2+, and Na+. Commercial alginates are extracted from three species of brown algae; Laminaria hyperborean, Ascophyllum nodosum and Macrocystis pyrifera, in which alginate comprises up to 40% of the dry weight [50]. Furthermore, bacterial alginates, not commercial products, have also been isolated from Azotobacter vinelandii and several Pseudomonas species [51].

Commercially available alginates today have M-content ranging between 30 and 70%. Alginates with even higher M-content, typically higher than 80%, have also been shown to be potent stimulators of immune cells such as human monocytes [52]. High-M alginate has also been used as an immunostimulant for enhancement of innate immune resistance in fish larvae and fry [53-56].

Ergosan is an algal based product and contains 1 % alginic acid extracted from Laminaria digitata. The first study on Ergosan in aquaculture, to the author's knowledge, was reported by Miles and coauthors on striped snakehead (Channa striata)

[57]. In a study with rainbow trout, Peddie et al. reported that a single injection of 1 mg of Ergosan significantly augmented the proportion of neutrophils in the peritoneal wall, increased the degree of phagocytosis, respiratory burst activity and expression of interleukin-1β (IL-1β), interleukin-8 (IL-8) and one of the two known isoforms of tumour necrosis factor-alpha (TNF-α) in peritoneal leucocytes one day post-injection [58]. However, humoral immune parameters were less responsive to intraperitoneal alginate administration with complement stimulation only evident in the 1 mg-treated group at 2 days post-injection. (Table 2) presents an overview of reports using alginates and Ergosan in fish studies.

Plant extracts

Some immunostimulants cannot be used because of various disadvantages, such as high cost and limited effectiveness upon parenteral administration. Numerous plants have on the other hand long been used in traditional medicine for the treatment and control of several diseases [59]. As herbs have little side effects and are easily degradable and abundantly available in farm areas, numerous investigations have investigated the effect of plant products on innate and adaptive immune response and to prevent and control fish and shellfish diseases [60].

Dügenci and co-authors investigated the effects of mistletoe, nettle and ginger on dietary intake of rainbow trout [61]. The diets contained lyophilized extracts of these plants at two inclusions levels, 0.1 and 1%, in a 3 week experiment. At the end of the experimental period, various parameters of innate defence mechanisms, including extracellular and intracellular respiratory burst activities, phagocytosis in blood leukocytes, total plasma protein level, specific growth rates and condition factors were examined. Inclusion of the plant materials increased the extracellular respiratory burst activity ($P<0.001$) compared to control. Furthermore, fish fed the diet containing 1% ginger roots exhibited a significant innate immune response. Phagocytosis and extracellular burst activities of blood leukocytes were significantly higher in this group compared to the control group. All plant extracts increased plasma protein level except for the 0.1% ginger supplemented diet. The highest level of plasma proteins was observed in the group fed with 1% ginger extract.

Oral administration of the medical plant, Eclipta alba, on the non-specific immune responses and disease resistance of tilapia (Oreochromis mossambicus) has been investigated [62]. The results indicated that E. alba administrated in the diet significantly enhanced the non-specific immune parameters tested. Furthermore, when tilapia was challenged with Aeromonas hydrophila mortality was significantly reduced in E. alba treated fish.

Lectins are proteins or glycoprotein substances, usually of plant origin, and they are sugar-binding proteins which are highly specific for their sugar moieties. Lectins are also known to play important roles in the immune system by recognizing carbohydrates that are found exclusively on pathogenic bacteria, or that are inaccessible to host cells [63]. Examples are the lectin complement activation pathway, and mannose binding lectin (MBL), also named mannose- or mannanbinding protein (MBP). MBL recognizes carbohydrate patterns, found on the surface of a large number of pathogenic micro-organisms, including bacteria, viruses, protozoa and fungi. Readers with special interest in lectins and immune response are referred to the recent comprehensive reviews [64-65].

In a recent paper, Galina and co-authors reviewed studies currently being carried out on herbs and herbal extracts that have been shown to modulate the immune system in fish and special attention was given to; Astragalus, Ganoderma, Lonicera and Chinese and Indian herbs [66]. Based on the results available per se Galina and co-authors suggested that herb extracts might have a potential application as an immunostimulant is fish, due to that they can easily be obtained, are as not expensive and act against a broad spectrum of pathogenic bacteria, and could be alternatives to vaccines, antibiotics or chemotherapeutic agents. In a more recent study, Soosean and co-authors reported that mangosteen (Garcinia mangostana) extracts as feed additive to African catfish (Clarias gariepinus) had no significant effect on growth parameters, feed conversion ratio and haemoglobin content. However, significantly higher red blood cells and white blood cell counts were recorded in fish fed the shoot extract [67]. In their study on kelp grouper (Epinephelus bruneus), Harikrishnan and co-authors evaluated the efficacy of dietary doses of Lactuca indica extracts on immunological parameters and disease resistance against Streptococcus iniae infection [68]. Based on their results, the authors suggested that supplementation of L. indica enhanced the immune system and the disease resistance against pathogenic infection. Enhanced immune response of tiger shrimp (Penaeus monodon) by the medical herb black nightshade (Solanum nigrum) and disease resistance against Vibrio harveyi were investigated by Harikrishnan and co-authors [69]. In this study the authors displayed improved immune response including respiratory bursts, PO-, SOD – and GPx activities and disease resistance by feeding tiger shrimp black nightshade at 0.1 and 1 % doses. This herb has also been shown to have inhibitory In vitro growth effect on A. hydrophila and improved survival and increased haematological parameters of spotted snakehead (Channa punctatus) infected with A. hydrophila [70].

Table 2: Bioactive alginate (high-M alginate) and Ergosan as immunostimulants in fish

Immunostimulant	Fish	Administration and dose	Length of administ-ration	Mechanism of action/results	References
High-M alginate	Atlantic halibut	Immersion	n.i	↑ survival	[214]
	Atlantic halibut	Bioencapsulated in *Artemia*	Different periods during initial feeding	↑ growth	[215]
	Atlantic cod and spotted wolffish	Diet	59 days 55 days	↑ growth Uptake of 125 I-labelled molecule in the stomach and intestine	[56]
	Turbot	Enriched in *Artemia*	1 week	↑ survival	[53]
Alginate	Atlantic salmon	Macrophage culture	n.i	↑ phagocytic activity and respiratory burst	Rokstad *et al.* (unpublished data) cited in Vadstein [54]
	Atlantic salmon	Diet	Six months	↑ lysozyme activity → survival, colour, taste consistency and grading	[216]
	Turbot	Diet	13 days	↑ protein synthesis and protein turnover → survival and larval size	[217]
Ergosan	Striped snakehead	i.p	14 days	↑ inhibition index of macrophages and serum on *in vitro* growth of *Aphanomyces invidans*	[57]
	Rainbow trout	i.p	1 day	↑ neutrophils and respiratory burst	[58]
	Sea bass	Diet (0.5 %)	60 days	↑ serum complement activity, serum lysozyme, gill and liver HSP	[162]
AquaVac Ergosan	Rainbow trout	Diet (0.5 %)	Three cycles (95 days in total)	↑ growth (95 days) and the growth related gene expression of IGFI, TRαmRNA and TRβ, and gene expression of IL-1β, IL 8 and TLR3 in spleen ↓ cortisol, HSP70 gene expression in liver, MYO gene expression	[218]

AquaVac Ergosan – extracts of *Laminaria digitata* and *Ascophyyllum nodosum*.

Symbols represent an increase (↑) in the specified response; no change (→); decrease (↓). n.i – no information given

Readers with special interest in the topic plant products on innate and adaptive immune response of fish and shellfish are referred to the recent review of Harikrishnan et al. [60].

MECHANISMS OF ACTION

Phagocytic Activity of Kidney Leucocytes

Phagocytic activity is the principle function of kidney leucocytes of teleosts. Monocytes, macrophages, dendritic cells, and neutrophils belong to the phagocytic leucocytes called phagocytes. In host defense, phagocytes respond to microbes in a sequential manner: active recruitment of the cells to the sites of infection through inflammatory responses, ingestion of microbes by the process of phagocytosis, and destruction of the ingested microbes. Phagocytosis is an active, energyrequiring engulfment process of large particles (>0.5 μm in diameter). The first step in phagocytosis is the recognition of microbe by the specific receptors expressed on the phagocyte such as TLRs (toll-like receptors). The bound microbes are ingested into vesicles forming phagosomes inside of the cells and the phagosomes are then fused with lysosomes that contain many different proteases. The internalized microbes will be killed there through proteolytic processes [71].

Macrophage Activation

Macrophage activation is initiated upon binding of the microbe to the cell. The recognition of microbes by phagocytes such as macrophages, neutrophils, and dendritic cells is selective and mediated by receptors on phagocytic cells. Upon microbe binding, the receptors cooperatively activate the cells through a receptor-mediated signal transduction to kill ingested microbes. The receptors responsible for microbe recognition include TLR (toll-like receptors), G protein-coupled receptors, antibody Fc and complement C3 receptors, and receptors for cytokines, mainly IFN-γ. Activated macrophages kill phagocytosed microbes by the action of microbicidal molecules in phagolysosomes of phagocytes. In addition, activated macrophages secrete cytokines such as TNF, IL-1, and IL-12, which cause inflammation and bridge to activate the adaptive immune system.

Respiratory Burst Activity

Respiratory burst (sometimes called oxidative burst) is the rapid release process of reactive oxygen species (ROS) such as superoxide anion, hydroxyl radical and hydrogen peroxide from activated macrophages and neutrophils. The respiratory burst, releasing important bactericidal ROS molecules, plays a crucial role to remove microbes from the body. The respiratory burst is mediated by the enzyme phagocyte oxidase, which converts molecular oxygen to ROS and is usually triggered by bacterial products or by inflammatory mediators. The respiratory burst activity is an indicator of the status of macrophage and neutrophil activation [71].

Acid Phosphatase

Acid phosphatase is a ubiquitous enzyme that removes phosphate groups from other molecules. The acid phosphatase is one of the lysosomal hydrolases which exist in phagolysosomes of macrophages (lysosomal acid phosphatases). When the macrophage is activated, the inside of phagosome becomes acidic. In an acidic environment, lysosomal hydrolases including acid phosphatase are activated to kill microbes. Acid phosphatase activity is an indicator of the microbicidal activity of macrophages and a high concentration of acid phosphatase in the blood is a sign of a chronic infection [71,72]. Recently, acid phosphatase activity tests are used for the diagnosis of prostate cancer and breast cancer in human.

Serum Complement Activity

The complement system is one of the major effector mechanisms of the humoral immunity as well as the adaptive immunity, which involves the clearing of pathogens from an organism. The complement system consists of serum and cell surface proteins that are activated by microbes (the alternative pathway) and antibodies (the classical pathway). Activation of the complement system involves the sequential proteolysis of proteins to generate new enzyme complexes with proteolytic activity, which is a characteristic of a proteolytic enzyme cascade. The products of the complement activation become covalently attached to microbes, antibodies bound to microbes, and to other antigens. The binding of complement products to the target molecules cause lysis of the microbe, facilitation of opsonization (the process of attaching IgG or complement fragments to microbial surfaces for phagocytosis) and phagocytosis, and inflammation [71].

Phenoloxidase Activity

Phenoloxidases are known to be involved in the innate immune response of invertebrates such as shrimps and crabs. Phenoloxidases are composed of tyrosinases, catecholases and laccases which are the terminal components of the prophenoloxidase (proPO) system, the modified complement system of several phyla of invertebrates [73]. The proPO system is present in the blood of a wide range of marine invertebrates, especially crustaceans. Phenoloxidases play a role in bactericidal activity and the enhancement of phagocytosis in crustaceans [74-76].

Serum Antibody Concentration

B lymphocytes recognize antigens such as microbes using their receptors (membrane-bound antibody molecules). The engagement of antigen receptors and other signals trigger the lymphocytes to expand their numbers and to differentiate into antibody-secreting mast cells, which secrete different classes of antibodies with distinct functions. The functionally different antibodies have the same antigen specificity, through which the antibodies recognize the same microbes despite their different classes. Antibodies bind to microbes and prevent them from infecting cells (neutralization). The binding of antibodies to microbes also promotes opsonization and phagocytosis of microbes to kill and mediate antibody-dependent cellular cytotoxicity (ADCC). In addition, antibodies trigger activation of the complement system (the classical pathway), which in turn causes to increase phagocytosis of microbes opsonized with complement fragments and inflammation. Serum antibody concentration is an indicator of B lymphocyte activation against specific microbes [71].

Leukocyte Number

Leukocytes (white blood cells) are immune cells involved in defending the body against both infectious microbes and foreign materials. Leucocytes consist of neutrophil, eosinophil, basophil, monocyte, and lymphocyte. They are all derived from a hematopoietic stem cell in the bone marrow. Neutrophils and monocytes are involved in the innate immune response as the first line of the defense system. Effector cells of the innate immune system circulate in the blood and migrate into tissues and kill microbes through phagocytosis. The innate immune cells also play a role in the inflammatory response by releasing cytokines. Lymphocytes including T cells and B cells are the only cells in the body capable of specifically recognizing and distinguishing different antigenic determinants (epitopes). When these adaptive immune effector cells recognize the microbes, the number of lymphocytes is significantly increased as a part of their effector functions. Although the number of the innate immune cells are also expanded upon activation, the level of increase is much lower than that of the adaptive immune cells. The number of leukocytes in the blood is often an indicator of the activation of the immune system or of leukemia [71].

Lysozyme Activity

Lysozyme is a cationic enzyme that attacks the β-1,4 glycosidic bond between N-acetylmuramic acid and N-acetylglucosamine in the peptidoglycan of bacterial cell walls. This activity causes the lysozyme to lyse certain gram-positive bacteria and even some gram-negative bacteria. Lysozyme activity has been detected in mucus, serum, organs and eggs of several fish species [77]. Lysozyme is produced mainly by macrophages and its production is induced in response to microbial components such as microbial lipopolysaccharide (LPS) [78] and many other immunostimulants.

Do Immunostimulants raise the level of Protection against Disease?

A large number of reviews have been published during the last 20 years concerning the advantages of immunostimulants in fish and the immune system [1,3,23,28,30,32,79-90]. In this section an attempt has been made to compile the recent developments in the field, mainly over the last five years, besides the earlier work done in the field of disease resistance in fish with exposure to immunostimulants.

Several types of β-glucan have been successfully used to enhance resistance of fish and crustacean against bacterial and viral infections. An overview is presented in (Table 3). Most pathogens used in these challenge experiments are bacteria reported in commercial aquaculture and include; A. hydrophila, Aeromonas salmonicida, Aphanomyces invadans, Edwardsiella ictaluri, Edwardsiella tarda, Photobacterium damselae ssp. piscicida, Vibrio (Listonella) anguillarum, Vibrio harvei, Yersinia ruckeri, Lactococcus garvieae, S.iniae and Streptococcus sp. as well as infectious hematopoietic necrosis virus (IHNV), white spot syndrome virus (WSSV) and parasites. Readers with special interest in the topic immune response and aquatic viruses are referred to the recent reviews [91,92].

Recognition of invasive pathogens is an essential step for the activation of the immune system. This is accomplished by binding of PRRs to PAMPs. The PRRs and PAMPs activate appropriate immune responses [93]. β-glucan binding proteins (β-GBP) are known as one of the most important PRRs in invertebrates [94] and fish [95]. The presence of these proteins in invertebrates and fish may indicate that the defence systems against β-glucan containing microorganisms are crucial for these organisms.

Effect of Immunostimulants on gut Microbiota

Immunostimulants seem to be valuable for the control of fish diseases. However, knowledge regarding the ability of immunostimulants to modulate the gut microbiota in a healthier way and decrease the infection pressure by improving the gut function is scarce. To our knowledge only two studies, Gildberg and Mikkelsen [46] and Skjermo et al. [96] have been done on the topic; immunostimulants and modulation of the gut microbiota. In their study on Atlantic cod (Gadus morhua) fry, Gildberg and Mikkelsen evaluated the effect of supplementation of LAB either alone or in combination with immunostimulating peptides [46]. After three weeks of feeding, the fish were challenged by bath exposure to V. anguillarum (107 ml- 1; 1hour). Twelve days after infection significantly (p<0.05) reduced cumulative mortality was recorded in fish fed diet supplemented with LAB, originally isolated from Atlantic salmon (Salmo salar) and with immunostimulating peptides. No synergistic or cumulative effects were achieved by combining LAB and immunostimulating peptides. Four weeks after infection similar cumulative mortality (80–84%) was observed in all groups. LAB seems to colonize the internal mucus layer of the cod fry pyloric caeca, and a significant number of the bacteria survived the passage of the whole gastrointestinal tract.

In a study with Atlantic cod larvae, Skjermo and co-authors [96] evaluated the effect of β-(1→3, 1→6)-glucans (chrysolaminaran) from the marine diatom Chaetoceros müllerі, a commercial yeast-glucan product and high-M alginate (high content of mannuronic acid isolated from Durvillaea antarctica), on the microbial community in larval gut and water. Total colony forming units (CFU) were evaluated on marine agar, and Vibrio - and Pseudomonas - like species on selective agars (TCBS and marine Pseudomonas agar with CFC - supplement). The larvae were rapidly colonised after hatching, but no or weak effects of the stimulants were observed on the bacterial colonisation rates or the bacterial community. The total CFU varied from 101 to 102 CFU per µg larva after initiation of the first feeding. Bacteria belonging to Pseudomonas appeared to increase throughout the period, whereas the level of Vibrio-like bacteria was low and stable. However, in this investigation the authors only focused on characterisation of Pseudomonas- and Vibrio-like bacteria and did not use molecular methods to evaluate the entire bacterial community.

Table 3: Immunostimulants and disease susceptibility in fish and shrimps

Resistance to pathogen	Agent	Species	Route of exposure	Length of administration	Mechanism of action/results	References
Aeromonas hydrophila	Yeast glucan	Tilapia and grass carp	i.p	14 days	↑ protection in both species and number of NTB-positive staining cells	[219]
	Glucan	Rohu	Oral	7 days	↑ phagocytotic activity, bactericidal activity, antibody titre and agglutinin level	[220]
	Glucan	Rohu	i.p	28 days	↑ phagocytotic activity, lysozyme activity, bactericidal activity, complement activity and resistance	[221]
	Garlic	Rohi	i.p	10 days	↑superoxide production, lysozyme activity and resistance	[222]
	Glucan	Catla	i.p	30 days	↑ antibody titre and macrophage activating factor	[223]
	Glucan from *S. cerevisiae*	Carp (25-30 g)	i.p (0, 100, 500 and 1000 µg fish^{-1})	14 days	↑ total leukocyte count and neutrophil and monocytes, increase survival in a challenge experiment with *Aeromonas hydrophila*	[224]
	Glucan from *S. cerevisiae*	Carp (25-30 g)	i.p (0, 100, 500 and 1000 µg fish^{-1})	14 days	↑ total leukocyte count and neutrophil and monocytes, superoxide anion production by kidney macrophages and resistance (500 µg) against *A. hydrophila*	[225]
	β-glucan in combination with LPS	Carp (25-30 g)	i.p, bathing and oral ad-ministration	14	↑ total blood leukocyte count and neutrophil and monocytes and resistance against *A. hydrophila* when i.p and oral administration were used	[226]
	Levamisole	carp	Diet (0, 125, 250 and 500 mg kg^{-1})	70 days	↑ resistance and erythrocyte count, haemoglobin content, haematocrit, total serum protein, albumin and globulin of fish fed 250 mg kg^{-1}	[227]

	β-glucan and *S. uvarum*	Carp	i.p	15 days	↑ resistance and serum lysozyme activity	[228]
	Glucan	Asian catfish	Oral	30 days	↑ antibody titre and protection	[229]
	Glucan	Blue gourami	i.p (5, 10, 20 and 40 mg kg^{-1})	22 days	↑ chemiluminescent response of head-kidney phagocytic cell isolated from fish i.p with a dose of 20 mg kg^{-1} and resistance	[230]
	Glucan	Indian major carp	i.p	1 week	↑superoxide production and phagocytic activity resistance unclear	[231]
	S. cerevisiae	zebrafish	i.p (0.5, 2 and 5 mg ml^{-1})	6 + 4 days	↑ resistance (5 mg), myelomonocytic cells in kidney and enhanced the ability of kidney cells to kill the pathogen	[232]
	S. cerevisiae	Nile tilapia	Oral (0, 0.25, 0.5, 1, 2 and 5 g kg^{-1}) and i.p	12 weeks + 10 days	12 weeks ↑ growth, feed utilization and protein turn-over (1-5 g kg^{-1}) and resistance (5 g kg^{-1})	[233]
	S. cerevisiae	Nile tilapia	Oral (2 % autolyzed yeast; 0.3 % cell wall)	70 days	↑thrombocytes and nonspecific acute inflammatory response ↓neutrophils, macrophages an lymphocytes	[234]
	Green tea (*Camellia sinensis*)	Nile tilapia	Oral (0, 0.125, 0.25, 0.5, 1 and 2 g kg^{-1})	12 weeks + 10 days (i.p challenge)	↑ growth and feed utilization (highest at 0.5 G kg-1), haematological and biochemical parameters (0.25-2 g kg^{-1}) and resistance with increasing inclusion levels	[235]
	Debaryomyces hansenii	Leopard grouper	Oral (10^6 yeast g^{-1})	5 weeks	↑ resistance and growth → immunological parameters after 4 weeks	[236]
	Macrogard	Tench	Oral (0, 0.5, 1 and 2 g kg^{-1})	1 month	↑ resistance and phagocytic activity (1 and 2 g kg^{-1})	[237]
	Schizophyllan	Two ornamental fish species	i.p (10^8 cells ml^{-1})	24 days	↑ resistance against both pathogens	[238]
A.hydrophila and *Pseudomonas fluorescens*	Glucan	Brook trout	i.p and 30 min immersion	4 weeks	↑ resistance (up to 7 days). By day 14 no resistance was reduced	[239]
Aeromonas salmonicida	Macrogard	Trout	Oral	1 week	↑ MPO activity, phagocytic activity, superoxide production and Ig level	[240]
Aeromonas salmonicida and *Vibrio salmonicida*	Laminaran	Atlantic salmon	i.p (23.6 mg kg^{-1})	35 days	↑resistance against *V. salmonicida* → resistance against *A. salmonicida*	[241]
Aphanomyces invadans	Ergosan	Striped snakehead	Injected intramuscularly	2 weeks	↑ inhibitory effect of Serum and macrophage activating factor	[57]
Edwardsiella ictaluri	Yeast glucan	Channel catfish	i.p	2 weeks	↑ resistance and phagocytic activity	[153]
	Glucan	Channel catfish	Oral	2 weeks	↑ macrophage and neutrophil migration → resistance	[242]
Edwardsiella tarda	Glucan	Carp	i.p	12 days	↑ resistance	[152]
	Glucan	Rohu	i.p	10 days	↑ bactericidal activity ↑resistance	[243]
	Glucan	Rohu	i.p	28 days	↑ phagocytotic activity, lysozyme activity, bactericidal activity, complement activity and resistance	[221]
	Glucan - injections	Rohu	i.p	28 days	↑ phagocytotic activity, lysozyme activity, bactericidal activity, complement activity and resistance	[222]

	Yeast glucan	Tilapia and grass carp	i.p	14 days	↑ protection in both species and number of NTB-positive staining cells	[219]
Pasteurella piscicida	Glucan	Yellowtail	i.p	10 days	→ resistance	[244]
Photobacterium damselae ssp. *piscicida*	Glucan	Gilthead sea bream	Immersion (10^7 cells l^{-1})	10 days	↑ resistance (10 g kg^{-1}) → resistance (5 g kg^{-1}) and phagocytic activity	[245]
Vibrio alginolyticus	Glucan	White shrimp	Oral	120 hours	↑ resistance, phenol-oxidase and respiratory burst	[246]
	Glucan	Fresh-water prawn	Bath exposure with glucan.	4 hours + 1 week immersion with the pathogen	↑ haemocyte lysate supernatant, lysozyme and phosphatase active-ties and increased resistance → total protein content	[247]
Vibrio anguillarum	Yeast glucan	Salmon	i.p	20 days	↑ resistance	[183]
	High-M alginate	Turbot	Immersion (10^5 cells ml^{-1})	1 week	↑ resistance	[53]
	High-M alginate	Atlantic halibut	Immersion ($5x10^5$ or $1.4x10^7$ cells ml^{-1})	Different periods during initial feeding	↑ resistance at the highest dose → resistance at lowest dose	[215]
	Cationic cod milt protein	Atlantic cod (5 g)	Bath challenge ($5x10^6$ cells ml^{-1})	4 weeks	↑ resistance	[248]
Vibrio anguill-arum and *Vibrio campbellii*	Glucan from *S. cerevisiae*	Gnoto-biotic *Artemia*	Immersion ($5x10^6$ cells ml^{-1})	6 days	↑ resistance against both pathogens	[249]
Vibrio campbellii	Glucan from *S. cerevisiae*	*Artemia*	Immersion ($5x10^6$ cells ml^{-1})	5 days	Protection against *Vibrio campbellii* when supplied in advance of the challenge	[250]
	Glucan from *S. cerevisiae*	*Artemia*	Immersion ($5x10^6$ cells ml^{-1})	6 days	↑ survival and resistance against *V. campbellii*	[251]
Vibrio damsela	Yeast glucan	Turbot	Yeast glucan was used as adjuvant	42 days	↑ activities of the immune parameters when glucan was injected after the bacterin	[252]
Vibrio harvei	Glucan	Croaker	i.p	8 weeks	↑ lysozyme activity, phagocytic activity and resistance (0.09%) → resistance (0.18%)	[253]
Vibrio vulnificus	Glucan	Tiger shrimp	Immerse	43 days	↑ resistance in shrimp treated with 0.5 and 1 mg/ ml glucan but no in groups treated with 0.25 and 2 mg/ ml glucan	[194]
Vibrio sp. 90-69B3	Yeast Culture	Pacific white shrimp	Oral	i.p after 4 weeks	↑ resistance	[254]
Yersinia ruckeri	Yeast glucan	Salmon	i.p	20 days	↑ resistance	[183]
	S. cerevisiae	Rainbow trout	i.p	14 days	↑ resistance, growth, lysozyme and complement activities	[255]
Lactococcus garviae	Alginate micro-particles	Trout	i.p	3 weeks	→ resistance	[256]
Streptococcus sp.	Glucan	Yellowtail	i.p	10 days	↑ resistance and serum complement ↑ lysozyme activity	[244]
	Sodium alginate	Grouper	i.p	6 days	↑ lysozyme activity, respiratory bursts, phagocytic activity and resistance	[257]
Streptococcus iniae	Glucan	Hybrid striped bass	Oral	3 weeks	→ resistance	[258]
	Glucan	Nile tilapia	Oral	14 weeks	↑ resistance when fish were fed 100 and 200 mg glucan	[259]
	Glucan	Shark	Oral	2 weeks	↑ resistance	[260]
Virus						
IHNV	Glucan	Rainbow trout	i.p	18 days	↑ resistance → respiratory bursts and TNF-α-expression	[39]
WSSV	β-1,3-glucan	*Penaeus monodon*	Oral (0 and 2g kg^{-1})	15 days	↑ resistance	[261]
	β-1,3-glucan	*Penaeus monodon*	i.p	20 days	↑ resistance and the Immune system	[262]
	Marine yeast (*Candida sake*)	White prawn	Oral (10 %)	40 + 10 days	↑ immunity index and resistance	[263]
	Marine yeast glucan	*Penaeus monodon*	Oral	21 days	↑ resistance, but higher molecular weight and lower degree of branching acts better	[264]
Parasite						
Loma morhua	Glucan	Atlantic cod	Oral	n.i	↑ lymphocyte density	[265]
Loma salmonae	β-1,3/1,6 glucan	Rainbow trout	i.p (0, 50, 100, 500 and 1000 μg)	10 weeks	During week 8-9 post exposure a significant reduction in no. of xenoma-positive fish was noticed.	[266]

Symbols represent an increase (↑) in the specified response; no change (→); decrease (↓); n.i – no information given.

i.p – intraperitoneal injection

n.i – no information given

IHNV – infectious hematopoietic necrosis virus

IHNV – infectious hematopoietic necrosis virus

WSSV - White spot syndrome virus

Hoseinifar and co-authors evaluated the effect of brewer's yeast (S.cerevisiae var. ellipsoideus) on gut microbiota of juvenile beluga (Huso huso) and displayed that juveniles fed 2 % S. cerevisiae var. ellipsoideus increased the levels of autochthonous LAB [97]. The authors suggested that this interesting finding might occur as a result of the provision of enzymes, RNA and free nucleotides, B-complex vitamins and/or amino acids by the dietary yeast. Whether supplementation of S. cerevisiae var. ellipsoideus improved disease resistance against pathogens was not investigated and merits further evaluation.

Nucleotides

Dietary nucleotides have attracted attention as key ingredients missing from nutritional formulae for many years. They are the building blocks of tissue RNA and DNA and of ATP, and their presence in breast milk has stimulated research in babies which has indicated that supplementation of infant formula milk leads to improved growth and reduced susceptibility to infection [108-110]. Nucleotide fortification of breast milk substitutes has been recommended to the U.S. Food and Drug Administration for approval [111]. There is increasing evidence that nucleotides administered intravenously or in the diet are capable of modifying immune responsiveness and recovery of organs that have undergone a metabolic or inflammatory insult.

It is generally accepted that nucleotides have essential physiological and biochemical functions including encoding and deciphering genetic information, mediating energy metabolism and cell signalling as well as serving as components of coenzymes, allosteric effectors and cellular agonists [112,113]. However, controversy has existed for many years over the roles of nucleotides administered exogenously. As neither overriding biochemical malfunctions nor classical signs of deficiency are developed in endothermic animal models, nucleotides have traditionally been considered to be non-essential nutrients. However, this opinion has been challenged by several research publications during the last decade which suggest that dietary nucleotide deficiency may impair liver, heart, intestine and immune functions [114]. The modulatory effects of dietary nucleotides on lymphocyte maturation, activation and proliferation, macrophage phagocytosis, immunoglobulin responses, gut microbiota as well as genetic expression of certain cytokines have been reported in endothermic animals [115,116]. Nucleotide supplementation has been one important aspect of research on clinical nutrition and functional food development for humans [109,110].

Although initial efforts in evaluation of dietary supplementation of nucleotides for fishes could be traced to the early 1970s, research at that time mainly focused on the possible chemo-attractive effects of these compounds [117-119]. However, the pioneer investigations by Burrels and co-authors resulted in increased attention on nucleotide supplementation for fishes as their studies indicated that dietary supplementation of nucleotides enhanced resistance of salmonids to viral, bacterial and parasitic infections as well as improved efficacy of vaccination and osmoregulation capacity [5,6]. To date, research related to nucleotide nutrition in fishes has to some extent showed consistent and encouraging beneficial results in fish health management (Table 4), although most of the suggested explanations put forward by the authors remain hypothetical. Systematic research on fishes is therefore needed.

Because increasing concerns of antibiotic use have resulted in a ban on subtherapeutic antibiotic usage in Europe and the potential for a ban in the US and other countries [120], research on immunonutrition for aquatic animals is becoming increasingly important [121]. Research on nucleotide nutrition in fish and shrimp is needed to provide insights concerning interactions between nutrition and physiological responses as well as provide practical solutions to reduce basic risks from infectious diseases for the aquaculture industry. Devresse hypothesized that nucleotides are a key nutrient for the shrimp immune system and supplementation of nucleotides or other nucleic acid-rich ingredients such as yeast or yeast extract may enhance disease resistance and growth of shrimp [122]. Although yeast products have been used in shrimp diet formulations, the role of yeast nucleotides remains largely unanswered. In their comprehensive review, Li and Gatlin summarized and evaluated knowledge of nucleotide nutrition in fishes as compared with that of terrestrial animals [123].

The roles of nucleotides and metabolites in fish diets have been sparingly studied for nearly 20 years. Beside possible involvement in diet palatability, fish feeding behaviour and biosynthesis of nonessential amino acids, exogenous nucleotides have shown promise most recently as dietary supplements to enhance immunity and disease resistance of fish produced in aquaculture. Research on dietary nucleotides in fishes has shown they may improve growth in early stages of development, enhance larval quality via brood stock fortification, alter intestinal structure, increase stress tolerance as well as modulate innate and adaptive immune responses. Fishes fed nucleotide supplemented diets generally have shown enhanced resistance to viral, bacterial and parasitic infection. Despite occasional inconsistency in physiological responses, dietary supplementation of nucleotides has shown rather consistent beneficial influences on various fish species. Although nucleotide nutrition research in fishes is in its infancy and many fundamental questions remain unanswered, observations thus far support the contention that nucleotides are conditionally or semi-essential nutrients for fishes, and further exploration of dietary supplementation of nucleotides for application in fish culture is warranted. Hypothesized reasons associated with these beneficial effects include dietary provision of physiologically required levels of nucleotides due to limited synthetic capacity of certain tissues (e.g. lymphoid), inadequate energetic expenditure for de novo synthesis, immunoendocrine interactions and modulation of gene expression patterns.

Table 4: Research on dietary supplementation of nucleotides on fishes

Nucleotide form	Dose and/or feeding regime	Length of administration	Species	Initial size	Effect*	Authors
Ascogen S	2 and 5 g kg^{-1} diet	16 weeks	Hybrid tilapia	21 days old	↑growth and survival	[267]
Ascogen P	5 g kg^{-1} diet, fixed ration approaching satiation daily	7 weeks	Hybrid striped bass	7.1; 9.1 g	↑ neutrophil oxidative radical production and survival after challenge with *S. iniae*	[268-269]
Ascogen	5 g kg^{-1} diet	120 days	Hybrid tilapia	30 days old	↑ antibody titer after vaccination and mitogenic response of lymphocyte	[270]
	0.62, 2.5 and 5 g kg^{-1} diet at 1% bw day^{-1}	37 days	Rainbow trout	163.4–169.7 g fish^{-1}	↑ growth	[134]
Nucleic acid	5.8 and 11.5 %	10 weeks	Sea bream	12.7 g	↑ growth, ornithine carbamyltransferase activity → body composition, hepatic glutamate dehydrogenase and uricase activities	[203]
Optimûn	2 g kg^{-1} diet, containing 0.03% NT, 2% bw day^{-1}	3 weeks	Rainbow trout	217 ± 62 g	↑ survival after challenge with *V. anguillarum*	[5]
	2 g kg^{-1} diet, containing 0.03% NT, 1% bw day^{-1}	2 weeks	Rainbow trout	53–55 g	↑ survival after challenge with infectious salmon anaemia virus	[5]
	2 g kg^{-1} diet, containing 0.03% NT, 2% bw day^{-1}	3 weeks	Coho salmon	100 g	↑ survival after challenge with *Piscirickettsia salmonis*	[5]
	2 g kg^{-1} diet, containing 0.03% NT, 2% bw day^{-1}	3 weeks	Atlantic salmon	60 g	↓ sea lice infection	[5]
	2 g kg^{-1} diet, containing 0.03% NT at 1.5% bw day^{-1}	3 weeks before vaccination and 5 weeks post-vaccination	Atlantic salmon	34.7 ± 9.6 g	↑ antibody titer ↓ mortality	[6]
	2 g kg^{-1} diet, containing 0.03% NT at 1.5% bw day^{-1}	8 weeks	Atlantic salmon	43 ± 3.0 g	↑ growth ↓ plasma chloride	[6]
	2 g kg^{-1} diet, containing 0.03% NT	10 weeks	Atlantic salmon	205 g	↑ intestinal fold	[6]
	NA	120 days	"all-female" rainbow trout	80–100 g	↑ B lymphocytes and resistance to IPN virus ↓ plasma cortisol	[271]
	2 g kg^{-1} diet, containing 0.03% NT to hand satiation daily	15 weeks	Turbot	120.9 ± 5.1 g	Altered immunogene expression in various tissues	[7]
	2 g kg^{-1} diet	6 weeks	Red drum	1 g	→ Growth, whole body composition and *in situ* challenge with *Amyloodinium ocellatum*	[126]
Ribonuclease-digested yeast RNA	15 mg fish^{-1}, by intubation	3 days	Common carp	100 g	↑ phagocytosis, respiratory burst, complement and lysozyme ↓ *A. hydrophila* infection	[273]
Aquagen	Oral	2 weeks	Shark	1.4 ± 0.2 g	↑ resistance after challenge with *S. iniae*	[260]
Nucleotide mixture	4 g kg^{-1} diet	5 weeks	Pacific white shrimp	ca. 0.83 g	↑ growth	[274]
	4 g kg^{-1} diet	4 weeks	Red drum	10.2 ± 0.2 g	↑ growth, neutrophil oxidative radical production and survival after challenge with *V. harveyi*	[275]

* - symbols represent an increase (↑), decrease (↓) or no change (→) in the specified response.

The numbers of scientific publications dealing with dietary nucleotides in fish has been relatively limited since the comprehensive review of Li and Gatlin [123] was published. The section below summarises the dietary effects of nucleotides on a number of fish species such as red drum (Sciaenops ocellatus), red-tail black shark (Epalzeorhynchos bicolor), Asian carp (Catla catla), barramundi (Lates calcarifer), cobia (Rachycentron canadum), grouper (Epinephelus malabaricus) and salmonids such as rainbow trout and Atlantic salmon published post 2006.

The use of dietary nucleotides in red drum Sciaenops ocellatus has also been evaluated on growth and feed utilisation [124]. Juvenile fish (ca. 122g) were fed purified a commercial nucleotide product 0.2% (Optimûn, Chemoforma, August, Switzerland) or nucleotides mixtures, which contained equal amounts of CMP (cytidine monophosphate), UMP (uridine monophosphate), AMP (adenosine monophosphate), IMP (inosine monophosphate) and GMP (guanosine monophosphate), at 0.03%, 0.1% and 0.3% of the diet for a period of four weeks [125]. Fish fed dietary nucleotides had showed a clear trend towards increased growth, measured as SGR at the end of the trial. Growth improvements ranged from 5-6% up to 8% higher than the control diet for the fish fed the nucleotide mixtures and commercial product respectively. A similar trend towards improved feed efficiencies was observed in all the groups fed diets containing added nucleotides. No diet effect on body composition was observed in contrast to previous observations. A clear trend of diet effect on neutrophil oxidative radical production was also observed that could suggest that excessive dose could inhibit immune responses. Fish were challenged against V.harveyi at the end of the four-week feeding period but no differences were observed in terms of survival when compared to the control diet. The authors reported on unexpected high mortalities within the 3 days postchallenge in contrast to previous experiments so it was concluded that further studies are needed in order to evaluate potential effect of dietary nucleotides on red drum resistance against V. harveyi [126].

Red drum was also the fish species selected by Cheng and co-authors to assess the effects of dietary nucleotides on intestinal morphology and immune responses [127]. The authors used around 7g fish to compare a commercial nucleotide product,

Ascogen P®(Canadian Byosystems Inc., Calgary, Alberta, Canada), at 0.5% and 1% inclusion levels against a control diet not enriched with nucleotides. Evaluation of immune response after six weeks of feeding showed increased extracellular superoxide anion production at the highest dose. In general intestinal morphology, measured as fold height, enterocytes height and microvillus height, in the pyloric caeca, proximal, mid- and distal intestine, was positively influenced by dietary nucleotides. Thus, fold height in the proximal intestine was increased by nucleotides as previously reported for Atlantic salmon [5,6]. Enterocyte height was increased in the pyloric caeca, proximal and distal intestine, not in the mid intestine, by nucleotides. The microvillus height was also increased in the pyloric caeca, proximal, mid and distal intestine.

Russo et al. examined the effects of commercial beta-glucans (Macroguard®, Biotec Pharmacon ASA, Tromsø, Norway) and nucleotide product (AquagenTM, NOVARTIS-Aqua Health Ltd., Charlottetown, Canada) on disease resistance of red-tail black shark (Epalzeorhynchos bicolor) against S.iniae infection [128]. Two trials were conducted using 1.4g fish feeding control, beta-glucan (0.1% inclusion) or nucleotide diet (0.2% product inclusion) for 24 days. In the first trial both vaccinated and non-vaccinated fish were used while in the second experiment nucleotide- or beta-glucan-enriched diets were fed only to vaccinated fish and results compared against a nonvaccinated fish fed control diet. Results of the first trial showed that both commercial products were able to reduce mortalities compared to the fish fed control diet in both vaccinated and non-vaccinated fish. In the second experiment vaccinated fish fed beta-glucans and nucleotide suffered lower mortalities than the vaccinated fish fed control diet and despite slightly lower mortality in the beta-glucan group compared to nucleotide group no statistical difference could be claimed. Growth performance was influenced significantly by dietary treatment in any of the trials conducted.

Jha et al. tested yeast nucleotides, in the form of RNA (Sisco Research Laboratories, India), at 0.4% and 0.8% inclusion levels on Catla catla fingerlings (ca. 8g start weight) to assess potential changes in haemato-immunological responses followed by a challenge trial against A. hydrophila after 60 days feeding [129]. Feeding dietary nucleotides increased leucocyte counts, total protein, globulins, and albumin: globulin ratio, lysozyme activity and respiratory burst activity compared to controls. When comparing differences between nucleotide doses only the respiratory burst activity was increased by augmenting the dietary nucleotides from 0.4% to 0.8% dose. Conversely, the highest survival was observed in the fish fed the highest dose of nucleotides. Fish fed 0.4% nucleotides also showed significantly higher survival than control fish.

At 30°C barramundi fed diets supplemented with nucleotides (Optimûn, Chemoforma, August, Switzerland) grew significantly better and had lower FCR than those fed the reference diet with no additional nucleotides added [130]. In the same study it is noted that the supplementation of dietary nucleotide at 30°C significantly improved the retention efficiencies of protein, energy, lysine and histidine compared to the reference diet. However at 37°C , under heat-stress conditions, despite dietary nucleotides reduced FCR an improvement in growth was not observed.

Salze et al. supplemented cobia diets with a nucleotide-rich yeast extract (Nupro, Alltech, Nicholasville, KY, USA) part of a zero-fishmeal feed [131]. The 104g fish fed the zero-fish meal diet, which was based on soy protein concentrate, worm meal and mannan-oligosaccharides, had similar growth as a control diet with 25.3% fish meal. However the experimental design tested was insufficient to separate the effects of the different raw materials and to evaluate whether cobia benefited from nucleotide supplementation.

Lin et al. reported the benefits of supplementing dietary nucleotides on growth and immune responses of grouper (Epinephelus malabaricus) juveniles [132]. The first experiment assessed different inclusion levels of a nucleotide mixture containing equal amounts of IMP, AMP, GMP, UMP and CMP. Weight gain and feed efficiency were the highest in the group of grouper fed 0.15% of the nucleotide mixture; with head kidney leucocyte superoxide anion production ratio and plasma total immunoglobulin concentration also being higher than the fish fed the control diet. In the second experiment a diet containing 0.15% nucleotide mixture was compared against a control diet and also diets containing 0.15% of IMP, AMP, GMP, UMP or CMP. From the second experiment it was concluded that juvenile grouper diets containing 0.15% AMP seemed to have better effects than other diets supplemented with different single-nucleotides. Overall conclusion showed that both growth and immune responses of juvenile grouper were enhanced in grouper diet with 0.15% nucleotide mixture.

Due to its economical impact as a high-value species, research on salmonids species and nucleotides has received lots of attention in the last years. Tahmasebi-Kohyani and co-authors conducted a dose response trial on rainbow trout fingerlings (ca. 23g start weight) in order to evaluate their effects on growth, humoral immune responses and resistance against S.iniae [133]. A basal diet was supplemented with Optimûn (Chemoforma, August, Switzerland) at 0.05%, 0.1%, 0.15% and 0.2% inclusion levels and fish were fed for eight weeks. Results showed that 0.05% dose was too low and thereafter increased growth, FCR, alternative complement activity, lysozyme activity and IgM as nucleotide dose was augmented, reaching highest values for each of them at the highest nucleotide dose. Similar results were observed at the end of the challenge trial and increased survival was observed with increasing dietary level of nucleotides from 0.05% to 0.2%. This is in agreement with previous work on rainbow trout by Adamek and co-authors that used a wider range of concentrations of the same commercial product, from 0% up to 0.5% and reported approximately 0.2% as the optimal dose (Figure 1)[134].

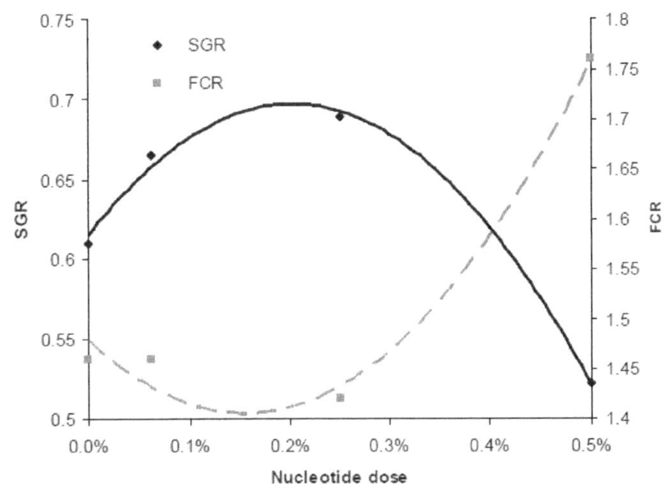

Figure 1: Growth and FCR data of rainbow trout fed diets with increasing doses of nucleotides. Adaption of data reported by Adamek et al. [135].

The efficacy of dietary nucleotides (Optimûn, Chemoforma, August, Switzerland) increasing survival (+39%) against Piscirickettsia salmonis was first reported by Burrells and co-authors on 100g coho salmon (Oncorhynchus kisutch) after feeding for three weeks before challenge [5]. The efficacy of dietary nucleotides against P. salmonis has also been confirmed in other salmonid species such as Atlantic salmon. González Vecino et al. reported increased protection against P. salmonis when Atlantic salmon were fed diets containing a combination of dietary nucleotides with a bacterial cell-wall extract (Sanictum®, Chemoforma, Augst, Switzerland) [135]. Thus, a synergistic effect of the combination of nucleotides and Sanictum® was observed since survival was significantly better than feeding nucleotides or bacterial cell-wall extract separately.

Dietary Nucleotides in Broodstock Diets

Fish reproduction, and egg and larval quality are affected by broodstock nutrition. Nucleotides, the building blocks of nucleic acids, are now considered as semiessential nutrients during periods of food deficiency, stress, rapid growth and immunological stress. González Vecino reported that since oogenesis is a process of intensive cell division with high nucleic acid formation and a concomitant high requirement for nucleotides, supplementation of broodstock diets with nucleotides was beneficial for fish [136]. The results showed that Atlantic halibut (Hippoglossus hippoglossus L.) and haddock (Melanogrammus aeglefinus L.), two cold water species with different life histories, had improved fecundity, egg quality, larval quality and survival when broodstock diets were supplemented with nucleotides. In addition it was observed that haddock larvae from broodfish fed nucleotides had a significantly better developed gut and first feeding success that those from broodstock fed diets not supplemented with nucleotides. Similar beneficial effects have been observed in salmon broodstock diets supplemented with nucleotides (González Vecino, unpublished data).

Dietary Nucleotides against sea Lice

Immune modulation of salmon can significantly affect the infection intensity of sea lice both Caligus spp. and Lepeophtheirus salmonis. Salmon with reduced or compromised immunity have higher infection with sea lice [137-139]. In addition it is known that during the attachment period sea lice release immunosupressive compounds, including prostaglandin E2 (PGE2) and leukotriene B4 (LTB4) (cytokines) creating a local immunosuppression at the infestation site allowing the lice to attach unhindered [139-144]. Salmon with enhanced immunity have lower infection with sea lice, thus a range of products such as alginates, ß-glucans and nucleotides have shown to reduce sea lice infections. Burrells et al. [5] reported a 38% reduction in sea lice (L. salmonis) infestation on 60g Atlantic salmon fed diets containing nucleotides for three weeks. This was in agreement with reports of 38% reduction on attached stages of L. salmonis on 700g Atlantic salmon after feeding nucleotides for 3 weeks [145]. In the same study Burrells et al. demonstrated that the combination of dietary nucleotides with anti-sea lice bath treatments increased the efficacy of the treatment by 53% [145]. Later work on Caligus rogercresseyi also reported increased efficacy of deltamethrin (bath treatment) and emamectin benzoate (oral treatment) when applied in combination with diets containing nucleotides Optimûn and bacterial cell-wall extract Sanictum® (Chemoforma, August, Switzerland) [146].

Sea lice are well adapted to develop resistance against medicines. Even within a fully sensitive population a number of sea lice will survive treatment. Wadsworth et al. reported that the use of pyrethroids at the recommended dose showed a survival of 20% of attached stages in a naïve population [147]. Although these lice survived the sub-lethal toxicity of the compound and were able to complete their life cycle, the pyrethroid still had a significant, negative effect upon the rate of development, dramatically increasing the time required to reach pre adult I stage. Increasing the time the lice takes to reach pre adults will expose them to the host immunity during this extended period. In their weakened state, it is unlikely the lice are deploying the full range of immune suppressive compounds as effectively as untreated lice. Therefore, it is recommended that immune modulating compounds such as dietary nucleotides are deployed against lice during treatments with anti-sea lice medicines. It is essential that attached lice surviving treatments are targeted to prevent the development of resistant populations.

Furthermore a sea lice infestation represents a substantial stress to the infected animal. An anti-sea lice bath treatment is in its nature a severe multi stress event including starvation, handling, crowding, hypoxia and exposure to a toxic pharmaceutical all within a short period of time. Even a presumably soft action such as feeding diets containing emamectin benzoate has been shown to trigger stress. Thus, Olsvik and co-authors reported that dietary treatment with Slice® (Schering-Plough Animal Health, Boxmeer, The Netherlands) in salmon triggers the expression of stress coding genes in the fish due to sheer metabolisation and degrading of emamectin benzoate in the liver [148]. Since dietary nucleotides have been shown to reduce levels of cortisol its use could also be suggested with emamectin benzoate treatment.

The use of dietary nucleotides as an immune modulating tool against sea lice should be utilised in an integrated pest management programme with rotation of compounds, resistance monitoring, coordinated treatments, cleaner fish as well as effective monitoring and control of other diseases.

CONCLUSIONS AND RECOMMENDATIONS

Immunostimulants

It is generally accepted that immunostimulants used in fish experiments induce beneficial effects such as disease protection due to increased cellular and humoral responses. However, precautions have to be taken regarding issues such as tolerance, non-wanted side effects such as immunosuppression using too high doses of immunostimulants or non-desirable effects caused by a prolonged use of such compounds. In future, it is hoped that following the development of genomic and proteomic tools for several fish species, many issues with special attention to immune response polarisation after receptor binding of immunostimulants will be unveiled.

Possible effect of immunostimulants on potential probiotics ability to adhere to intestinal mucus should be given high priority in future studies. New and vital information on this topic is needed as the gastrointestinal tract is a potential port of entry for pathogenic bacteria.

Both prebiotics Bio-MOS® and β-1.3 glucan are two commercially available products, derived from the cell wall of S.cerevisiae, in which the use of Bio-MOS® in aquaculture has increased during the last years [149]. However, as less information per se in available on the effect of Bio-MOS® and β-1.3 glucan in aquatic animals [150], this topic should be given high priority in future studies.

Nucleotides

Based on available knowledge and experience within nucleotides, demonstrating that nucleotide supplementation improves the composition of the gut microbiota in formula-fed infants should encourage bacteriologists working with fish to investigate interaction between nucleotides and gut microbiota. Furthermore, we recommend that the following topics should be given high priority in future; (1) in broodstock during oogenesis to improve fecundity, egg and larval quality, (2) during sea water transfer: feeding pre- and post- seawater transfer provides fish with a better osmoregulation and adaptation to marine environment, (3) combination with vaccines: feeding nucleotides during (pre- and post-) the vaccination period modulates the immune response and reduces the negative side effects caused by vaccines (e.g. growth suppression), (4) improved immune response and protection against pathogens such as; V. anguillarum, A. salmonicida, IPN, ISA, rickettsia-like bacteria, Flavobacterium, Moritella viscosus and (5) protection against sea lice and improved (higher) efficacy of pyrethroids when used combined with nucleotide feed.

ACKNOWLEDGEMENTS

This review is an extended revision of the section immunostimulants and nucleotides in the Ad-hoc report "A risk-assessment on the use of plant ingredients in diets for carnivorous fish" financed by the Norwegian Scientific Committee of Food Safety. The report is available at: www.vkm.no

REFERENCES

1. Sakai M (1999) Current research status of fish immunostimulants. Aquaculture 172: 63-92.
2. Seljelid R (1990) Immunomodulators – medicine for the 90-ies? In: Wadström T, et al. (Eds.), Workshop on Pathogenesis of Wound and Biomaterial-Associated Infections. Aug. 30–Sep. 01, 1989, Springer Verlag, New York, pp. 107–113.
3. Bricknell I, Dalmo RA (2005) The use of immunostimulants in fish larval aquaculture. Fish Shellfish Immunol 19: 457-472.
4. Volman JJ, Ramakers JD, Plat J (2008) Dietary modulation of immune function by β-glucans. Physiol Behavior 94: 276-284.
5. Burrells C, Williams PD, Forno PF (2001) Dietary nucleotides: a novel supplement in fish feeds. 1. Effects on resistance to disease in salmonids. Aquaculture 199: 159–169.
6. Burrells C, Williams PD, Southgate PJ, Wadsworth SL (2001) Dietary nucleotides: a novel supplement in fish feeds. 2. Effects on vaccination, salt water transfer, growth rates and physiology of Atlantic salmon (Salmo salar L.). Aquaculture 199: 171–184.
7.
8. Low C, Wadsworth S, Burrells C, Secombes CJ (2003) Expression of immune genes in turbot (Scophthalmus maximus) fed a nucleotide-supplemented diet. Aquaculture 221: 23-40.
9. Nayak SK (2010) Probiotics and immunity: A fish perspective. Fish Shellfish Immunol 29: 2-14.
10. Panigrahi A, Kiron V, Puangkaew J, Kobayashi T, Satoh S, et al. (2005) The viability of probiotic bacteria as a factor influencing the immune response in rainbow trout Oncorhynchus mykiss. Aquaculture 243: 241-254.
11. Gomez GD, Balcazar JL (2007) A review on the interactions between gut microbiota and innate immunity of fish. FEMS Immunol Med Microbiol 52: 145- 154.
12. Franca FM, de Carla Dias D, Teixeira PC, Marcantonio AS, De Stefani MV, et al. (2008) Effect of the probiotic Bacillus subtilis on growing, survival and fisiology in the bullfrog (Rana catesbeiana). Boletim Do Instituto de Pesca 34: 403-412.
13. Shida K, Nanno M (2008) Probiotics and immunology. Separating the wheat from the chaff. Trends Immunol 29: 565-573.
14. Borchers, AT, Selmi C, Meyers FJ, Keen CL, Gershwin ME (2009) Borchers, AT, Selmi C, Meyers FJ, Keen CL, Gershwin ME (2009) Probiotics and immunity. J Gastroenterol 44: 26-46 J Gastroenterol 44: 26-46.
15. Panigrahi A, Azad IS, Das BK, Dandpat J, Das G, et al. (2009) Probiotic induced immunomodulation. Investigation into the cellular and molecular mechanism involved. Res J Biotechnol 4: 7-13.
16. Zanello G, Meurens F, Mustapha B, Salmon H (2009) Saccharomyces boulardii effects on gastrointestinal diseases. Curr Iss Mol Biol 11: 47-58.
17. Dimitroglou A, Merrifield DL, Carnevali O, Picchietti S, Avella M, et al. (2010) Microbial manipulations to improve fish health and production – a Mediterranean perspective. Fish Shellfish Immunol 30: 1-16.

18. Ganguly S, Paul I, Mukhopadhayay SK (2010) Application and effectiveness of immunostimulants, probiotics, and prebiotics in aquaculture: A review. The Israeli J Aquacult – Bamidgeh 62: 130-138.
19. Kaila M, Isolauri E, Soppi E, Virtanen E, Laine S, et al. (1992) Enhanchement of the circulating antibody secreting cell response in human diarrhea by a human Lactobacillus strain. Ped Res 32: 141-144.
20. Panigrahi A, Kiron V, Kobayashi T, Puangkaew J, Satoh S, et al. (2004) Immune responses in rainbow trout Oncorhynchus mykiss induced by a potential probiotic bacteria Lactobacillus rhamnosus JCM 1136. Vet Immunol Immunopathol 102: 379–388.
21. Panigrahi A, Kiron V, Satoh S, Hirono I, Kobayashi T, et al. (2007) Immune modulation and expression of cytokine genes in rainbow trout Oncorhynchus mykiss upon probiotic feeding. Dev Comp Immunol 31: 372-382.
22. Salinas I, Abelli L, Bertoni F, Picchietti S, Roque A, et al. (2008) Monospecies and multispecies probiotic formulations produce different systemic and local immunostimulatory effects in gilthead seabream (Sparus aurata L.). Fish Shellfish Immunol 25: 114-123.
23. Oelschlaeger TA (2010) Mechanisms of probiotic actions – A review. Int J Med Microbiol 300: 57-62.
24. Suchner U, Kuhn KS, Fürst P (2000) The scientific basis of immunonutrition. Proc Nutr Soc 59: 553-563.
25. Landolt ML (1989) The relationship between diet and the immune responses of fish. Aquaculture 79: 193-206.
26. Anderson DP (1992) Immunostimulants, adjuvants, and vaccine carriers in fish: application to aquaculture. Ann Rev Fish Dis 281-307.
27. Blazer VS (1992) Nutrition and disease resistance in fish. Ann Rev Fish Dis 2: 309-323.
28. Waagbø R (1994) The impact of nutritional factors on the immune system in Atlantic salmon Salmo salar L.: A review. Aquacult Fish Manag 25: 175-197.
29. Raa J (1996) The use of immunostimulatory substances in fish and shellfish farming. Rev Fish Sci 4: 229-288.
30. Galeotti M (1998) Some aspects of the application of immunostimulants and critical review of methods for their evaluation. J Appl Ichthyol 14: 189-199.
31. Gannam AL, Schrock RM (1999) Immunostimulants in fish diets. J Appl Aquacult 9: 53-89.
32. Sohn KS, Kim MK, Kim JD, Han IK (2000) The role of immunostimulants in monogastric animal and fish – review. Asian-Australasian J Anim Sci 13: 1178- 1187.
33. Galindo-Villegas J, Hosokawa H (2004) Immunostimulants: Towards temporary prevention of diseases in marine fish. In: Cruz Suarez LE, Ricque MD, Nieto Lopez MG, Villareal D, Scholz Y, Gonzalez M (Eds.), Advances en Nutricion. Acuicola VII Memorias del VII Simposium Internationale de Nutricion Acuícola, 16-19 279-319.
34. Trushenski JT, Kasper CS, Kohler CC (2006) Challenges and opportunities in finfish nutrition. North Am J Aquacult 68: 122-140.
35. Panigrahi A, Azad IS (2007) Microbial intervention for better fish health in aquaculture: the Indian scenario. Fish Physiol Biochem 33: 429-440.
36. Sahoo PK (2007) Role of immunostimulants in disease resistance of fish. CAB Reviews: Perspectives in Agriculture. Vet SciNutrNat Res 2: no. 045.
37. Galli C, Calder PC (2009) Effects of fat and fatty acid intake on inflammatory and immune responses. A critical review. Ann Nutr Metab 55: 123-139.
38. Verlhac TV (2010) Nutrition and immunity. Aquaculture Research 41: 356-372.
39. Zekovic DB, Kwiatowski S (2005) Natural and modified (1->3)-beta-D-glucans in health promotion and disease alleviation. Crit Rev Biotechnol 25: 205-230.
40. Sealey WM, Barrows FT, Hang A, Johansen KA, Overturf K, et al. (2008) Evaluation of the ability of barley genotypes containing different amounts of β-glucan to alter growth and disease resistance of rainbow trout Oncorhynchus mykiss. Anim Feed Sci Technol 141: 115-128.
41. Soltanian S, Stuyven E, Cox E, Sorgeloos P, Bossier P (2009) Beta-glucans as immunostimulants in vertebrates and invertebrates.Crit Rev Microbiol 35: 109-138.
42. Supamattaya K, Pongmaneerat J, Klowklieng T (2000) The effect of β-glucan (MacroGard®) on growth performance, immune response and disease resistance in black tiger shrimp, Penaeus monodon Fabricius. Songklanakarin J Sci Technol 22: 677-688.
43. Ferreira IMPLVO, Pinho O, Vieira E, Tavarela JG (2010) Brewer`s Saccharomyces yeast biomass: characteristics and potential applications. Trend Food Sci Technol 21: 77-84.
44. Schrøder MB, Villena AJ, Jørgensen TØ (1998) Ontogeny of lymphoid organs and immunoglobulin producing cells in Atlantic cod (Gadus morhua L.). Dev Comp Immunol 22: 507–517.
45. Grøntvedt RN, Espelid S (2003) Immunoglobulin producing cells in the spotted wolffish (Anarhichas minor Olafsen): localization in adults and during juvenile development. Dev Comp Immunol 27: 569–578.
46. Gildberg A, Johansen A, Bøgwald J (1995) Growth and survival of Atlantic salmon (Salmo salar) fry given diets supplemented with fish protein hydrolysate and lactic acid bacteria during a challenge trial with Aeromonas salmonicida. Aquaculture 138: 23–34.
47. Gildberg A, Mikkelsen H (1998) Effects of supplementing the feed to Atlantic cod (Gadus morhua) fry with lactic acid bacteria and immuno-stimulating peptides during a challenge trial with Vibrio anguillarum. Aquaculture 167: 103- 113.
48. Nikoskelainen S, Ouwehand A, Salminen S, Bylund G (2001) Protection of rainbow trout (Oncorhynchus mykiss) from furunculosis by Lactobacillus rhamnosus. Aquaculture 198: 229–236.
49. Remminghorst U, Rehm B (2006) Bacterial alginates: from biosynthesis to applications. Biotechnol Lett 28: 1701-1712.
50. Haug A, Myklestad S, Larsen B, Smidsrød O (1967)

Correlation between chemical structure and physical properties of alginates.Acta Chemica Scand 21: 768–778.

51. Smidsrød O, Skjåk-Bræk G (1990) Alginate as immobilization material for cells. Trend Biotechnol 8: 71-78.

52. Remminghorst U, Rehm B (2006) Remminghorst U, Rehm B (2006) Bacterial alginates: from biosynthesis to applications. Biotechnol Lett 28: 1701-1712 Biotechnol Lett 28: 1701-1712.

53. Skjåk-Bræk G, Flo T, Halaas Ø, Espevik T (2000) Immune stimulating properties of de-equatorially β(1–4) linked poly-uronides. Bioactive Carbohydrate Polymers. Kluwer Academic Publishers, 85–93.

54. Skjermo J, Defoort T, Dehasque M, Espevik T, Olsen Y, et al. (1995) Immunostimulation of juvenile turbot (Scophthalmus maximus L.) using an alginate with high mannuronic acid content administrated via the live food organism Artemia. Fish Shellfish Immunol 5: 531-534.

55. Vadstein O (1997) The use of immunostimulation of marine larviculture: possibilities and challenges. Aquaculture 155: 401-417.

56. Skjermo J, Vadstein, O (1999) Techniques for microbial control in the intensive rearing of marine larvae. Aquaculture 177: 333–343.

57. Vollstad D, Bøgwald J, Gaserød O, Dalmo RA (2006) Influence of high-M alginate on the growth and survival of Atlantic cod (Gadus morhua L.) and spotted wolffish (Anarhichas minor Olafsen) fry. Fish Shellfish Immunol 20: 548–561.

58. Miles DJC, Polchana J, Lilley JH, Kanchanakhan S, Thompson KD, et al. (2001) Immunostimulation of striped snakehead Channa striata against epizootic ulcerative syndrome. Aquaculture 195: 1-15.

59. Peddie S, Zou J, Secombes CJ (2002) Immunostimulation in the rainbow trout (Oncorhynchus mykiss) following intraperitoneal administration of Ergosan. Vet Immunol Immunopathol 86: 101–113.

60. Duke JA (1987) CRC Handbook of Medicinal Herbs (5th ed.). CRC Press, Boca Raton, FL.

61. Harikrishnan R, Balasundaram C, Heo MS (2011) Impact of plant products on innate and adaptive immune system of cultured finfish and shellfish. Aquaculture 317: 1-15.

62. Dügenci SK, Arda N, Candan A (2003) Some medicinal plants as immunostimulants for fish. J Ethnopharmacol 88: 99-106.

63. Christybapita D, Divyagnaneswari M, Michael RD (2007) Oral administration of Eclipta alba leaf aqueous extract enhances the non-specific immune responses and disease resistance of Oreochromis mossambicus. Fish Shellfish Immunol 23: 840-852.

64. Dam TK, Brewer CF (2010) Lectins as pattern recognition molecules: The effects epitope density in innate immunity. Glycobiol 20: 270-279.

65. Boshra H, Li J, Sunyer JO (2006) Recent advances on the complement system of teleost fish. Fish Shellfish Immunol 20: 239-262.

66. van Kooyk Y, Rabinovich GA (2008) Protein-glycan interactions in the control of innate and adaptive immune responses. Nature Immunol 9: 593-601.

67. Galina J, Yin G, Ardo, Jenny Z (2009) The use of immunostimulating herbs in fish. An overview of research. Fish Physiol Biochem 35: 669-676.

68. Soosean C, Marimuthu K, Sudhakaran S, Xavier R (2010) Effect of mangosteen (Garcinia mangostana L.) extracts as a feed additive on growth and haematological parameters of African catfish (Clarias gariepinus). Eur Rev Med Pharmacol Sci 14: 605-611.

69. Harikrishnan R, Kim JS, Kim MC, Balasundaram C, Heo MS (2011) Lactuca indica extract as feed additive enhances immunological parameters and disease resistance in Epinephelus bruneus to Streptococcus iniae. Aquaculture 318: 43-47.

70. Harikrishnan R, Balasundaram C, Jawahar S, Heo MS (2011) Solanum nigrum enhancement of the immune response and disease resistance of tiger shrimp, Penaeus monodon against Vibrio harveyi. Aquaculture 318: 67-73.

71. Rajendiran A, Natarajan E, Subramanian P (2008) Control of Aeromonas hydrophila infection in spotted snakehead, Channa punctatus, by Solanum nigrum L., a medical plant. J World Aquacult Soc 39: 375-383.

72. Abbas AK, Lichtman AH, Pillai S (2010) Cellular and molecular immunology (updated 6/E). Saunders Elsevier Philadelphia, PA, USA

73. Møller M, Rath MF, Klein DC (2006) The perivascular phagocyte of the mouse pineal gland. Chronobiol Int 23: 393-401.

74. Beck G, Habicht GS (1996) Immunity and the invertebrates. Sci Am November 60~66.

75. Smith, VJ, Soderhall K (1991) A comparison of phenoloxidase activity in the blood of marine invertebrates. Dev Comp Immunol 15: 251-261.

76. Hellio C, Bado-Nilles A, Gagnaire B, Renault T, Thomas-Guyon H (2007) Demonstration of a trae phenoloxidase activity and activation of a ProPO cascade in pacific oyster, Crassostrea gigas (Thunberg) In vitro. Fish Shellfish Immunol 22: 433-440.

77. Liu H, Jiravanichpaisal P, Cerenius L, Lee BL, Söderhöll I, et al. (2007) Phenoloxidase is an important component of the defense against aeromonas hydrophila infection in a crustacean, Pacifastacus leniusculus. J Biol Chem 282: 33593-33598.

78. Alexander JB, Ingram GA (1992) Noncellular nonspecific defence mechanisms of fish. Ann Rev Fish Dis 2: 249–280.

79. Goethe R, Phi-van L (1998) Posttranscriptional lipopolysaccharide regulation of the lysozyme gene at processing of the primary transcript in myelomonocytic HD11 cells. J Immunol 160: 4970–4978.

80. Robertsen B, Engstad R, Jørgensen JB (1994) β-glucans as immunostimulators in fish. In: Stolen, JS, Fletcher, TC (Eds.), Modulators of Fish Immune Responses. SOS Publication, Fair Haven, NJ, USA, Vol. 1, pp. 83-99.

81. Robertsen B (1999) Modulation of the non-specific defence of fish by structurally conserved microbial

polymers. Fish Shellfish Immunol 9: 269-290.

82. Sahoo PK, Mukherjee SC (1999) Immunostimulants in aquaculture. In: Mohapatra, BC, Ingole, PG, Bharad, GM (Eds.), Aquaculture with Special Reference to Vidarbha. Maharashtra State, Akola, India, pp. 282-293.

83. Philip R, Sahoo PK, Sreekumar PK, Anas A, Mukherjee SC, et al. (2003) Immunostimulants – source, diversity and mode of application. In: Aquaculture Medicine. Bright Singh IS, Somnath Pai, S, Philip, R, Mohandas, A (ds.) p 135- 147 Centre for Fish Disease Diagnosis and Management, Cochin University of Science and Technology (CUSAT), Cochin, India.

84. Peddie S, Secombes CJ (2005) An overview of fish immunostimulant research. Fish Vet J 8: 1-31.

85. Tassakka ACMAR, Sakai M (2005) Current research on the immunostimulatory effects of CpG oligodeoxynucleotides in fish.Aquaculture 246: 25-36.

86. Magnadóttir B (2010) Immunological control of fish diseases. Mar Biotechnol 12: 361-379.

87. Smith VJ, Brown JH, Hauton C (2003) Immunostimulation in crustaceans: does it really protect against infections? Fish Shellfish Immunol 15: 71-90.

88. Magnadóttir B, Gudmundsdottir BK, Lange S, Steinarsson A, Oddgeirsson M, et al. (2006) Immunostimulation of larvae and juveniles of cod, Gadus morhua L. J Fish Dis 29: 147-155.

89. Whyte SK (2007) The innate immune response of fish – A review of current knowledge. Fish Shellfish Immunol 23: 1127-1151.

90. Dalmo RA, Bøgwald J (2008) β-glucans as conductors of immune symphonies. Fish Shellfish Immunol 25: 384-396.

91. Rombout J, Abelli L, Picchietti S, Scapigliati G, Kiron V (2010) Teleost intestinal immunology. Fish Shellfish Immunol early view.Doi:10.1016/j.fsi.2010.09.001.

92. Walker PJ, Mohan CV (2009) Viral disease emergence in shrimp aquaculture: origins, impact and the effectiveness of health management strategies. Rev Aquacult 1: 125-154.

93. Workenhe ST, Rise ML, Kibenge MJT, Kibenge FSB (2010) The fight between the teleost fish immune response and aquatic viruses. Mol Immunol 47: 2525-2536.

94. Palm NW, Medzhitov R (2009) Pattern recognition receptors and control of adaptive immunity. Immunol Reviews 227: 221-233.

95. Johansson MW, Keysera P, Sritunyalucksanaa K, Söderhäll K (2000) Crustacean haemocytes and haematopoiesis. Aquaculture 191: 45-52

96. Magnadóttir B (2006) Innate immunity of fish (overview). Fish Shellfish Immunol 20: 137-151

97. Skjermo J, Størseth TR, Hansen K, Handå A, Øie G (2006) Evaluation of β-(1→3, 1→6)-glucans and high-M alginate used as immunostimulatory dietary supplement during first feeding and weaning of Atlantic cod (Gadus morhua L.). Aquaculture 261: 1088-1101.

98. Hoseinifar SH, Mirvaghefi AM, Merrifield DL (2011) The effect of dietary inactive brewer`s yeast Saccharomyces cerevisiae var. ellipsoideus on the growth, physiological responses and gut microbiota of juvenile beluga (Huso huso). Aquaculture 318: 90-94.

99. Ringø E, Jutfelt F, Kanapathippillai P, Bakken Y, Sundell K, et al. (2004) Damaging effect of the fish pathogen Aeromonas salmonicida ssp. salmonicida on intestinal enterocytes of Atlantic salmon (Salmo salar L.). Cell Tiss Res 318: 305-311.

100. Birkbeck TH, Ringø E (2005) Pathogenesis and the gastrointestinal tract of growing fish. In: Holzapfel W, Naughton P (Eds.), Microbial Ecology in Growing Animals. Elsevier, Edinburgh, UK, pp 208-234.

101. Ringø E, Myklebust R, Mayhew TM, Olsen RE (2007). Bacterial translocation and pathogenesis in the digestive tract of larvae and fry. Aquaculture 268: 251-264.

102. Ringø E, Salinas I, Olsen RE, Nyhaug A, Myklebust R, et al. (2007) Histological changes in Atlantic salmon (Salmo salar L.) intestine following In vitro exposure to pathogenic and probiotic bacterial strains. Cell Tiss Res 328: 109-116.

103. Salinas I, Myklebust R, Esteban MA, Olsen RE, Meseguer J, Ringø E (2008) In vitro studies of Lactobacillus delbrueckii ssp. lactis in Atlantic salmon (Salmo salar L.) foregut: tissue responses and evidence of protection against Aeromonas salmonicida ssp. salmonicida epithelial damage. Vet Microbiol 128: 167-177.

104. Sugita H, Mizuki H, Itoi S (2008) Prevalence of a fish pathogen, Listonella anguillarum in the intestinal tract of fish collected off the coast of Japan. Aquacult Res 39: 103-105.

105. Nayak SK (2010) Role of gastrointestinal microbiota in fish. Aquacult Res 41: 1553-1573.

106. Ringø E, Løvmo L, Kristiansen M, Bakken M, Salinas I, et al. (2010) Lactic acid bacteria vs. pathogens in the gastrointestinal tract of fish: a review. Aquacult Res 41: 451-467.

107. Harikrishan R, Balasundaram C (2005) Modern trends in Aeromonas hydrophila disease management with fish. Rev Fish Sci 13: 281-320.

108. Liu Y, Zhou Z, Yao B, Shi P, He S, et al. (2008) The effect of intraperitoneal injection of immunostimulatory substances on allochthonous gut microbiota of Atlantic salmon (Salmo salar L.) determined using Denaturing Gradient Gel Electrophoresis (DGGE). Aquacult Res 39: 635-646.

109. Grimble GK (1996) Why are dietary nucleotides essential nutrients? Br J Nutr 76: 475–478.

110. Grimble GK, Westwood OMR (2001) Nucleotides as immunomodulators in clinical nutrition. Curr Op Clin Nutr Metabolic Care 4: 57-64.

111. Gutierrez-Castellon P, Mora-Magana I, Diaz-Garcia L, Jiménez-Gutierrez C, Ramirez-Mayanas J, et al. (2007) Immune response to nucleotidesupplemented infant formulae: systematic review and meta-analysis. Br J Nutr 98: S64-S67.

112. Aggett P, Leach JL, Rueda R, MacLean WC (2003) Innovation in infant formula development: a reassessment of ribonucleotides in 2002. Nutrition 19: 375–384.
113. Carver JD, Walker WA (1995) The role of nucleotides in human nutrition. J Nutr Biochem 6: 58–72.
114. Cosgrove M (1998) Nucleotides. Nutrition 14: 748–751.
115. Grimble GK, Westwood OMR (2000) Nucleotides In: German JB, Keen CL (Eds.), Nutrition and Immunology: Principles and Practice. Humana Press Inc., Totowa, NJ, USA, pp. 135–144.
116. Gil A (2002) Modulation of the immune response mediated by dietary nucleotides. Eur J Clin Nutr 56: S1–S4.
117. Singhal A, Macfarlane G, Macfarlane S, Lanigan J, Kennedy K, et al. (2008) Dietary nucleotides and fecal microbiota in formula-fed infants: a randomized controlled trial. Am J Clin Nutr 87: 1785-1792.
118. Mackie AM (1973) The chemical basis of food detection in the lobster Homarus gammarus. Mar Biol 21: 103–108.
119. Kiyohara S, Hidaka I, Tamura T (1975) Gustatory response in the puffer-II. Single fiber analysis. Bull Jap Soc Sci Fish 41: 383–391.
120. Mackie AM, Adron JW (1978) Identification of inosine and inosine 5′-monophosphate as the gustatory feeding stimulants for the turbot, Scophthalmus maximus. Comp Biochem Physiol - A 60: 79–83.
121. Patterson JA, Burkholder KM (2003) Application of prebiotics and probiotics in poultry production. Poult Sci 82: 627–631.
122. Gatlin DM (2002) Nutrition and fish health. In: Halver JE, Hardy RW (Eds.), Fish Nutrition, Academic Press. San Diego, CA, USA, pp. 671–702.
123. Devresse B (2000) Nucleotides–a key nutrient for shrimp immune system. Feed Mix 8: 20–22.
124. Li P, Gatlin DM (2006) Nucleotide nutrition in fish: Current knowledge and future applications. Aquaculture 251: 141-152.
125. Li P, Gatlin DM, Neill WH (2007) Dietary supplementation of a purified nucleotide mixture transiently enhanced growth and feed utilization of juvenile red drum, Sciaenops ocellatus. J World Aquacult Soc 38: 281-286.
126. Li P, Burr GS, Goff J, Whiteman KW, Davis KB, et al. (2005) A preliminary study on the effects of dietary supplementation of brewers yeast and nucleotides, singularly or in combination, on juvenile red drum (Sciaenops ocellatus). Aquacult Res 36: 1120-1127.
127. Liu P, Chuang CWH, Lee KK (2003) Infectious gastroenteritis caused by Vibrio harveyi (V. carchariae) in cultured red drum, Sciaenops ocellatus. J Appl Ichthyol 19: 59–61.
128. Cheng, Z, Buentello A, Gatlin DM (2011) Dietary nucleotides influence immune responses and intestinal morphology of red drum Sciaenops ocellatus. Fish Shellfish Immunol 30: 143-147.
129. Russo R, Yanong RPE, Mitchell H (2007) Dietary beta-glucans and nucleotides enhance resistance of red-tail black shark (Epalzeorhynchos bicolor, fam. Cyprinidae) to Streptococcus iniae infection. J World Aquacult Soc 37: 298- 306
130. Jha AK, Pal AK, Sahu NP, Kumar S, Mukherjee SC (2007) Haematoimmunological responses to dietary yeast RNA, ω-3 fatty acid and β-carotene in Catla catla juveniles. Fish Shellfish Immunol 23: 917-927.
131. Glencross B, Rutherford N (2010) Dietary strategies to improve the growth and feed utilization of barramundi, Lates calcarifer under high water temperature conditions. Aquacult Nutr 16:4 343-350.
132. Salze G, McLean E, Battle PR, Schwarz MH, Craig SR (2010) Use of soy protein concentrate and novel ingredients in the total elimination of fish meal and fish oil in diets for juvenile cobia, Rachycentron canadum. Aquaculture 298: 294-299.
133. Lin YH, Wang H, Shiau SY (2009) Dietary nucleotide supplementation enhances growth and immune responses of grouper, Epinephelus malabaricus. Aquacult Nutr 15: 117-122.
134. Tahmasebi-Kohyani A, Keyvanshokooh S, Nematollahi A, Mahmoudi N, Pasha-Zanoosi H (2011) Dietary administration of nucleotides to enhance growth, humoral immune responses, and disease resistance of the rainbow trout (Oncorhynchus mykiss) fingerlings. Fish Shellfish Immunol 30: 189-193.
135. Adamek Z, Hamackova J, Kouril J, Vachta R, Stibranyiova I (1996) Effect of Ascogen probiotics supplementation on farming success in rainbow trout (Oncorhynchus mykiss) and wels (Silurus glais) under conditions of intensive culture. Krmiva (Zagreb) 38: 11–20.
136. González VJL, Wadsworth S, González J, Ruohonen K (2010) The combined used of nucleotides and immune-stimulant (Sanictum®) in feed against salmon rickettsia syndrome. Poster presentation at the 14th International Symposium on Fish Nutrition and Feeding, Qingdao, China Book of abstracts page 356.
137. González Vecino JL (2005) Nucleotide enhancement of diets, fish reproduction and egg quality. PhD Thesis. Open University, UHI Millennium Institute, Scottish Association for Marine Science, UK.
138. MacKinnon BM (1998) Host factors important in sea lice infections. ICES J Mar Sci 55. 188–192.
139. Johnson S, Margolis L (1993) Efficacy of invermectin for the control of the salmon louse Lepeophtheirus salmonis on Atlantic salmon. Dis Aquatic Org 17: 101-105.
140. Fast MD, Johnson S, Eddy T, Pinto D, Ross N (2008) Cortisol response and immune-related effects of Atlantic salmon (Salmo salar Linnaeus) subjected to short- and long term stress. Fish Shellfish Immunol 24: 194–204.
141. Fast MD, Burka JF, Johnson SC, Ross NW (2003) Enzymes released from Lepeophtheirus salmonis in response to mucus from different salmonids. J Parasitol 89: 7-13.
142. Fast MD, Ross NW, Craft CA, Locke SJ, MacKinnon SL, et al. (2004) Lepeophtheirus salmonis: characterization of prostaglandin E-2 in secretory products of the salmon

louse by Rp-Hplc and mass spectrometry. Exp Parasitol 107: 5-13.

143. Fast MD, Ross NW, Muise DM, Johnson SC (2006) Differential gene expression in Atlantic salmon, Salmo salar, infected with Lepeophtheirus salmonis (Copepoda: Caligidae). J Aquatic Health 18: 116–127.
144. Fast MD, Johnson S, Eddy T, Pinto D, Ross N (2007) Lepeophtheirus salmonis secretory / excretory products and their effects on Atlantic salmon immune gene regulation. Parasite Immunol 29: 179–189.
145. Wagner G, Fast M, Johnson SC (2008) Physiology and immunology of Lepeophtheirus salmonis infections of salmonids. Trend Parasitol 24: 176- 183.
146. Burrells C, Williams PD, Southgate PJ, Wadsworth SL (2001) The effects of a nucleotide-enriched diet on experimental infestation with sea lice (Lepeophtheirus salmonis) and on re-infestation rates following anti-louse bath treatment with cyperthin. SCI Conference, Aberdeen.
147. González J, Troncoso J (2009) Reducing Caligus rogercresseyi infestation in Salmo salar fed a diet supplemented with microbial based nucleotides and immunostimulants. Oral presentation at World Aquaculture 2009, Veracruz, Mexico. Book of abstracts p25.
148. Wadsworth S, Lygren B (2009) Nukleotider reduserer påslag av lus og forhindrer resistens. Norsk Fiskeoppdrett, Special sea lice Issue June 09, 50- 53.
149. Olsvik PA, Lie KK, Mykkeltvedt E, Samuelsen OB, Petersen K, et al. (2008) Pharmakinetics and transcripcional effects of the anti-salmon lice drug emamectin benzoate in Atlantic salmon (Salmo salar L). BMC Pharmacol 8:16.
150. Ringø E, Olsen RE, Gifstad TØ, Dalmo RA, Amlund H, et al. (2010) Prebiotics in aquaculture: a review. Aquacult Nutr 16: 117-136.
151. Hai NV, Fotedar R (2009) Comparison of the effect of the prebiotics (Bio- MOS® and β-1,3-D-glucan) and the customised probiotics (Pseudomonas synxantha and P. aeruginosa) on the culture of juvenile western king prawns (Penaeus latisulcatus Kishinouye, 1896). Aquaculture 289: 310-316.
152. George M, Abraham TE (2006) Pylyionic hydrocolloids for the intestinal delivery of protein drugs: Alginate and chitosan – a review.J Cont Rel 114: 1-14.
153. Yano T, Mangindaan REP, Matsuyama H (1989) Enhancement of the resistance of carp Cyprinus carpio to experimental Edwardsiella tarda infection, by some β-1,3 glucans. Nipp Sui Gakk 55: 1815-1819.
154. Chen D, Ainsworth, AJ (1992) Glucan administration potentiates immune defence mechanisms of channel catfish, Ictalurus punctatus Rafinesque. J Fish Dis 15: 295-304.
155. Sang HM, Fotedar R (2010) Effects of dietary β – 1.3 – glucan on the growth, survival, physiological and immune response of marron, Cherax tenuimanus (smith, 1912). Fish Shellfish Immunol 28: 957-960.
156. Strand HK, Dalmo RA (1997) Absorption of immunomodulating β(1,3)-glucan in yolk sac larvae of Atlantic halibut, Hippoglossus hippoglossus (L). J Fish Dis 20: 41-49.
157. Kim Y-s, Ke F, Zhang Q-Y (2009) Effect of β-glucan on activity of antioxidant enzymes and Mx gene expression in virus infected grass carp. Fish Shellfish Immunol 27: 336-340.
158. Nikl L, Albright LJ, Evelyn TPT (1991) Influence of seven immunostimulants on the immune response of coho salmon to Aeromonas salmonicida. Dis Aqua Org 12: 7-12.
159. Jeney G, Galeotti M, Volpatti D, Jeney Z, Anderson DP (1997) Prevention of stress in rainbow trout (Oncorhynchus mykiss) fed diets containing different doses of glucan. Aquaculture 154: 1-15.
160. Ogier de Baulny M, Quentel C, Fournier V, Lamour F, Le Gouvello R (1996) Effect of long-term oral administration of β-glucan as an immunostimulant or an adjuvant on non-specific parameters of the immune response of turbot Scophthalmus maximus. Dis Aqua Org 26: 139-147
161. Djorddjevic B, Skugor S, Jørgensen SM, Øverland M, Mydland LT, et al. (2009) Modulation of splenic immune responses to bacterial lipopolysaccharide in rainbow trout (Oncorhynchus mykiss) fed lentinan, a beta-glucan from mushroom Lentinula edodes.Fish Shellfish Immunol 26: 201-209.
162. Bagni M, Archetti L, Amadori M, Marino G (2000) Effects of long-term oral administration diet on innate immunity in sea bass (Dicentrarchus labrax). J Vet Med B 47: 745-751.
163. Bagni M, Romano N, Finoia MG, Abelli L, Scapigliati G, et al. (2005) Shortand long-term effects of a dietary yeast beta-glucan (Macrogard) and alginic acid (Ergosan) preparation on immune response in sea bass (Dicentrarchus labrax). Fish Shellfish Immunol 18: 311-325.
164. Bonaldo A, Thompson KD, Manfrin A, Adams A, Murano E, et al. (2007) The influence of dietary beta-glucans on the adaptive and innate immune responses of European sea bass (Dicentrarchus labrax) vaccinated against vibriosis. Italian J Anim Sci 6: 151-164.
165. Cook MT, Hayball PJ, Hutchinson W, Nowak B, Hayball JD (2003) Administration of a commercial immunostimulant preparation, EcoActiva™, as a feed supplement enhances macrophage respiratory burst and the growth rate of snapper (Pagrus auratus, Sparidae (Bloch and Schneider)) in winter. Fish Shellfish Immunol 14: 333-345.
166. Wang WS, Hung SW, Lin YH, Tu CY, Wong ML, et al. (2007) The effect of five different glycans on innate immune responses by phagocytes of hybrid tilapia and Japanese eels, Anguilla japonica. J Aquatic Anim Health 19: 49-59.
167. Cain KD, Grabowski L, Reilly J, Lytwyn M (2003) Immunomodulatory effects of a bacterial-derived β-1, 3 glucan administered to tilapia (Oreochromis nilotocus L) in a Spirulina-based diet. Aquacult Res 34: 1241-1244.
168. Galindo-Villegas J, Masumoto T, Hosokawa H (2002)

Immunostimulants and disease resistance in Japanese flounder, Paralichthys olivaceus. In: Proceedings of the 5th Korean-Japan Biannual Joint Meeting, Kunsan University, Kunsan, Korea, p 26.

169. Welker T, Lim C, Yildirim-Aksoy M, Shelby R, Klesius PH (2007) Immune response and resistance to stress and Edwardsiella ictaluri challenge in channel catfish, Ictalurus punctatus, fed diets containing commercial wholecell yeast or yeast subcomponents. J World Aquacult Soc 38: 24-35.

170. Das BK, Debnath CD, Patnaik P, Swain DK, Kumar K, et al. (2009) Effect of β-glucan on immunity and survival of early stage of Anabas testudineus (Bloch). Fish Shellfish Immunol 27: 678-683.

171. Campa-Cordova AI, Hernandez-Saavedra NY, de Philippis R, Ascencio F (2002) Generation of superoxide anion and SOD activity in haemocytes and muscle of American White shrimp (Litopenaeus vannamei) as a response to β-glucan and sulphated polysaccharide. Fish Shellfish Immunol 12: 353-366.

172. Anas, A, Lowmann DW, Williams D, Millen S, Pai SS, et al. (2009) Alkali insoluble glucan extracted from Acremonium diospyri is a more potent immunostimulant in the Indian White Shrimp, Fenneropenaeus indicus than soluble glucan. Aquacult Res 40: 1320-1327.

173. Bai N, Zhang W, Mai K, Wang X, Xu W, et al. (2010) Effects of discontinous administration of ß-glucan and glycyrrhizin on the growth and immunity of white shrimp Litopenaeus vannamei. Aquaculture 306: 218-224.

174. Costa MM, Novoa B, Figueras A (2008) Influence of ß-glucans on the immune responses of carpet shell clam (Ruditapes decussates) and Mediterranean mussel (Mytilus galloprovincialis). Fish Shellfish Immunol 24: 498-505.

175. Lin T, Xing J, Jiang J, Tang X, Zhan W (2011) ß-glucan-stimulated activation of six enzymes in the haemocytes of the scallop Chlamys farreri at different water temperatures. Aquaculture 315: 213-221.

176. Cook MT, Hayball PJ, Hutchinson W, Nowak B, Hayball JD (2001) The efficacy of a commercial ß-glucan preparation, EcoActivaT, on stimulating respiratory burst activity of head-kidney macrophages from pink snapper (Pagrus auratus), Sparidae. Fish Shellfish Immunol 11: 661-672.

177. Paulsen SM, Engstad RE, Robertsen B (2001) Enhanced lysozyme production in Atlantic salmon (Salmo salar L) macrophages treated with yeast ß-glucan and bacterial lipopolysaccharide. Fish Shellfish Immunol 11: 23-37.

178. Kiseleva MI, Balabanova LA, Rasskazov VA, Zvyagintseeva TN (2008) Effect of 1,3;1,6-ß-D-glucans on developing sea urchin embryos. Mar Biotechnol 10: 466-470.

179. Jørgensen JB, Lunde H, Robertsen B (1993) Peritoneal and head kidney cell response to intraperitoneal injected yeast glucan in Atlantic salmon, Salmo salar L. J Fish Dis 16: 313-325.

180. Refstie S, Baeverfjord G, Seim RP, Elvebø O (2010) Effects of dietary yeast cell wall ß-glucans and MOS on performance, gut health, and salmon lice resistance in Atlantic salmon (Salmo salar) fed sunflower and soybean meal. Aquaculture 305: 109-116

181. Jørgensen, JB, Sharp GJE, Secombes CJ, Robertsen B (1993) Effect of a yeast cell wall glucan on the bactericidal activity of rainbow trout macrophages. Fish Shellfish Immunol 3: 267-277.

182. Chang CF, Chen HY, Su MS, Liao IC (2000) Immunomodulation of dietary ß-1,3-glucan in the brooders of the black tiger shrimp Penaeus monodon. Fish Shellfish Immunol 10: 505-514.

183. Jeney G, Anderson DP (1993) Glucan injection or bath exposure given alone or in combination with bacterin enhance the non-specific defence mechanism in rainbow trout (Oncorhynchus mykiss). Aquaculture 116: 315-329.

184. Robertsen B, Rørstad G, Engstad R, Raa J (1990) Enhancement of nonspecific disease resistance in Atlantic salmon, Salmo salar L, by a glucan from Saccharomyces cerevisiae cell walls. J Fish Dis 13: 391-400.

185. Engstad RE, Robertsen B, Frivold, E (1992) Yeast glucan induces increase in lysozyme and complement-mediated haemolytic activity in Atlantic salmon blood. Fish Shellfish Immunol 2: 287-297.

186. Rørstad G, Aasjord P M, Robertsen B (1993) Adjuvant effect of a yeast glucan in vaccines against furunculosis in Atlantic salmon (Salmo salar L). Fish Shellfish Immunol 3: 179-190.

187. Aakre R, Wergeland HI, Aasjord PM, Endresen C (1994) Enhanced antibody response in Atlantic salmon (Salmo salar L) to Aeromonas salmonicida cell wall antigens using a bacterin containing ß-1,3-M-glucan as adjuvant. Fish Shellfish Immunol 4: 47-61.

188. Dalmo RA, Bøgwald J, Ingebrigtsen K, Seljelid R (1996) The immunomodulatory effect of laminaran [beta(1,3)-D-glucan] on Atlantic salmon, Salmo salar L, anterior kidney leucocytes after intraperitoneal, peroral and peranal administration. J Fish Dis 19: 449-457.

189. Thompson KD, Cachos A, Inglis V (1995) Immunomodulating effects of glucans and oxytetracycline in rainbow trout, Oncorhynchus mykiss, on serum lysozyme and protection. In: Shariff M, Subasighe RP, Arthur JR (Eds.), Diseases in Asian Aquaculture Vol. 11. Fish health Section, Asian Fisheries Society, Manila, Philippines, pp. 433-439.

190. Barnes ME, Durben DJ, Reeves SG, Sanders R (2006) Dietary yeast culture supplementation improves initial rearing of McConaughy strain rainbow trout. Aquacult Nutr 12: 388-394.

191. Pal D, Joardar SN, Roy B (2007) Immunostimulatory effects of a yeast (Saccharomyces cerevisiae) cell wall feed supplement on rohu (Labeo rohita), an Indian major carp. The Israeli J Aquacult - Bamidgeh 59: 175-181.

192. Pal D, Joardar SN, Roy B (2007) Yeast cell wall preparation from Saccharomyces cerevisiae enhances immune effector cells of Indian major carp, Catla catla (Hamilton). Indian J Fish 56: 33-38.

193. Watanabe AL, Macedo Viegas EM, Goncalves LU (2010) Levels of yeast and its by-products on pacu

juveniles feeding. Revista Bras Zoo 39: 447-453.

194. Palic D, Andreasen CB, Herolt DM, Menzel BW, Roth JA (2006) Immunomodulatory effects of ß-glucan on neutrophil function in fathead minnows (Pimephales promelas Rafinesque, 1820). Dev Comp Immunol 30: 817-830.

195. Sung HH, Kou GH, Song YL (1994) Vibriosis resistance induced by glucan treatment in tiger shrimp (Penaeus monodon). Fish Pathol 29: 11-17.

196. Thanardkit P, Khunrae P, Suphantharika M, Verduyn C (2002) Glucan from brewer's yeast: preparation, analysis and use as a potential immunostimulant in shrimp feed. World J Microbiol Biotechnol 18: 527-539.

197. Chotikachinda R, Lapjatupon W, Chaisilapasung S, Sangsue D, Tantikitti C (2008) Effect of inactive yeast cell wall on growth performance, survival rate and immune parameters in Pacific white shrimp, (Litopenaeus vannamei). Songklanakarin J Sci Technol 22: 677-688.

198. Verlhac V, Gabaudan J, Obach A, Schüep W, Hole R (1996) Influence of dietary glucan and vitamin C on non-specific and specific immune responses of rainbow trout (Oncorhynchus mykiss). Aquaculture 143: 123-133.

199. Oliva-Teles A, Goncalves P (2001) Partial replacement of fishmeal by brewers yeast (Saccaromyces cerevisiae) in diets for sea bass (Dicentrarchus labrax) juveniles. Aquaculture 202: 269-278.

200. Rodríguez A, Cuesta A, Ortuño J, Esteban MA, Meseguer J (2003) Immunostimulant properties of a cell wall-modified whole Saccharomyces cerevisiae strain administered by diet to seabream (Sparus aurata L). Vet Immunol Immunopathol 96: 183–192.

201. Rodríguez A, Esteban MA, Meseguer J (2003) Phagocytosis and peroxidise release by seabream (Sparus aurata L) leucocytes in response to yeast cells. The Anatom Rec Part A 272 A: 415–423.

202. Cuesta A, Meseguer J, Esteban MA (2004) Total serum immunoglobulin M levels are affected by immunomodulators in seabream (Sparus aurata L). Vet Immunol Immunopathol 101: 203-210.

203. Esteban MA, Rodriguez A, Meseguer J (2004) Glucan receptor but not mannose receptor is involved in the phagocytosis of Saccharomyces cerevisiae by seabream (Sparus aurata L) blood leucocytes. Fish Shellfish Immunol 16: 447-451.

204. Oliva-Teles A, Guedes MJ, Vachot C, Kaushik, SJ (2006) The effect of nucleic acids on growth, ureagenesis and nitrogen extretion of gilthead sea bream Sparus aurata juveniles. Aquaculture 253: 608-617.

205. Ortuño J, Cuesta A, Rodríguez A, Esteban MA, Meseguer J (2002) Oral administration of yeast, Saccharomyces cerevisiae, enhances the cellular innate immune response of gilthead seabream (Sparus aurata L). Vet Immunol Immunopathol 85: 41–50.

206. Salnur S, Gultepe N, Hossu B (2009) Replacement of fish meal by yeast (Saccharomyces cerevisiae): Effects on digestibility and blood parameters of gilthead sea bream (Sparus aurata). J Anim Vet Adv 8: 2557-2561.

207. Lara-Flores M, Olvera-Novoa MA, Guzman-Mendez BE, Lopez-Madrid W (2003) Use of the bacteria Streptococcus faecium and Lactobacillus acidophilus, and the yeast Saccharomyces cerevisiae as growth promotors in Nile tilapia (Oreochromis niloticus). Aquaculture 216: 193-201.

208. Zerai DB, Fitzsimmons KM, Collier RJ, Duff GC (2008) Evaluation of brewer`s waste as partial replacement of fish meal protein in Nile tilapia, Oreochromis niloticus, diets. J World Aquacult Soc 39: 556-564.

209. Zaki M, Nour AE, Gaber M, Labib E (2010) Brewers dried yeast as a good alternative ingredient in Nile tilapia, (Oreochromis niloticus) feeds. Proc 2010 Intern Conf Agricult Anim 338-342.

210. Li P, Gatlin DM (2003) Evaluation of brewers yeast (Saccharomyces cervisiae) as a feed supplement for hybrid striped bass (Morone chrysops×M. saxatilis). Aquaculture 219: 681–692.

211. Li P, Gatlin DM (2004) Dietary brewers yeast and the prebiotic GroBiotic™ AE influence growth performance, immune responses and resistance of hybrid striped bass (Morone chrysops×M. saxatilis) to Streptococcus iniae infection. Aquaculture 231: 445–456.

212. Li P, Gatlin DM (2005) Evaluation of the prebiotic GroBiotic®-A and brewers yeast as dietary supplements for sub-adult hybrid striped bass (Morone chrysops×M. saxatilis) challenged in situ with Mycobacterium marinum. Aquaculture 248: 197-2005.

213. Ozòrio ROA, Turini BGS, Moro GV, Oliveira LST, Portz L, et al. (2010) Growth, nitrogen gain and indispensable amino acid retention of pacu (Piaractus mesopotamicus, Holmberg 1887) fed different brewers yeast (Saccharomyces cerevisiae) levels.Aquacult Nutr 16: 276-283.

214. Abdel-Tawwab M, Mousa MAA, Mohammed MA (2010) Use of live baker`s yeast, Saccharomyces cerevisiae, in practical diet to enhance the growth performance of galilee tilapia, Sarotherodon galilaeus (L), and its resistance to environmental copper toxicity. J World Aquacult Soc 41: 214-223.

215. Vadstein O, Øie G, Olsen Y, Salvesen I, Skjermo J, et al. (1993) A strategy to obtain microbial control during larval development of marine fish. In: Reinertsen H, Dahle LA, Jørgensen L, Tvinnereim K (Eds.), Fish Farming Technology AA Balkema Publishers, Rotterdam, pp. 69-75.

216. Skjermo J, Berg Ø (2004) High-M alginate immunostimulation of Atlantic halibut (Hippoglossus hippoglossus) larvae using Artemia for delivery, increases resistance against vibriosis. Aquaculture 238: 107-113.

217. Gabrielsen BO, Austreng E (1998) Growth, product quality and immune status of Atlantic salmon, Salmo salar L., fed wet feed with alginate. Aquacult Res 29: 397-401.

218. Conceicao LEC, Skjermo J, Skjåk-Bræk G, Verreth JAJ (2001) Effect of an immunostimulating alginate on protein turnover of turbot (Scophthalmus maximus L.) larvae. Fish Physiol Biochem 24: 207–212.

219. Gioacchini G, Lombardo F, Avella MA, Olivotto I,

Carnevali O (2010) Welfare improvements using algínico acid in rainbow trout (Oncorhynchus mykiss) juveniles. Chem Ecol 26: 111-121.

220. Wang W, Wang D (1997) Enhancement of the resistance of tilapia and grass carp to experimental Aeromonas hydrophila and Edwardsiella tarda infections by several polysaccharides. Comp Immunol Microbiol Inf Dis 20: 261-270.

221. Sahoo PK, Mukherjee SC (2001) Effect of dietary β-1, 3 glucan on immune responses and disease resistance of healthy and aflatoxin B1-induced immunocompromised rohu (Labeo rohita Hamilton). Fish Shellfish Immunol 11: 683–695.

222. Misra CK, Das BK, Mukherjee SC, Pattnaik P (2006) Effect of long term administration of dietary β-glucan on immunity, growth and survival of Labeo rohita fingerlings. Aquaculture 255: 82-94.

223. Misra CK, Das BK, Mukherjee SC, Pattnaik P (2006) Effect of multiple injections of β-glucan on non-specific immune response and disease resistance in Labeo rohita fingerlings. Fish Shellfish Immunol 20: 305-319.

224. Kamilya D, Maiti TK, Joardar SN, Mal BC (2006) Adjuvant effect of mushroom glucan and bovine lactoferrin upon Aeromonas hydrophila vaccination in catla, Catla catla (Hamilton). J Fish Dis 29: 331-337.

225. Selvaraj V, Sampath K, Sekar V (2005) Use of glucan from Saccharomyces cerevisiae as an immunostimulant in carp: Impact on hematology, phagocyte function, and infection with Aeromonas hydrophila. The Israeli J Aquacult – Bamidgeh 57: 39-48.

226. Selvaraj V, Sampath K, Sekar V (2005) Administration of yeast glucan enhanced survival and some non-specific and specific immune parameters in carp (Cyprinus carpio) infected with Aeromonas hydrophila. Fish Shellfish Immunol 19: 293-306.

227. Selvaraj V, Sampath K, Sekar V (2006) Adjuvant and immunostimulatory effects of β-glucan administration in combination with lipopolysaccharide enhances survival and some immune parameters in carp challenge with Aeromonas hydrophila. Vet Immunol Immunopathol 114: 15-24.

228. Moqsood S, Samoon MH, Singh P (2009) Immunomodulatory and growth promoting effect of dietary levamisole in Cyprinus carpio fingerlings against the challenge of Aeromonas hydrophila. Tur J Fish Aqua Sci 9: 111-120.

229. Gopalakannan A, Arul V (2010) Enhancement of the innate immune system and disease-resistant activity in Cyprinus carpio by oral administration of β-glucan and whole yeast. Aquacult Res 41: 884-892.

230. Kumari J, Sahoo PK (2006) Dietary immunostimulants influence specific immune response and resistance of healthy and immunocompromised Asian catfish Clarias batrachus to Aeromonas hydrophila infection. Dis Aqua Org 70: 63-70.

231. Samuel M, Lam TJ, Sin YM (1996) Effect of laminaran [β (1-3)-D-glucan] on the protective immunity of blue gourami, Trichogaster trichopterus against Aeromonas hydrophila. Fish Shellfish Immunol 6: 443-454.

232. Kamilya D, Joardar SN, Mal BC, Maiti TK (2008) Effects of a glucan from the edible mushroom (Pleurotus florida) as an immunostimulant in farmed Indian major carp (Catla catia). Israeli J Aquacult - Bamidgeh 60: 37-45.

233. Rodriguez I, Chammoro R, Novoa B, Figueras A (2009) β-glucan administration enhances disease resistance and some innate immune responses in zebrafish (Danio rerio). Fish Shellfish Immunol 27: 369-373.

234. Abdel-Tawwab M, Abdel-Rahman AM, Ismael NEM (2008) Evaluation of commercial live bakers`yeast, Saccharomyces cerevisiaeas growth and immunity promoter for fry Nile tilapia, Oreochromis niloticus (L.) challenged in situ with Aeromonas hydrophila. Aquaculture 280, 185-189.

235. Reque VR, de Moraes JRE, de Andrade Belo MA, de Moraes FR (2010) Inflammation induced by inactivated Aeromonas hydrophila in Nile tilapia fed diets supplemented with Saccharomyces cerevisiae. Aquaculture 300: 37-42.

236. Abdel-Tawwab M, Ahmad MH, Seden MEA, Sakr SFM (2010) Use of green tea, Camellia sinensis L., in practical diet for growth and protection of Nile tilapia, Oreochromis niloticus (L.), against Aeromonas hydrophila infection. J World Aquacult Soc 41: 208-213.

237. Reyes-Becerril M, Tovar-Ramirez D, Ascenico-valle F, Civera-Cerecedo R, Gracia-lopez V, et al. (2010) Effects of dietary supplementation with probiotic live yeast Debaryomyces hansenii on the immune and antioxidant systems of leooard grouper Mycteroperca rosacea infected with Aeromonas hydrophila. Aquacult Res early view.

238. Siwicki AK, Zakes Z, Terech-Majewska E, Kazun K, Lepa A, et al. (2010) Dietary Macrogard reduces Aeromonas hydrophila mortality in tench (Tinca tinca) through the activation of cellular and humoral defence mechanisms. Rev Fish Biol Fish 20: 435-439.

239. Türnau D, Schmidt H, Kürzinger H, Böhm KH (2000) Potency testing of β-glucan immunostimulating effect in food for ornamental fish. Bull Eur Ass Fish Pathol 20: 143-147.

240. Anderson DP, Siwick AK (1992) Duration of protection against Aeromonas salmonicida in brook trout immunostimulated with glucan or chitosan by injection or immersion. Prog Fish Cult 56: 258-261.

241. Siwicki AK, Anderson DP, Rumsey GL (1994) Dietary intake of immunostimulants by rainbow trout affects non-specific immunity and protection against furunculosis. Vet Immunol Immunopathol 41: 125–139.

242. Dalmo RA, Martinsen B, Horsberg TE, Ramstad A, Seljelid R, (1998) Prophylactic effect of β (1,3-D-glucan (laminaran) against experimental Aeromonas salmonicida and Vibrio salmonicida infections. J Fish Dis 21: 459- 462.

243. Duncan PL, Klesius, PH (1996) Dietary immunostimulants enhance nonspecific immune response in channel catfish but not resistance to Edwardsiella ictaluri. J Aquatic Anim Health 8: 241-248.

244. Sahoo PK, Mukherjee SC (2002) The effect of dietary

immunomodulation upon Edwardsiella tarda vaccination in healthy and immunocompromised Indian major carp (Labeo rohita). Fish Shellfish Immunol 12: 1–16.

245. Matsuyama H, Mangindaan REP, Yano T (1992) Protective effect of schizophyllan and scleroglucan against Streptococcus sp. infection in yellowtail (Seriola quinqueradiata). Aquaculture 101: 197-203.
246. Couso N, Castro R, Magarinos B, Obach A, Lamas J (2003) Effect of oral administration of glucans on the resistance of gilthead seabream to pasteurellosis. Aquaculture 219: 99-109.
247. Chang J, Zhang W, Mai K, Liufu Z, Ai Q (2011) Effects of dietary β-glucan and glycyrrhizin on non-specific immunity and disease resistance of white shrimp, Litopenaeus vannamei (Boone) challenged with Vibrio alginolyticus. Aquacult Res 42: 1101-1109.
248.
249. Misra CK, Das BK, Pradhan J, Pattnaik P, Sethi S, et al. (2004) Changes in lysosomal enzyme activity and protection against Vibrio infection in Macrobrachium rosenbergii (De Man) post larvae after bath immunostimulation with β-glucan. Fish Shellfish Immunol 17: 389-395.
250. Pedersen GM, Gildberg A, Olsen RL (2004) Effects of including cationic proteins from cod milt in the feed to Atlantic cod (Gadus morhua) fry during a challenge trial with Vibrio anguillarum. Aquaculture 233: 31-43.
251. Marques A, Dhont J, Sorgeloos P, Bossier P (2006) Immunostimulatory nature of β-glucans and baker's yeast in gnotobiotic Artemia challenge tests. Fish Shellfish Immunol 20: 682-692.
252. Soltanian S, Thai TQ, Dhont J, Sorgeloos P, Bossier P (2007) The protective effect against Vibrio campbellii in Artemia nauplii by pure β-glucan and isogenic yeast cells differing in β-glucan and chitin content operated with a source-dependent time lag. Fish Shellfish Immunol 23: 1003-1004.
253. Soltanian S, Francois JM, Dhont J, Arnouts S, Sorgeloos P, et al. (2007) Enhanched disease resistance in Artemia by application of commercial β-glucans sources and chitin in a gnotobiotic Artemia challenge test. Fish Shellfish Immunol 23: 1304-1314.
254. Figueras A, Santarém MM, Novoa B (1998) Influence of the sequence of administration of β-glucan and a Vibrio damsela vaccine on immune response of turbot (Scophthalmus maximus L.). Vet Immunol Immunopathol 64: 59-68.
255. Ai Q, Mai K, Zhang L, Tan B, Zhang W (2007) Effects of dietary β-1,3 glucan on innate immune response of large yellow croaker, Pseudosciaena crocea. Fish Shellfish Immunol 22: 394–402.
256. Burgents JE, Burnett KG, Burnet LE (2004) Disease resistance of pacific white shrimp, Litopenaeus vannamei, following the dietary administration of a yeast culture supplement. Aquaculture 231: 1-8.
257. Tukmechi A, Andani HRR, Manaffar R, Sheikhzadeh N (2011) Dietary administration of beta-mercapto-ethanol treated Saccharomyces cerevisiae enhanced the growth, innate immune response and disease resistance of the rainbow trout, Oncorhynchus mykiss. Fish Shellfish Immunol 30: 923-928.
258. Romalde JL, Luzardo-Alaverez A, Ravelo C, Toranzo AE, Blanco-Mendez J (2004) Oral immunization using alginate microparticles as a useful strategy for booster vaccination against fish lactococcosis. Aquaculture 236: 119-129.
259. Chiu ST, Tsai RT, Hsu JP, Liu CH, Cheng W (2008) Dietary sodium alginate administration to enhance the non-specific immune responses, and disease resistance of the juvenile grouper Epinephelus fuscoguttatus. Aquaculture 277: 66-72.
260. Jaramillo F, Gatlin DM (2004) Comparison of purified and practical diets supplemented with or without beta-glucan and selenium on resistance of hybrid striped bass Morone chrysops female x M. saxatilis male to Streptococcus iniae infection. J World Aquacult Soc 35: 245-252.
261. Whittington R, Lim C, Klesius, PH (2005) Effect of dietary β-glucan levels on the growth response and efficacy of Streptococcus iniae vaccine in Nile tilapia, Oreochromis niloticus. Aquaculture 248: 217-225.
262. Russo R, Yanong RPE, Mitchell H (2006) Dietary beta-glucans and nucleotides enhance resistance of red-tail black shark (Epalzeorhynchos bicolor, fam. Cyprinidae) to Streptococcus iniae infection. J World Aquacult Soc 37: 298- 306.
263. Chang CF, Su MS, Chen HY, Lo CF, Kou GH (1999) Effect of dietary β-1,3- glucan on resistance to white spot syndrome virus (WSSV) in postlarval and juvenile Penaeus monodon. Dis Aqua Org 36: 163-168.
264. Chang CF, Su MS, Chen HY, Liao IC (2003) Dietary β-1,3-glucan effectively improves immunity and survival of Penaeus monodon challenged with white spot syndrome virus. Fish Shellfish Immunol 15: 297-310.
265. Sajeevan TP, Lowman DW, Williams DL, Selven S, Anas A, et al. (2009) Marine yeast diet confers better protection than its cell wall component (1- 3)- β-D-glucan as an immunostimulant in Fenneropenaeus indicus. Aquacult Res 40: 1723-1730.
266. Sukumaran V, Lowman DW, Sajeevan T, Philip R (2010) Marine yeast glucans confer better protection than that of baker's yeast in Penaeus monodon against white spot syndrome virus infection. Aquacult Res 41: 1799-1805.
267. Jenkins CS, Barker DE (2004) The efficacy of oral immunostimulants in enhancing resistance against microsporidiosis in juvenile Atlantic cod (Gadus morhua L.). Aquacult Ass Can 9: 97-100.
268. Guselle NJ, markham RFJ, Speare DJ (2006) Intraperitoneal administration of B-1,3/1,6-glucan to rainbow trout, Oncorhynchus mykiss (Walbaum), protects against Loma salmonae. J Fish Dis 29: 375-381.
269. Ramadan A, Atef M (1991) Effect of the biogenic performance enhancer (Ascogen "S") on growth rate of tilapia fish. Acta Vet Scand 87: S304–S306.
270. Li P, Lewis DH, Gatlin DM (2004) A Dietary oligonucleotide from yeast RNA influences immune

responses and resistance of hybrid striped bass (Morone chrysops × Morone saxatilis) to Streptococcus iniae infection. Fish Shellfish Immunol 16: 561–569.

271. Li P, Wang X, Gatlin DM (2004) Excessive dietary levamisole suppresses growth performance of hybrid striped bass (Morone chrysops × M. saxatilis), and elevated levamisole In vitro impairs macrophage function. Aquacult Res 35: 1380–1383.

272. Ramadan A, Afifi NA, Moustafa M, Samy AM (1994) The effect of ascogen on the immune response of tilapia fish to Aeromonas hydrophila vaccine. Fish Shellfish Immunol 5: 159–165.

273. Lionardi M, Sandino AM, Klempau A (2003) Effect of a nucleotide-enriched diet on the immune system, plasma cortisol levels and resistance to infectious pancreatic necrosis (IPN) in juvenile rainbow trout (Oncorhynchus mykiss). Bull Eur Ass Fish Pathol 23: 52–59.

274. Sahu S, Das BK, Mishra BK, Pradhan J, Sarangi N (2007) Effect of Allium sativum on the immunity and survival of Labeo rohita infected with Aeromonas hydrophila. J Appl Ichthyol 23: 80-86.

275. Sakai M, Taniguchi K, Mamoto K, Ogawa H, Tabata M (2001) Immunostimulant effects of nucleotide isolated from yeast RNA on carp, Cyprinus carpio L. J Fish Dis 24: 433–438.

276. Li P, Lawrence A, Castille FI, Gatlin DM (2007) Preliminary evaluation of a purified nucleotide mixture as a dietary supplement for pacific white shrimp Litopenaeus vannamei (Boone). Aquacult Res 38: 887-890.

277. Li P, Gatlin DM, Neill WH (2007) Dietary supplementation of a purified nucleotide mixture transiently enhanced growth and feed utilization of juvenile red drum, Sciaenops ocellatus. J World Aquacult Soc 38: 281-286.

INDEX